Surveys of Modern Mathematics
Volume I

Analytic Methods in Algebraic Geometry

by Jean-Pierre Demailly

International Press
www.intlpress.com

高等教育出版社
HIGHER EDUCATION PRESS

Surveys of Modern Mathematics, Volume I
Analytic Methods in Algebraic Geometry

Jean-Pierre Demailly
Institut Fourier, Laboratoire de Mathématiques, Saint-Martin d'Hères, France

2010 Mathematics Subject Classification. 14-XX.

ISBN: 978-1-57146-234-3

Printed in the United States of America.

16 15 14 13 12 1 2 3 4 5 6 7 8 9

Surveys of Modern Mathematics

Preface

The main purpose of this book is to describe analytic techniques which are useful to study questions such as linear series, multiplier ideals and vanishing theorems for algebraic vector bundles. One century after the ground-breaking work of Riemann on geometric aspects of function theory, the general progress achieved in differential geometry and global analysis on manifolds resulted into major advances in the theory of algebraic and analytic varieties of arbitrary dimension. One central unifying concept is positivity, which can be viewed either in algebraic terms (positivity of divisors and algebraic cycles), or in more analytic terms (plurisubharmonicity, Hermitian connections with positive curvature). In this direction, one of the most basic results is Kodaira's vanishing theorem for positive vector bundles (1953–1954), which is a deep consequence of the Bochner technique and the theory of harmonic forms initiated by Hodge during the 1940's. This method quickly led Kodaira to the well-known embedding theorem for projective varieties, a far reaching extension of Riemann's characterization of abelian varieties. Further refinements of the Bochner technique led ten years later to the theory of L^2 estimates for the Cauchy-Riemann operator, in the hands of Kohn, Andreotti-Vesentini and Hörmander among others. Not only can vanishing theorems be proved or reproved in that manner, but perhaps more importantly, extremely precise information of a quantitative nature can be obtained about solutions of $\overline{\partial}$-equations, their zeroes, poles and growth at infinity.

We try to present here a condensed exposition of these techniques, assuming that the reader is already somewhat acquainted with the basic concepts pertaining to sheaf theory, cohomology and complex differential geometry. In the final chapter, we address very recent questions and open problems, e.g. results related to the finiteness of the canonical ring and the abundance conjecture, as well as results describing the geometric structure of Kähler varieties and their positive cones.

This book is an expansion of lectures given by the author at the Park City Mathematics Institute in 2008 and was published partly in Analytic and Algebraic Geometry, edited by Jeff McNeal and Mircea Mustaţă, It is a volume in the Park City Mathematics Series, a co-publication of the Park City Mathematics Institute and the American Mathematical Society.

Jean-Pierre Demailly
Université de Grenoble I, Institut Fourier
Saint-Martin d'Hères
April 21, 2010

Contents

Introduction

This introduction will serve as a general guide for reading the various parts of this text. The first three chapters briefly introduce basic materials concerning complex differential geometry, Dolbeault cohomology, plurisubharmonic functions, positive currents and holomorphic vector bundles. They are mainly intended to fix notation. Although the most important concepts are redefined, readers will probably need to already possess some related background in complex analysis and complex differential geometry — whereas the expert readers should be able to quickly proceed further.

The heart of the subject starts with the Bochner technique in Chapter 4, leading to fundamental L^2 existence theorems for solutions of $\overline{\partial}$-equations in Chapter 5. What makes the theory extremely flexible is the possibility to formulate existence theorems with a wide assortment of different L^2 norms, namely norms of the form $\int_X |f|^2 e^{-2\varphi}$ where φ is a plurisubharmonic or strictly plurisubharmonic function on the given manifold or variety X. Here, the weight φ need not be smooth, and on the contrary, it is extremely important to allow weights which have logarithmic poles of the form $\varphi(z) = c \log \sum |g_j|^2$, where $c > 0$ and (g_j) is a collection of holomorphic functions possessing a common zero set $Z \subset X$. Following Nadel [Nad89], one defines the *multiplier ideal sheaf* $\mathcal{I}(\varphi)$ to be the sheaf of germs of holomorphic functions f such that $|f|^2 e^{-2\varphi}$ is locally summable. Then $\mathcal{I}(\varphi)$ is a coherent algebraic sheaf over X and $H^q(X, K_X \otimes L \otimes \mathcal{I}(\varphi)) = 0$ for all $q \geqslant 1$ if the curvature of L is positive as a current. This important result can be seen as a generalization of the Kawamata-Viehweg vanishing theorem [Kaw82, Vie82], which is one of the cornerstones of higher dimensional algebraic geometry, especially in relation with Mori's minimal model program.

In the dictionary between analytic geometry and algebraic geometry, the ideal $\mathcal{I}(\varphi)$ plays a very important role, since it directly converts an analytic object into an algebraic one, and, simultaneously, takes care of the singularities in a very efficient way. Another analytic tool used to deal with singularities is the theory of positive currents introduced by Lelong [Lel57]. Currents can be seen as generalizations of algebraic cycles, and many classical results of intersection theory still apply to currents. The concept of Lelong number of a current is the analytic analogue of the concept of multiplicity of a germ of algebraic variety. Intersections of cycles correspond to wedge products of currents (whenever these products are defined).

Besides the Kodaira-Nakano vanishing theorem, one of the most basic "effective result" expected to hold in algebraic geometry is expressed in the following conjecture of Fujita [Fuj87]: if L is an ample (i.e. positive) line bundle on a projective n-dimensional algebraic variety X, then $K_X + (n+1)L$ is generated by sections and $K_X + (n+2)L$ is very ample. In the last two decades, a lot of efforts have been brought for the solution of this conjecture — but reaching the expected optimal bounds will probably require new ideas. The first major results are the proof of the Fujita conjecture in the case of surfaces by Reider [Rei88] (the case of curves is easy and has been known since a very long time), and the numerical criterion for the very ampleness of $2K_X + L$ given in [Dem93b], obtained by means of analytic techniques and Monge-Ampère equations with isolated singularities. Alternative algebraic techniques were developed slightly later by Kollár [Kol92], Ein-Lazarsfeld [EL93], Fujita [Fuj93], Siu [Siu95, 96], Kawamata [Kaw97] and Helmke [Hel97]. We will explain here Siu's method because it is technically the simplest method; one of the results obtained by this method is the following effective result: $2K_X + mL$ is very ample for $m \geqslant 2 + \binom{3n+1}{n}$. The basic idea is to apply the Kawamata-Viehweg vanishing theorem, and to combine this with the Riemann-Roch formula in order to produce sections through a clever induction procedure on the dimension of the base loci of the linear systems involved.

Although Siu's result is certainly not optimal, it is sufficient to obtain a nice constructive proof of *Matsusaka's big theorem* [Siu93, Dem96]. The result states that there is an effective value m_0 depending only on the intersection numbers L^n and $L^{n-1} \cdot K_X$, such that mL is very ample for $m \geqslant m_0$. The basic idea is to combine results on the very ampleness of $2K_X + mL$ together with the theory of holomorphic Morse inequalities [Dem85b]. The Morse inequalities are used to construct sections of $m'L - K_X$ for m' large. Again this step can be made algebraic (following suggestions by F. Catanese and R. Lazarsfeld), but the analytic formulation apparently has a wider range of applicability.

In the subsequent chapters, we pursue the study of L^2 estimates, in relation with the Nullstellenstatz and with the extension problem. Skoda [Sko72b, 78] showed that the division problem $f = \sum g_j h_j$ can be solved holomorphically with very precise L^2 estimates, provided that the L^2 norm of $|f|\,|g|^{-p}$ is finite for some sufficiently large exponent p ($p > n = \dim X$ is enough). Skoda's estimates have a nice interpretation in terms of local algebra, and they lead to precise qualitative and quantitative estimates in connection with the Bézout problem. Another very important result is the L^2 extension theorem by Ohsawa-Takegoshi [OT87, Ohs88], which has also been generalized later by Manivel [Man93]. The main statement is that every L^2 section f of a suitably positive line bundle defined on a subavariety $Y \subset X$ can be extended to a L^2 section $\tilde{f}$ defined over the whole of X. The positivity condition can be understood in terms of the canonical sheaf and normal bundle to the subvariety. The extension theorem turns out to have an incredible amount of important consequences: among them, let us mention for instance Siu's theorem [Siu74] on the analyticity of Lelong numbers, the basic approximation theorem of closed positive $(1,1)$-currents by divisors, the subadditivity property $\mathcal{I}(\varphi+\psi) \subset \mathcal{I}(\varphi).\mathcal{I}(\psi)$ of multiplier ideals [DEL00], the restriction formula $\mathcal{I}(\varphi_{|Y}) \subset \mathcal{I}(\varphi)_{|Y}$, ... A suitable combination of these results yields another important result of Fujita [Fuj94] on approximate Zariski decomposition, as we show in Chapter 14.

In Chapter 15, we show how subadditivity can be used to derive an "equisingular" approximation theorem for (almost) plurisubharmonic functions: any such function can be approximated by a sequence of (almost) plurisubharmonic functions which are smooth outside an analytic set, and which define the same multiplier ideal sheaves. From this, we derive a generalized version of the *hard Lefschetz theorem* for cohomology with values in a pseudo-effective line bundle; namely, the Lefschetz map is surjective when the cohomology groups are twisted by the relevant multiplier ideal sheaves.

Chapter 16 explains the proof of Siu's theorem on the invariance of plurigenera, according to a beautiful approach developped by Mihai Păun [Pău07]. The proofs consists of an iterative process based on the Ohsawa-Takegoshi theorem, and a very clever limiting argument for currents.

Chapters 17 and 18 are devoted to the study of positive cones in Kähler or projective geometry. Recent "algebro-analytic" characterizations of the Kähler cone [DP04] and the pseudo-effective cone of divisors [BDPP04] are explained in detail. This leads to a discussion of the important concepts of volume and mobile intersections, following S. Boucksom's PhD work [Bou02]. As a consequence, we show that a projective algebraic manifold has a pseudo-effective canonical line bundle if and only if it is not uniruled.

Chapter 19 presents further important ideas of H. Tsuji, later refined by Berndtsson and Păun, concerning the so-called "super-canonical metrics", and their interpretation in terms of the invariance of plurigenera and of the abundance conjecture. In the concluding Chapter 20, we state Păun's version of the Shokurov-Hacon-McKernan-Siu non vanishing theorem and give an account of the very recent approach of the proof of the finiteness of the canonical ring by Birkar-Păun [BiP09], based on the ideas of Hacon-McKernan and Siu.

Chapter 1

Preliminary Material: Cohomology, Currents

1.A. Dolbeault Cohomology and Sheaf Cohomology

Let X be a $\mathbb{C}$-analytic manifold of dimension n. We denote by $\Lambda^{p,q}T^*_X$ the bundle of differential forms of bidegree (p,q) on X, i.e., differential forms which can be written as

$$u = \sum_{|I|=p,\,|J|=q} u_{I,J} dz_I \wedge d\overline{z}_J.$$

Here $(z_1, \ldots, z_n)$ denote arbitrary local holomorphic coordinates on X, $I = (i_1, \ldots, i_p)$, $J = (j_1, \ldots, j_q)$ are multi-indices (increasing sequences of integers in the range $[1, \ldots, n]$, of lengths $|I| = p$, $|J| = q$), and

$$dz_I := dz_{i_1} \wedge \cdots \wedge dz_{i_p}, \qquad d\overline{z}_J := d\overline{z}_{j_1} \wedge \cdots \wedge d\overline{z}_{j_q}.$$

Let $\mathscr{E}^{p,q}$ be the sheaf of germs of complex valued differential (p,q)-forms with $\mathscr{C}^\infty$ coefficients. Recall that the exterior derivative d splits as $d = d' + d''$ where

$$d'u = \sum_{|I|=p,\,|J|=q,1\leqslant k\leqslant n} \frac{\partial u_{I,J}}{\partial z_k} dz_k \wedge dz_I \wedge d\overline{z}_J,$$

$$d''u = \sum_{|I|=p,\,|J|=q,1\leqslant k\leqslant n} \frac{\partial u_{I,J}}{\partial \overline{z}_k} d\overline{z}_k \wedge dz_I \wedge d\overline{z}_J$$

are of type $(p+1,q)$, $(p,q+1)$ respectively. The well-known Dolbeault-Grothendieck lemma asserts that any d''-closed form of type (p,q) with $q > 0$ is locally d''-exact (this is the analogue for d'' of the usual Poincaré lemma for d, see e.g. [Hör66]). In other words, the complex of sheaves $(\mathscr{E}^{p,\bullet}, d'')$ is exact in degree $q > 0$; in degree $q = 0$, $\operatorname{Ker} d''$ is the sheaf Ω^p_X of germs of holomorphic forms of degree p on X.

More generally, if F is a holomorphic vector bundle of rank r over X, there is a natural d'' operator acting on the space $\mathscr{C}^\infty(X, \Lambda^{p,q}T^*_X \otimes F)$ of smooth (p,q)-forms with values in F; if $s = \sum_{1\leqslant\lambda\leqslant r} s_\lambda e_\lambda$ is a (p,q)-form expressed in terms of a local holomorphic frame of F, we simply define $d''s := \sum d''s_\lambda \otimes e_\lambda$, observing that the holomorphic transition matrices involved in changes of holomorphic frames do not affect the computation of d''. It is then clear that the Dolbeault-Grothendieck lemma still holds for F-valued forms. For every integer $p = 0, 1, \ldots, n$, the *Dolbeault Cohomology* groups $H^{p,q}(X, F)$ are defined to be the cohomology groups of the complex of global (p, q) forms (graded by q):

$$H^{p,q}(X,F) = H^q\big(\mathscr{C}^\infty(X, \Lambda^{p,\bullet}T^*_X \otimes F)\big). \tag{1.1}$$

Now, let us recall the following fundamental result from sheaf theory (De Rham-Weil isomorphism theorem): let $(\mathscr{L}^\bullet, d)$ be a resolution of a sheaf $\mathscr{A}$ by acyclic sheaves, i.e. a complex of sheaves $(\mathscr{L}^\bullet, \delta)$ such that there is an exact sequence of sheaves

$$0 \longrightarrow \mathscr{A} \xrightarrow{\ j\ } \mathscr{L}^0 \xrightarrow{\ \delta^0\ } \mathscr{L}^1 \longrightarrow \cdots \longrightarrow \mathscr{L}^q \xrightarrow{\ \delta^q\ } \mathscr{L}^{q+1} \longrightarrow \cdots,$$

and $H^s(X, \mathscr{L}^q) = 0$ for all $q \geqslant 0$ and $s \geqslant 1$. Then there is a functorial isomorphism

$$H^q\big(\Gamma(X, \mathscr{L}^\bullet)\big) \longrightarrow H^q(X, \mathscr{A}). \tag{1.2}$$

We apply this to the following situation: let $\mathscr{E}(F)^{p,q}$ be the sheaf of germs of $\mathscr{C}^\infty$ sections of $\Lambda^{p,q}T^*_X \otimes F$. Then $(\mathscr{E}(F)^{p,\bullet}, d'')$ is a resolution of the locally free $\mathscr{O}_X$-module $\Omega^p_X \otimes \mathscr{O}(F)$ (Dolbeault-Grothendieck lemma), and the sheaves $\mathscr{E}(F)^{p,q}$ are acyclic as modules over the soft sheaf of rings $\mathscr{C}^\infty$. Hence by (1.2) we get

(1.3) Dolbeault Isomorphism Theorem (1953). *For every holomorphic vector bundle F on X, there is a canonical isomorphism:*

$$H^{p,q}(X, F) \simeq H^q(X, \Omega^p_X \otimes \mathscr{O}(F)).$$

If X is projective algebraic and F is an algebraic vector bundle, Serre's GAGA theorem [Ser56] shows that the algebraic sheaf cohomology group $H^q(X, \Omega^p_X \otimes \mathscr{O}(F))$ computed with algebraic sections over Zariski open sets is actually isomorphic to the analytic cohomology group. These results are the most basic tools to attack algebraic problems via analytic methods. Another important tool is the theory of plurisubharmonic functions and positive currents originated by K. Oka and P. Lelong in the decades 1940–1960.

1.B. Plurisubharmonic Functions

Plurisubharmonic functions have been introduced independently by Lelong and Oka in the study of holomorphic convexity. We refer to [Lel67, 69] for more details.

(1.4) Definition. *A function $u : \Omega \longrightarrow [-\infty, +\infty[$ defined on an open subset $\Omega \subset \mathbb{C}^n$ is said to be plurisubharmonic (psh for short) if*

(a) *u is upper semicontinuous ;*

(b) *for every complex line* $L \subset \mathbb{C}^n$, $u_{\restriction \Omega \cap L}$ *is subharmonic on* $\Omega \cap L$, *that is, for all* $a \in \Omega$ *and* $\xi \in \mathbb{C}^n$ *with* $|\xi| < d(a, \complement\Omega)$, *the function* u *satisfies the mean value inequality:*

$$u(a) \leqslant \frac{1}{2\pi} \int_0^{2\pi} u(a + e^{i\theta}\,\xi)\,d\theta.$$

The set of psh functions on Ω *is denoted by* $\mathrm{Psh}(\Omega)$.

We list below the most basic properties of psh functions. They all follow easily from the definition.

(1.5) Basic Properties.

(a) Every function $u \in \mathrm{Psh}(\Omega)$ is subharmonic, namely it satisfies the mean value inequality on Euclidean balls or spheres:

$$u(a) \leqslant \frac{1}{\pi^n r^{2n}/n!} \int_{B(a,r)} u(z)\,d\lambda(z)$$

for every $a \in \Omega$ and $r < d(a, \complement\Omega)$. Either $u \equiv -\infty$ or $u \in L^1_{\mathrm{loc}}$ on every connected component of Ω.

(b) For any decreasing sequence of psh functions $u_k \in \mathrm{Psh}(\Omega)$, the limit $u = \lim u_k$ is psh on Ω.

(c) Let $u \in \mathrm{Psh}(\Omega)$ be such that $u \not\equiv -\infty$ on every connected component of Ω. If (ρ_ε) is a family of smoothing kernels, then $u * \rho_\varepsilon$ is $\mathscr{C}^\infty$ and psh on

$$\Omega_\varepsilon = \big\{x \in \Omega\,;\, d(x, \complement\Omega) > \varepsilon\big\},$$

the family $(u * \rho_\varepsilon)$ is increasing in ε and $\lim_{\varepsilon \to 0} u * \rho_\varepsilon = u$.

(d) Let $u_1, \ldots, u_p \in \mathrm{Psh}(\Omega)$ and $\chi : \mathbb{R}^p \longrightarrow \mathbb{R}$ be a convex function such that $\chi(t_1, \ldots, t_p)$ is increasing in each t_j. Then $\chi(u_1, \ldots, u_p)$ is psh on Ω. In particular $u_1 + \cdots + u_p$, $\max\{u_1, \ldots, u_p\}$, $\log(e^{u_1} + \cdots + e^{u_p})$ are psh on Ω. □

(1.6) Lemma. *A function* $u \in C^2(\Omega, \mathbb{R})$ *is psh on* Ω *if and only if the Hermitian form:*

$$Hu(a)(\xi) = \sum_{1 \leqslant j,k \leqslant n} \partial^2 u/\partial z_j \partial \overline{z}_k(a)\,\xi_j \overline{\xi}_k$$

is semi-positive at every point $a \in \Omega$.

Proof. This is an easy consequence of the following standard formula:

$$\frac{1}{2\pi} \int_0^{2\pi} u(a + e^{i\theta}\,\xi)\,d\theta - u(a) = \frac{2}{\pi} \int_0^1 \frac{dt}{t} \int_{|\zeta| < t} Hu(a + \zeta\xi)(\xi)\,d\lambda(\zeta),$$

where $d\lambda$ is the Lebesgue measure on $\mathbb{C}$. Lemma 1.6 is a strong evidence that plurisubharmonicity is the natural complex analogue of linear convexity. □

For non smooth functions, a similar characterization of plurisubharmonicity can be obtained by means of a regularization process.

(1.7) Theorem. *If $u \in \mathrm{Psh}(\Omega)$, $u \not\equiv -\infty$ on every connected component of Ω, then for all $\xi \in \mathbb{C}^n$*

$$Hu(\xi) = \sum_{1 \leqslant j,k \leqslant n} \frac{\partial^2 u}{\partial z_j \partial \overline{z}_k} \xi_j \overline{\xi}_k \in \mathcal{D}'(\Omega)$$

is a positive measure. Conversely, if $v \in \mathcal{D}'(\Omega)$ is such that $Hv(\xi)$ is a positive measure for every $\xi \in \mathbb{C}^n$, there exists a unique function $u \in \mathrm{Psh}(\Omega)$ which is locally integrable on Ω and such that v is the distribution associated to u.

In order to get a better geometric insight of this notion, we assume more generally that u is a function on a complex n-dimensional manifold X. If $\Phi : X \to Y$ is a holomorphic mapping and if $v \in C^2(Y, \mathbb{R})$, we have $d'd''(v \circ \Phi) = \Phi^* d'd''v$, hence

$$H(v \circ \Phi)(a, \xi) = Hv\big(\Phi(a), \Phi'(a)\xi\big).$$

In particular Hu, viewed as a Hermitian form on T_X, does not depend on the choice of coordinates $(z_1, \ldots, z_n)$. Therefore, the notion of psh function makes sense on any complex manifold. More generally, we have

(1.8) Proposition. *If $\Phi : X \longrightarrow Y$ is a holomorphic map and $v \in \mathrm{Psh}(Y)$, then $v \circ \Phi \in \mathrm{Psh}(X)$.*

(1.9) Example. It is a standard fact that $\log|z|$ is psh (i.e. subharmonic) on $\mathbb{C}$. Thus $\log|f| \in \mathrm{Psh}(X)$ for every holomorphic function $f \in H^0(X, \mathcal{O}_X)$. More generally

$$\log\big(|f_1|^{\alpha_1} + \cdots + |f_q|^{\alpha_q}\big) \in \mathrm{Psh}(X)$$

for every $f_j \in H^0(X, \mathcal{O}_X)$ and $\alpha_j \geqslant 0$ (apply Property 1.5 (d) with $u_j = \alpha_j \log|f_j|$). We will be especially interested in the singularities obtained at points of the zero variety $f_1 = \cdots = f_q = 0$, when the α_j are rational numbers. □

(1.10) Definition. *A psh function $u \in \mathrm{Psh}(X)$ will be said to have analytic singularities if u can be written locally as*

$$u = \frac{\alpha}{2} \log\big(|f_1|^2 + \cdots + |f_N|^2\big) + v,$$

where $\alpha \in \mathbb{R}_+$, v is a locally bounded function and the f_j are holomorphic functions. If X is algebraic, we say that u has algebraic singularities if u can be written as above on sufficiently small Zariski open sets, with $\alpha \in \mathbb{Q}_+$ and f_j algebraic.

We then introduce the ideal $\mathcal{J} = \mathcal{J}(u/\alpha)$ of germs of holomorphic functions h such that $|h| \leqslant Ce^{u/\alpha}$ for some constant C, i.e.

$$|h| \leqslant C\big(|f_1| + \cdots + |f_N|\big).$$

This is a globally defined ideal sheaf on X, locally equal to the integral closure $\overline{\mathcal{J}}$ of the ideal sheaf $\mathcal{J} = (f_1, \ldots, f_N)$, thus $\mathcal{J}$ is coherent on X. If $(g_1, \ldots, g_{N'})$ are local generators of $\mathcal{J}$, we still have

$$u = \frac{\alpha}{2} \log\big(|g_1|^2 + \cdots + |g_{N'}|^2\big) + O(1).$$

If X is projective algebraic and u has analytic singularities with $\alpha \in \mathbb{Q}_+$, then u automatically has algebraic singularities. From an algebraic point of view, the singularities of u are in 1:1 correspondence with the "algebraic data" $(\mathcal{J}, \alpha)$. Later on, we will see another important method for associating an ideal sheaf to a psh function.

(1.11) Exercise. Show that the above definition of the integral closure of an ideal $\mathcal{J}$ is equivalent to the following more algebraic definition: $\overline{\mathcal{J}}$ consists of all germs h satisfying an integral equation:

$$h^d + a_1 h^{d-1} + \cdots + a_{d-1} h + a_d = 0, \qquad a_k \in \mathcal{J}^k.$$

Hint. One inclusion is clear. To prove the other inclusion, consider the normalization of the blow-up of X along the (non necessarily reduced) zero variety $V(\mathcal{J})$. □

1.C. Positive Currents

The reader can consult [Fed69] for a more thorough treatment of current theory. Let us first recall a few basic definitions. A *current* of degree q on an oriented differentiable manifold M is simply a differential q-form Θ with distribution coefficients. The space of currents of degree q over M will be denoted by $\mathcal{D}'^q(M)$. Alternatively, a current of degree q can be seen as an element Θ in the dual space $\mathcal{D}'_p(M) := \big(\mathcal{D}^p(M)\big)'$ of the space $\mathcal{D}^p(M)$ of smooth differential forms of degree $p = \dim M - q$ with compact support; the duality pairing is given by

(1.12) $$\langle \Theta, \alpha \rangle = \int_M \Theta \wedge \alpha, \quad \alpha \in \mathcal{D}^p(M).$$

A basic example is the *current of integration* $[S]$ over a compact oriented submanifold S of M:

(1.13) $$\langle [S], \alpha \rangle = \int_S \alpha, \quad \deg\alpha = p = \dim_{\mathbb{R}} S.$$

Then $[S]$ is a current with measure coefficients, and Stokes' formula shows that $d[S] = (-1)^{q-1}[\partial S]$, in particular $d[S] = 0$ if S has no boundary. Because of this example, the integer p is said to be the dimension of Θ when $\Theta \in \mathcal{D}'_p(M)$. The current Θ is said to be *closed* if $d\Theta = 0$.

On a complex manifold X, we have similar notions of bidegree and bidimension; as in the real case, we denote by

$$\mathscr{D}'^{p,q}(X) = \mathscr{D}'_{n-p,n-q}(X), \qquad n = \dim X,$$

the space of currents of bidegree (p,q) and bidimension $(n-p,n-q)$ on X. According to [Lel57], a current Θ of bidimension (p,p) is said to be (*weakly*) *positive* if for every choice of smooth $(1,0)$-forms $\alpha_1, \ldots, \alpha_p$ on X the distribution

(1.14) $\qquad \Theta \wedge \mathrm{i}\alpha_1 \wedge \overline{\alpha}_1 \wedge \cdots \wedge \mathrm{i}\alpha_p \wedge \overline{\alpha}_p$ is a positive measure.

(1.15) Exercise. If Θ is positive, show that the coefficients $\Theta_{I,J}$ of Θ are complex measures, and that, up to constants, they are dominated by the trace measure:

$$\sigma_\Theta = \Theta \wedge \frac{1}{p!}\beta^p = 2^{-p}\sum \Theta_{I,I}, \qquad \beta = \frac{\mathrm{i}}{2}d'd''|z|^2 = \frac{\mathrm{i}}{2}\sum_{1\leqslant j\leqslant n} dz_j \wedge d\overline{z}_j,$$

which is a positive measure.
Hint. Observe that $\sum \Theta_{I,I}$ is invariant by unitary changes of coordinates and that the (p,p)-forms $\mathrm{i}\alpha_1 \wedge \overline{\alpha}_1 \wedge \cdots \wedge \mathrm{i}\alpha_p \wedge \overline{\alpha}_p$ generate $\Lambda^{p,p}T^*_{\mathbb{C}^n}$ as a $\mathbb{C}$-vector space. □

A current $\Theta = \mathrm{i}\sum_{1\leqslant j,k\leqslant n} \Theta_{jk} dz_j \wedge d\overline{z}_k$ of bidegree $(1,1)$ is easily seen to be positive if and only if the complex measure $\sum \lambda_j \overline{\lambda}_k \Theta_{jk}$ is a positive measure for every n-tuple $(\lambda_1, \ldots, \lambda_n) \in \mathbb{C}^n$.

(1.16) Example. If u is a (not identically $-\infty$) psh function on X, we can associate with u a (closed) positive current $\Theta = \mathrm{i}\partial\overline{\partial}u$ of bidegree $(1,1)$. Conversely, each closed positive current of bidegree $(1,1)$ can be written under this form on any open subset $\Omega \subset X$ such that $H^2_{DR}(\Omega,\mathbb{R}) = H^1(\Omega,\mathcal{O}) = 0$, e.g. on small coordinate balls (exercise to the reader). □

It is not difficult to show that a product $\Theta_1 \wedge \cdots \wedge \Theta_q$ of positive currents of bidegree $(1,1)$ is positive whenever the product is well defined (this is certainly the case if all Θ_j but one at most are smooth; much finer conditions will be discussed in Chapter 2).

We now discuss another very important example of closed positive current. In fact, with every closed analytic set $A \subset X$ of pure dimension p is associated a current of integration $[A]$ such that:

(1.17) $$\langle [A], \alpha \rangle = \int_{A_{\mathrm{reg}}} \alpha, \qquad \alpha \in \mathscr{D}^{p,p}(X),$$

obtained by integrating over the regular points of A. In order to show that (1.17) is a correct definition of a current on X, one must show that A_{reg} has locally finite area in a neighborhood of A_{sing}. This result, due to [Lel57] is shown as follows.

Suppose that 0 is a singular point of A. By the local parametrization theorem for analytic sets, there is a linear change of coordinates on $\mathbb{C}^n$ such that all projections

$$\pi_I : (z_1, \ldots, z_n) \mapsto (z_{i_1}, \ldots, z_{i_p})$$

define a finite ramified covering of the intersection $A \cap \Delta$ with a small polydisk Δ in $\mathbb{C}^n$ onto a small polydisk Δ_I in $\mathbb{C}^p$. Let n_I be the sheet number. Then the p-dimensional area of $A \cap \Delta$ is bounded above by the sum of the areas of its projections counted with multiplicities, i.e.

$$\mathrm{Area}(A \cap \Delta) \leqslant \sum n_I \mathrm{Vol}(\Delta_I).$$

The fact that $[A]$ is positive is also easy. In fact

$$i\alpha_1 \wedge \overline{\alpha}_1 \wedge \cdots \wedge i\alpha_p \wedge \overline{\alpha}_p = |\det(\alpha_{jk})|^2\, iw_1 \wedge \overline{w}_1 \wedge \cdots \wedge iw_p \wedge \overline{w}_p$$

if $\alpha_j = \sum \alpha_{jk} dw_k$ in terms of local coordinates $(w_1, \ldots, w_p)$ on A_{reg}. This shows that all such forms are $\geqslant 0$ in the canonical orientation defined by $iw_1 \wedge \overline{w}_1 \wedge \cdots \wedge iw_p \wedge \overline{w}_p$. More importantly, Lelong [Lel57] has shown that $[A]$ is d-closed in X, even at points of A_{sing}. This last result can be seen today as a consequence of the Skoda-El Mir extension theorem. For this we need the following definition: a *complete pluripolar* set is a set E such that there is an open covering (Ω_j) of X and psh functions u_j on Ω_j with $E \cap \Omega_j = u_j^{-1}(-\infty)$. Any (closed) analytic set is of course complete pluripolar (take u_j as in Example 1.9).

(1.18) Theorem (Skoda [Sko82], El Mir [EM84], Sibony [Sib85]). *Let E be a closed complete pluripolar set in X, and let Θ be a closed positive current on $X \smallsetminus E$ such that the coefficients $\Theta_{I,J}$ of Θ are measures with locally finite mass near E. Then the trivial extension $\tilde{\Theta}$ obtained by extending the measures $\Theta_{I,J}$ by 0 on E is still closed on X.*

Lelong's result $d[A] = 0$ is obtained by applying the Skoda-El Mir theorem to $\Theta = [A_{\mathrm{reg}}]$ on $X \smallsetminus A_{\mathrm{sing}}$.

Proof of Theorem 1.18. The statement is local on X, so we may work on a small open set Ω such that $E \cap \Omega = v^{-1}(-\infty)$, $v \in \mathrm{Psh}(\Omega)$. Let $\chi : \mathbb{R} \to \mathbb{R}$ be a convex increasing function such that $\chi(t) = 0$ for $t \leqslant -1$ and $\chi(0) = 1$. By shrinking Ω and putting $v_k = \chi(k^{-1} v * \rho_{\varepsilon_k})$ with $\varepsilon_k \to 0$ fast, we get a sequence of functions $v_k \in \mathrm{Psh}(\Omega) \cap \mathscr{C}^\infty(\Omega)$ such that $0 \leqslant v_k \leqslant 1$, $v_k = 0$ in a neighborhood of $E \cap \Omega$ and $\lim v_k(x) = 1$ at every point of $\Omega \smallsetminus E$. Let $\theta \in \mathscr{C}^\infty([0,1])$ be a function such that $\theta = 0$ on $[0, 1/3]$, $\theta = 1$ on $[2/3, 1]$ and $0 \leqslant \theta \leqslant 1$. Then $\theta \circ v_k = 0$ near $E \cap \Omega$ and $\theta \circ v_k \to 1$ on $\Omega \smallsetminus E$. Therefore $\tilde{\Theta} = \lim_{k \to +\infty} (\theta \circ v_k)\Theta$ and

$$d'\tilde{\Theta} = \lim_{k \to +\infty} \Theta \wedge d'(\theta \circ v_k)$$

in the weak topology of currents. It is therefore sufficient to verify that $\Theta \wedge d'(\theta \circ v_k)$ converges weakly to 0 (note that $d''\tilde{\Theta}$ is conjugate to $d'\tilde{\Theta}$, thus $d''\tilde{\Theta}$ will also vanish).

Assume first that $\Theta \in \mathcal{D}'^{n-1,n-1}(X)$. Then $\Theta \wedge d'(\theta \circ v_k) \in \mathcal{D}'^{n,n-1}(\Omega)$, and we have to show that

$$\langle \Theta \wedge d'(\theta \circ v_k), \overline{\alpha}\rangle = \langle \Theta, \theta'(v_k) d' v_k \wedge \overline{\alpha}\rangle \underset{k\to+\infty}{\longrightarrow} 0, \qquad \forall \alpha \in \mathcal{D}^{1,0}(\Omega).$$

As $\gamma \mapsto \langle \Theta, \mathrm{i}\gamma \wedge \overline{\gamma}\rangle$ is a non-negative Hermitian form on $\mathcal{D}^{1,0}(\Omega)$, the Cauchy-Schwarz inequality yields

$$\big|\langle \Theta, \mathrm{i}\beta \wedge \overline{\gamma}\rangle\big|^2 \leqslant \langle \Theta, \mathrm{i}\beta \wedge \overline{\beta}\rangle \ \langle \Theta, \mathrm{i}\gamma \wedge \overline{\gamma}\rangle, \qquad \forall \beta, \gamma \in \mathcal{D}^{1,0}(\Omega).$$

Let $\psi \in \mathcal{D}(\Omega)$, $0 \leqslant \psi \leqslant 1$, be equal to 1 in a neighborhood of $\operatorname{Supp}\alpha$. We find

$$\big|\langle \Theta, \theta'(v_k) d' v_k \wedge \overline{\alpha}\rangle\big|^2 \leqslant \langle \Theta, \psi \mathrm{i} d' v_k \wedge d'' v_k\rangle \ \langle \Theta, \theta'(v_k)^2 \mathrm{i}\alpha \wedge \overline{\alpha}\rangle.$$

By hypothesis $\int_{\Omega \smallsetminus E} \Theta \wedge \mathrm{i}\alpha \wedge \overline{\alpha} < +\infty$ and $\theta'(v_k)$ converges everywhere to 0 on Ω, thus $\langle \Theta, \theta'(v_k)^2 \mathrm{i}\alpha \wedge \overline{\alpha}\rangle$ converges to 0 by Lebesgue's dominated convergence theorem. On the other hand,

$$\begin{aligned} \mathrm{i}d'd''v_k^2 &= 2v_k\, \mathrm{i}d'd''v_k + 2\mathrm{i}d'v_k \wedge d''v_k \geqslant 2\mathrm{i}d'v_k \wedge d''v_k, \\ &2\langle \Theta, \psi \mathrm{i} d' v_k \wedge d'' v_k\rangle \leqslant \langle \Theta, \psi \mathrm{i} d'd'' v_k^2\rangle. \end{aligned}$$

As $\psi \in \mathcal{D}(\Omega)$, $v_k = 0$ near E and $d\Theta = 0$ on $\Omega \smallsetminus E$, an integration by parts yields

$$\langle \Theta, \psi \mathrm{i} d'd'' v_k^2\rangle = \langle \Theta, v_k^2 \mathrm{i} d'd''\psi\rangle \leqslant C \int_{\Omega \smallsetminus E} \|\Theta\| < +\infty,$$

where C is a bound for the coefficients of $\mathrm{i}d'd''\psi$. Thus $\langle \Theta, \psi \mathrm{i} d' v_k \wedge d'' v_k\rangle$ is bounded, and the proof is complete when $\Theta \in \mathcal{D}'^{n-1,n-1}$.

In the general case $\Theta \in \mathcal{D}'^{p,p}$, $p < n$, we simply apply the result already proved to all positive currents $\Theta \wedge \gamma \in \mathcal{D}'^{n-1,n-1}$ where $\gamma = \mathrm{i}\gamma_1 \wedge \overline{\gamma}_1 \wedge \cdots \wedge \mathrm{i}\gamma_{n-p-1} \wedge \overline{\gamma}_{n-p-1}$ runs over a basis of forms of $\Lambda^{n-p-1,n-p-1} T^*_\Omega$ with constant coefficients. Then we get $d(\widetilde{\Theta} \wedge \gamma) = d\widetilde{\Theta} \wedge \gamma = 0$ for all such γ, hence $d\widetilde{\Theta} = 0$. □

(1.19) Corollary. *Let Θ be a closed positive current on X and let E be a complete pluripolar set. Then $\mathbb{1}_E \Theta$ and $\mathbb{1}_{X \smallsetminus E}\Theta$ are closed positive currents. In fact, $\widetilde{\Theta} = \mathbb{1}_{X \smallsetminus E}\Theta$ is the trivial extension of $\Theta_{\restriction X \smallsetminus E}$ to X, and $\mathbb{1}_E \Theta = \Theta - \widetilde{\Theta}$.*

As mentioned above, any current $\Theta = \mathrm{i}d'd''u$ associated with a psh function u is a closed positive $(1,1)$-current. In the special case $u = \log|f|$ where $f \in H^0(X, \mathcal{O}_X)$ is a non zero holomorphic function, we have the important

(1.20) Lelong-Poincaré Equation. *Let $f \in H^0(X, \mathcal{O}_X)$ be a non zero holomorphic function, $Z_f = \sum m_j Z_j$, $m_j \in \mathbb{N}$, the zero divisor of f and $[Z_f] = \sum m_j [Z_j]$ the associated current of integration. Then*

$$\frac{\mathrm{i}}{\pi} \partial\overline{\partial} \log|f| = [Z_f].$$

Proof (sketch). It is clear that $\mathrm{i}d'd''\log|f| = 0$ in a neighborhood of every point $x \notin \mathrm{Supp}(Z_f) = \bigcup Z_j$, so it is enough to check the equation in a neighborhood of every point of $\mathrm{Supp}(Z_f)$. Let A be the set of singular points of $\mathrm{Supp}(Z_f)$, i.e. the union of the pairwise intersections $Z_j \cap Z_k$ and of the singular loci $Z_{j,\mathrm{sing}}$; we thus have $\dim A \leqslant n-2$. In a neighborhood of any point $x \in \mathrm{Supp}(Z_f) \smallsetminus A$ there are local coordinates $(z_1, \ldots, z_n)$ such that $f(z) = z_1^{m_j}$ where m_j is the multiplicity of f along the component Z_j which contains x and $z_1 = 0$ is an equation for Z_j near x. Hence

$$\frac{\mathrm{i}}{\pi}d'd''\log|f| = m_j\frac{\mathrm{i}}{\pi}d'd''\log|z_1| = m_j[Z_j]$$

in a neighborhood of x, as desired (the identity comes from the standard formula $\frac{\mathrm{i}}{\pi}d'd''\log|z| =$ Dirac measure δ_0 in $\mathbb{C}$). This shows that the equation holds on $X \smallsetminus A$. Hence the difference $\frac{\mathrm{i}}{\pi}d'd''\log|f| - [Z_f]$ is a closed current of degree 2 with measure coefficients, whose support is contained in A. By Exercise 1.21, this current must be 0, for A has too small dimension to carry its support (A is stratified by submanifolds of real codimension $\geqslant 4$). □

(1.21) Exercise. Let Θ be a current of degree q on a real manifold M, such that both Θ and $d\Theta$ have measure coefficients ("normal current"). Suppose that $\mathrm{Supp}\,\Theta$ is contained in a real submanifold A with $\mathrm{codim}_{\mathbb{R}} A > q$. Show that $\Theta = 0$. *Hint:* Let $m = \dim_{\mathbb{R}} M$ and let $(x_1, \cdots, x_m)$ be a coordinate system in a neighborhood Ω of a point $a \in A$ such that $A \cap \Omega = \{x_1 = \ldots = x_k = 0\}$, $k > q$. Observe that $x_j\Theta = x_j d\Theta = 0$ for $1 \leqslant j \leqslant k$, thanks to the hypothesis on supports and on the normality of Θ, hence $dx_j \wedge \Theta = d(x_j\Theta) - x_j d\Theta = 0$, $1 \leqslant j \leqslant k$. Infer from this that all coefficients in $\Theta = \sum_{|I|=q}\Theta_I dx_I$ vanish. □

We now recall a few basic facts of slicing theory (the reader will profitably consult [Fed69] and [Siu74] for further developments). Let $\sigma : M \to M'$ be a submersion of smooth differentiable manifolds and let Θ be a *locally flat* current on M, that is, a current which can be written locally as $\Theta = U + dV$ where U, V have L^1_{loc} coefficients. It is a standard fact (see Federer) that every current Θ such that both Θ and $d\Theta$ have measure coefficients is locally flat; in particular, closed positive currents are locally flat. Then, for almost every $x' \in M'$, there is a well defined *slice* $\Theta_{x'}$, which is the current on the fiber $\sigma^{-1}(x')$ defined by

$$\Theta_{x'} = U_{\restriction\sigma^{-1}(x')} + dV_{\restriction\sigma^{-1}(x')}.$$

The restrictions of U, V to the fibers exist for almost all x' by the Fubini theorem. The slices $\Theta_{x'}$ are currents on the fibers with the same degree as Θ (thus of dimension $\dim\Theta - \dim(\text{fibers})$). Of course, every slice $\Theta_{x'}$ coincides with the usual restriction of Θ to the fiber if Θ has smooth coefficients. By using a regularization $\Theta_\varepsilon = \Theta * \rho_\varepsilon$, it is easy to show that the slices of a closed positive current are again closed and positive: in fact $U_{\varepsilon,x'}$ and $V_{\varepsilon,x'}$ converge to $U_{x'}$ and $V_{x'}$ in $L^1_{\mathrm{loc}}(\sigma^{-1}(x'))$, thus $\Theta_{\varepsilon,x'}$ converges weakly to $\Theta_{x'}$ for almost every x'. Now, the basic slicing formula is

$$(1.22)\qquad \int_M \Theta \wedge \alpha \wedge \sigma^*\beta = \int_{x' \in M'}\Big(\int_{x'' \in \sigma^{-1}(x')}\Theta_{x'}(x'') \wedge \alpha_{\restriction\sigma^{-1}(x')}(x'')\Big)\beta(x')$$

for every smooth form α on M and β on M', such that α has compact support and $\deg\alpha = \dim M - \dim M' - \deg\Theta$, $\deg\beta = \dim M'$. This is an easy consequence of the usual Fubini theorem applied to U and V in the decomposition $\Theta = U + dV$, if we identify locally σ with a projection map:

$$M = M' \times M'' \to M', \qquad x = (x', x'') \mapsto x',$$

and use a partition of unity on the support of α.

To conclude this section, we discuss De Rham and Dolbeault cohomology theory in the context of currents. A basic observation is that the Poincaré and Dolbeault-Grothendieck lemmas still hold for currents. Namely, if $(\mathscr{D}'^q, d)$ and $(\mathscr{D}'(F)^{p,q}, d'')$ denote the complex of sheaves of degree q currents (resp. of (p,q)-currents with values in a holomorphic vector bundle F), we still have De Rham and Dolbeault sheaf resolutions:

$$0 \to \mathbb{R} \to \mathscr{D}'^\bullet, \qquad 0 \to \Omega^p_X \otimes \mathscr{O}(F) \to \mathscr{D}'(F)^{p,\bullet}.$$

Hence we get canonical isomorphisms

$$\begin{aligned} H^q_{\mathrm{DR}}(M,\mathbb{R}) &= H^q\big((\Gamma(M,\mathscr{D}'^\bullet), d)\big),\\ H^{p,q}(X,F) &= H^q\big((\Gamma(X,\mathscr{D}'(F)^{p,\bullet}), d'')\big). \end{aligned} \tag{1.23}$$

In other words, we can attach a cohomology class $\{\Theta\} \in H^q_{\mathrm{DR}}(M,\mathbb{R})$ to any closed current Θ of degree q, resp. a cohomology class $\{\Theta\} \in H^{p,q}(X,F)$ to any d''-closed current of bidegree (p,q). Replacing if necessary every current by a smooth representative in the same cohomology class, we see that there is a well defined cup product given by the wedge product of differential forms:

$$\begin{aligned} H^{q_1}(M,\mathbb{R}) \times \cdots \times H^{q_m}(M,\mathbb{R}) &\longrightarrow H^{q_1+\cdots+q_m}(M,\mathbb{R}),\\ (\{\Theta_1\}, \ldots, \{\Theta_1\}) &\longmapsto \{\Theta_1\} \wedge \cdots \wedge \{\Theta_m\}. \end{aligned}$$

In particular, if M is a compact oriented variety and $q_1 + \cdots + q_m = \dim M$, there is a well defined intersection number:

$$\{\Theta_1\} \cdot \{\Theta_2\} \cdot \cdots \cdot \{\Theta_m\} = \int_M \{\Theta_1\} \wedge \cdots \wedge \{\Theta_m\}.$$

However, as we will see in the next section, the pointwise product $\Theta_1 \wedge \cdots \wedge \Theta_m$ need not exist in general.

Chapter 2

Lelong numbers and Intersection Theory

Lelong numbers were historically defined around 1960 as density numbers of positive currents, and were quickly realized to be natural generalizations of the concept of multiplicity in algebraic geometric. As emphasized e.g. in [Dem82a, 85a, 87], they then started to be viewed rather as a special case of a general intersection theory for closed positive currents. We will adopt here this viewpoint.

2.A. Multiplication of Currents and Monge-Ampère Operators

Let X be a n-dimensional complex manifold. We set

$$d^c = \frac{1}{2i\pi}(d' - d'').$$

It follows in particular that d^c is a real operator, i.e. $\overline{d^c u} = d^c \overline{u}$, and that $dd^c = \frac{i}{\pi} d'd''$. Although not quite standard, the $1/2i\pi$ normalization is very convenient for many purposes, since we may then forget the factor π or 2π almost everywhere (e.g. in the Lelong-Poincaré Equation (1.20)).

Let u be a psh function and let Θ be a closed positive current on X. Our desire is to define the wedge product $dd^c u \wedge \Theta$ even when neither u nor Θ are smooth. In general, this product does not make sense because $dd^c u$ and Θ have measure coefficients and measures cannot be multiplied; see Kiselman [Kis84] for interesting counterexamples. Even in the algebraic setting considered here, multiplication of currents is not always possible: suppose e.g. that $\Theta = [D]$ is the exceptional divisor of a blow-up in a surface; then $D \cdot D = -1$ cannot be the cohomology class of a closed positive current $[D]^2$. Assume however that u is a *locally bounded* psh function. Then the current $u\Theta$ is well defined since u is a locally bounded Borel function and Θ has measure coefficients. According to Bedford-Taylor [BT82] we define

$$dd^c u \wedge \Theta = dd^c(u\Theta)$$

where $dd^c(\ )$ is taken in the sense of distribution theory.

(2.1) Proposition. *If u is a locally bounded psh function, the wedge product $dd^c u \wedge \Theta$ is again a closed positive current.*

Proof. The result is local. Use a convolution $u_\nu = u * \rho_{1/\nu}$ to get a decreasing sequence of smooth psh functions converging to u. Then write

$$dd^c(u\Theta) = \lim_{\nu\to+\infty} dd^c(u_\nu\Theta) = \lim_{\nu\to+\infty} dd^c u_\nu \wedge \Theta$$

as a weak limit of closed positive currents. Observe that $u_\nu\Theta$ converges weakly to $u\Theta$ by Lebesgue's monotone convergence theorem. □

More generally, if $u_1, \ldots, u_m$ are locally bounded psh functions, we can define

$$dd^c u_1 \wedge dd^c u_1 \cdots \wedge dd^c u_m \wedge \Theta = dd^c\big(u_1 dd^c u_2 \wedge \cdots \wedge dd^c u_m \wedge \Theta\big)$$

by induction on m. Chern, Levine and Nirenberg [CLN69] noticed the following useful inequality. Define the *mass* of a current Θ on a compact set K to be

$$||\Theta||_K = \int_K \sum_{I,J} |\Theta_{I,J}|$$

whenever K is contained in a coordinate patch and $\Theta = \sum \Theta_{I,J} dz_I \wedge d\overline{z}_J$. Up to seminorm equivalence, this does not depend on the choice of coordinates. If K is not contained in a coordinate patch, we use a partition of unity to define a suitable seminorm $||\Theta||_K$. If $\Theta \geqslant 0$, Exercise 1.15 shows that the mass is controlled by the trace measure, i.e. $||\Theta||_K \leqslant C \int_K \Theta \wedge \beta^p$.

(2.2) Chern-Levine-Nirenberg Inequality. *For all compact subsets K, L of X with $L \subset K^\circ$, there exists a constant $C_{K,L} \geqslant 0$ such that*

$$||dd^c u_1 \wedge \cdots \wedge dd^c u_m \wedge \Theta||_L \leqslant C_{K,L}\ ||u_1||_{L^\infty(K)} \cdots ||u_m||_{L^\infty(K)}\ ||\Theta||_K$$

Proof. By induction, it is sufficient to prove the result for $m = 1$ and $u_1 = u$. There is a covering of L by a family of open balls $B'_j \subset\subset B_j \subset K$ contained in coordinate patches of X. Let (p, p) be the bidimension of Θ, let $\beta = \frac{\mathrm{i}}{2} d'd''|z|^2$, and let $\chi \in \mathcal{D}(B_j)$ be equal to 1 on $\overline{B}'_j$. Then

$$||dd^c u \wedge \Theta||_{L\cap\overline{B}'_j} \leqslant C \int_{\overline{B}'_j} dd^c u \wedge \Theta \wedge \beta^{p-1} \leqslant C \int_{B_j} \chi\, dd^c u \wedge \Theta \wedge \beta^{p-1}.$$

As Θ and β are closed, an integration by parts yields

$$||dd^c u \wedge \Theta||_{L\cap\overline{B}'_j} \leqslant C \int_{B_j} u\,\Theta \wedge dd^c\chi \wedge \beta^{p-1} \leqslant C'||u||_{L^\infty(K)}||\Theta||_K$$

where C' is equal to C multiplied by a bound for the coefficients of the smooth form $dd^c\chi \wedge \beta^{p-1}$. □

Various examples (cf. [Kis84]) show however that products of $(1,1)$-currents $dd^c u_j$ cannot be defined in a reasonable way for arbitrary psh functions u_j. However, functions u_j with $-\infty$ poles can be admitted if the polar sets are sufficiently small.

(2.3) Proposition. *Let u be a psh function on X, and let Θ be a closed positive current of bidimension (p,p). Suppose that u is locally bounded on $X \smallsetminus A$, where A is an analytic subset of X of dimension $< p$ at each point. Then $dd^c u \wedge \Theta$ can be defined in such a way that $dd^c u \wedge \Theta = \lim_{\nu \to +\infty} dd^c u_\nu \wedge \Theta$ in the weak topology of currents, for any decreasing sequence $(u_\nu)_{\nu \geqslant 0}$ of psh functions converging to u.*

Proof. When u is locally bounded everywhere, we have $\lim u_\nu \Theta = u\,\Theta$ by the monotone convergence theorem and the result follows from the continuity of dd^c with respect to the weak topology.

First assume that A is discrete. Since our results are local, we may suppose that X is a ball $B(0,R) \subset \mathbb{C}^n$ and that $A = \{0\}$. For every $s \leqslant 0$, the function $u^{\geqslant s} = \max(u,s)$ is locally bounded on X, so the product $\Theta \wedge dd^c u^{\geqslant s}$ is well defined. For $|s|$ large, the function $u^{\geqslant s}$ differs from u only in a small neighborhood of the origin, at which u may have a $-\infty$ pole. Let γ be a $(p-1,p-1)$-form with constant coefficients and set $s(r) = \liminf_{|z| \to r-0} u(z)$. By Stokes' formula, we see that the integral

$$I(s) := \int_{B(0,r)} dd^c u^{\geqslant s} \wedge \Theta \wedge \gamma \tag{2.4}$$

does not depend on s when $s < s(r)$, for the difference $I(s) - I(s')$ of two such integrals involves the dd^c of a current $(u^{\geqslant s} - u^{\geqslant s'}) \wedge \Theta \wedge \gamma$ with compact support in $B(0,r)$. Taking $\gamma = (dd^c|z|^2)^{p-1}$, we see that the current $dd^c u \wedge \Theta$ has finite mass on $B(0,r) \smallsetminus \{0\}$ and we can define $\langle \mathbb{1}_{\{0\}}(dd^c u \wedge \Theta), \gamma \rangle$ to be the limit of the integrals (2.4) as r tends to zero and $s < s(r)$. In this case, the weak convergence statement is easily deduced from the locally bounded case discussed above.

In the case where $0 < \dim A < p$, we use a slicing technique to reduce the situation to the discrete case. Set $q = p-1$. There are linear coordinates $(z_1, \ldots, z_n)$ centered at any point of A, such that 0 is an isolated point of $A \cap (\{0\} \times \mathbb{C}^{n-q})$. Then there are small balls $B' = B(0,r')$ in $\mathbb{C}^q$, $B'' = B(0,r'')$ in $\mathbb{C}^{n-q}$ such that $A \cap (\overline{B}' \times \partial B'') = \emptyset$, and the projection map

$$\pi : \mathbb{C}^n \to \mathbb{C}^q, \qquad z = (z_1, \ldots, z_n) \mapsto z' = (z_1, \ldots, z_q)$$

defines a finite proper mapping $A \cap (B' \times B'') \to B'$. These properties are preserved if we slightly change the direction of projection. Take sufficiently many projections π_m associated to coordinate systems $(z_1^m, \ldots, z_n^m)$, $1 \leqslant m \leqslant N$, in such a way that the family of (q,q)-forms

$$\mathrm{i}\, dz_1^m \wedge d\overline{z}_1^m \wedge \cdots \wedge \mathrm{i}\, dz_q^m \wedge d\overline{z}_q^m$$

defines a basis of the space of (q,q)-forms. Expressing any compactly supported smooth (q,q)-form in such a basis, we see that we need only define

$$(2.5)\quad \int_{B'\times B''} dd^c u\wedge\Theta\wedge f(z',z'')\,\mathrm{i}\,dz_1\wedge d\overline{z}_1\wedge\cdots\wedge\mathrm{i}\,dz_q\wedge d\overline{z}_q$$
$$=\int_{B'}\Big\{\int_{B''} f(z',\bullet)\,dd^c u(z',\bullet)\wedge\Theta(z',\bullet)\Big\}\mathrm{i}\,dz_1\wedge d\overline{z}_1\wedge\cdots\wedge\mathrm{i}\,dz_q\wedge d\overline{z}_q$$

where f is a test function with compact support in $B'\times B''$, and $\Theta(z',\bullet)$ denotes the slice of Θ on the fiber $\{z'\}\times B''$ of the projection $\pi:\mathbb{C}^n\to\mathbb{C}^q$. Each integral $\int_{B''}$ in the right hand side of (2.5) makes sense since the slices $(\{z'\}\times B'')\cap A$ are discrete. Moreover, the double integral $\int_{B'}\int_{B''}$ is convergent. Indeed, observe that u is bounded on any compact cylinder:

$$K_{\delta,\varepsilon}=\overline{B}\big((1-\delta)r'\big)\times\Big(\overline{B}(r'')\smallsetminus\overline{B}\big((1-\varepsilon)r''\big)\Big)$$

disjoint from A. Take $\varepsilon\ll\delta\ll 1$ so small that

$$\operatorname{Supp} f\subset\overline{B}\big((1-\delta)r'\big)\times\overline{B}\big((1-\varepsilon)r''\big).$$

For all $z'\in\overline{B}((1-\delta)r')$, the proof of the Chern-Levine-Nirenberg inequality (2.2) with a cut-off function $\chi(z'')$ equal to 1 on $B((1-\varepsilon)r'')$ and with support in $B((1-\varepsilon/2)r'')$ shows that

$$\int_{B((1-\varepsilon)r'')} dd^c u(z',\bullet)\wedge\Theta(z',\bullet)$$
$$\leqslant C_\varepsilon\|u\|_{L^\infty(K_{\delta,\varepsilon})}\int_{z''\in B((1-\varepsilon/2)r'')}\Theta(z',z'')\wedge dd^c|z''|^2.$$

This implies that the double integral is convergent. Now replace u everywhere by u_ν and observe that $\lim_{\nu\to+\infty}\int_{B''}$ is the expected integral for every z' such that $\Theta(z',\bullet)$ exists (apply the discrete case already proven). Moreover, the Chern-Levine-Nirenberg inequality yields uniform bounds for all functions u_ν, hence Lebesgue's dominated convergence theorem can be applied to $\int_{B'}$. We conclude from this that the sequence of integrals (2.5) converges when $u_\nu\downarrow u$, as expected. □

(2.6) Remark. In the above proof, the fact that A is an analytic set does not play an essential role. The main point is just that the slices $(\{z'\}\times B'')\cap A$ consist of isolated points for generic choices of coordinates (z',z''). In fact, the proof even works if the slices are totally discontinuous, in particular if they are of zero Hausdorff measure $\mathcal{H}_1$. It follows that Proposition 2.3 still holds whenever A is a closed set such that $\mathcal{H}_{2p-1}(A)=0$. □

2.B. Lelong Numbers

The concept of Lelong number is an analytic analogue of the algebraic notion of multiplicity. It is a very useful technique to extend results of the intersection theory of algebraic cycles to currents. Lelong numbers have been introduced for the first time by Lelong in [Lel57]. See also [Lel69; Siu74; Dem82a, 85a, 87] for further developments.

Let us first recall a few definitions. Let Θ be a closed positive current of bidimension (p,p) on a coordinate open set $\Omega \subset \mathbb{C}^n$ of a complex manifold X. The Lelong number of Θ at a point $x \in \Omega$ is defined to be the limit

$$\nu(\Theta,x) = \lim_{r \to 0+} \nu(\Theta,x,r), \qquad \text{where } \nu(\Theta,x,r) = \frac{\sigma_\Theta(B(x,r))}{\pi^p r^{2p}/p!}$$

measures the ratio of the area of Θ in the ball $B(x,r)$ to the area of the ball of radius r in $\mathbb{C}^p$. As $\sigma_\Theta = \Theta \wedge \frac{1}{p!}(\pi dd^c|z|^2)^p$ by Excercise 1.15, we also get

$$\nu(\Theta,x,r) = \frac{1}{r^{2p}} \int_{B(x,r)} \Theta(z) \wedge (dd^c|z|^2)^p. \tag{2.7}$$

The main results concerning Lelong numbers are summarized in the following theorems, due respectively to Lelong, Thie and Siu.

(2.8) Theorem ([Lel57]).

(a) *For every positive current Θ, the ratio $\nu(\Theta,x,r)$ is a nonnegative increasing function of r, in particular the limit $\nu(\Theta,x)$ as $r \to 0+$ always exists.*

(b) *If $\Theta = dd^c u$ is the bidegree $(1,1)$-current associated with a psh function u, then*

$$\nu(\Theta,x) = \sup\big\{\gamma \geqslant 0\,;\ u(z) \leqslant \gamma \log|z-x| + O(1) \ \text{ at } x\big\}.$$

In particular, if $u = \log|f|$ with $f \in H^0(X,\mathcal{O}_X)$ and $\Theta = dd^c u = [Z_f]$, we have

$$\nu([Z_f],x) = \mathrm{ord}_x(f) = \max\{m \in \mathbb{N}\,;\ D^\alpha f(x) = 0,\ |\alpha| < m\}.$$

(2.9) Theorem ([Thi67]). *In the case where Θ is a current of integration $[A]$ over an analytic subvariety A, the Lelong number $\nu([A],x)$ coincides with the multiplicity of A at x (defined e.g. as the sheet number in the ramified covering obtained by taking a generic linear projection of the germ (A,x) onto a p-dimensional linear subspace through x in any coordinate patch Ω).*

(2.10) Theorem ([Siu74]). *Let Θ be a closed positive current of bidimension (p,p) on the complex manifold X.*

(a) *The Lelong number $\nu(\Theta,x)$ is invariant by holomorphic changes of local coordinates.*

(b) *For every $c > 0$, the set $E_c(\Theta) = \{x \in X\,;\, \nu(\Theta, x) \geqslant c\}$ is a closed analytic subset of X of dimension $\leqslant p$.*

The most important result is Theorem 2.10 (b), which was initially proved as a (very deep) consequence of Hörmander's L^2 estimates (Chapter 5); Kiselman [Kis78] later found a much simpler proof based on his Legendre transformation for plurisubharmonic functions; however, there is now an even more direct route relying on the Ohsawa-Takegoshi L^2 extension theorem (cf. Corollary 13.3 below). The early proofs of the other results were also rather intricate in spite of their rather simple nature. We reproduce below a sketch of elementary arguments based on the use of a more general and more flexible notion of Lelong number introduced in [Dem87]. Let φ be a continuous psh function with an isolated $-\infty$ pole at x, e.g. a function of the form $\varphi(z) = \log \sum_{1\leqslant j\leqslant N} |g_j(z)|^{\gamma_j}$, $\gamma_j > 0$, where $(g_1, \ldots, g_N)$ is an ideal of germs of holomorphic functions in $\mathcal{O}_x$ with $g^{-1}(0) = \{x\}$. The *generalized Lelong number* $\nu(\Theta, \varphi)$ of Θ with respect to the weight φ is simply defined to be the mass of the measure $\Theta \wedge (dd^c\varphi)^p$ carried by the point x (the measure $\Theta \wedge (dd^c\varphi)^p$ is always well defined thanks to Proposition 2.3). This number can also be seen as the limit $\nu(\Theta, \varphi) = \lim_{t\to-\infty} \nu(\Theta, \varphi, t)$, where

$$\nu(\Theta, \varphi, t) = \int_{\varphi(z)<t} \Theta \wedge (dd^c\varphi)^p. \tag{2.11}$$

The relation with our earlier definition of Lelong numbers (as well as part (a) of Theorem 2.8) comes from the identity

$$\nu(\Theta, x, r) = \nu(\Theta, \varphi, \log r), \qquad \varphi(z) = \log|z - x|, \tag{2.12}$$

in particular $\nu(\Theta, x) = \nu(\Theta, \log|\bullet - x|)$. This equality is in turn a consequence of the following general formula, applied to $\chi(t) = e^{2t}$ and $t = \log r$:

$$\int_{\varphi(z)<t} \Theta \wedge (dd^c\chi\circ\varphi)^p = \chi'(t-0)^p \int_{\varphi(z)<t} \Theta \wedge (dd^c\varphi)^p, \tag{2.13}$$

where χ is an arbitrary convex increasing function. To prove the formula, we use a regularization and thus suppose that Θ, φ and χ are smooth, and that t is a non critical value of φ. Then Stokes' formula shows that the integrals on the left and right hand side of (2.13) are equal respectively to

$$\int_{\varphi(z)=t} \Theta \wedge \big(dd^c\chi\circ\varphi\big)^{p-1} \wedge d^c(\chi\circ\varphi), \qquad \int_{\varphi(z)=t} \Theta \wedge \big(dd^c\varphi\big)^{p-1} \wedge d^c\varphi,$$

and the differential form of bidegree $(p-1, p)$ appearing in the integrand of the first integral is equal to $(\chi'\circ\varphi)^p\,(dd^c\varphi)^{p-1} \wedge d^c\varphi$. The expected formula follows. Part (b) of Theorem 2.8 is a consequence of the Jensen-Lelong formula, whose proof is left as an exercise to the reader.

(2.14) Jensen-Lelong Formula. *Let u be any psh function on X. Then u is integrable with respect to the measure $\mu_r = (dd^c\varphi)^{n-1} \wedge d^c\varphi$ supported by the pseudo-sphere $\{\varphi(z) = r\}$ and*

$$\mu_r(u) = \int_{\{\varphi<r\}} u(dd^c\varphi)^n + \int_{-\infty}^{r} \nu(dd^c u, \varphi, t)\, dt.$$

In our case, we set $\varphi(z) = \log|z-x|$. Then $(dd^c\varphi)^n = \delta_x$ and μ_r is just the unitary invariant mean value measure on the sphere $S(x, e^r)$. For $r < r_0$, Formula 2.14 implies

$$\mu_r(u) - \mu_{r_0}(u) = \int_{r_0}^{r} \nu(dd^c u, x, t) \sim (r - r_0)\nu(dd^c u, x) \qquad \text{as } r \to -\infty.$$

From this, using the Harnack inequality for subharmonic functions, we get

$$\liminf_{z\to x} \frac{u(z)}{\log|z-x|} = \lim_{r\to-\infty} \frac{\mu_r(u)}{r} = \nu(dd^c u, x).$$

These equalities imply statement 2.8 (b).

Next, we show that the Lelong numbers $\nu(T,\varphi)$ only depend on the asymptotic behavior of φ near the polar set $\varphi^{-1}(-\infty)$. In a precise way:

(2.15) Comparison Theorem. *Let Θ be a closed positive current on X, and let $\varphi, \psi : X \to [-\infty, +\infty[$ be continuous psh functions with isolated poles at some point $x \in X$. Assume*

$$\ell := \limsup_{z\to x} \frac{\psi(z)}{\varphi(z)} < +\infty.$$

Then $\nu(\Theta,\psi) \leqslant \ell^p \nu(\Theta,\varphi)$, and the equality holds if $\ell = \lim \psi/\varphi$.

Proof. (2.12) shows that $\nu(\Theta, \lambda\varphi) = \lambda^p \nu(\Theta,\varphi)$ for every positive constant λ. It is thus sufficient to verify the inequality $\nu(\Theta,\psi) \leqslant \nu(\Theta,\varphi)$ under the hypothesis $\limsup \psi/\varphi < 1$. For any $c > 0$, consider the psh function

$$u_c = \max(\psi - c, \varphi).$$

Fix $r \ll 0$. For $c > 0$ large enough, we have $u_c = \varphi$ on a neighborhood of $\varphi^{-1}(r)$ and Stokes' formula gives

$$\nu(\Theta,\varphi,r) = \nu(\Theta,u_c,r) \geqslant \nu(\Theta,u_c).$$

On the other hand, the hypothesis $\limsup \psi/\varphi < 1$ implies that there exists $t_0 < 0$ such that $u_c = \psi - c$ on $\{u_c < t_0\}$. We thus get

$$\nu(\Theta,u_c) = \nu(\Theta,\psi - c) = \nu(\Theta,\psi),$$

hence $\nu(\Theta,\psi) \leqslant \nu(\Theta,\varphi)$. The equality case is obtained by reversing the roles of φ and ψ and observing that $\lim \varphi/\psi = 1/l$. □

Part (a) of Theorem 2.10 follows immediately from Theorem 2.15 by considering the weights $\varphi(z) = \log|\tau(z) - \tau(x)|$, $\psi(z) = \log|\tau'(z) - \tau'(x)|$ associated to coordinates systems $\tau(z) = (z_1, \ldots, z_n)$, $\tau'(z) = (z'_1, \ldots, z'_n)$ in a neighborhood of x. Another application is a direct simple proof of Thie's Theorem 2.9 when $\Theta = [A]$ is the current of integration over an analytic set $A \subset X$ of pure dimension p. For this, we have to observe that Theorem 2.15 still holds provided that x is an isolated point in $\mathrm{Supp}(\Theta) \cap \varphi^{-1}(-\infty)$ and $\mathrm{Supp}(\Theta) \cap \psi^{-1}(-\infty)$ (even though x is not isolated in $\varphi^{-1}(-\infty)$ or $\psi^{-1}(-\infty)$), under the weaker assumption that $\limsup_{\mathrm{Supp}(\Theta) \ni z \to x} \psi(z)/\varphi(z) = \ell$. The reason for this is that all integrals involve currents supported on $\mathrm{Supp}(\Theta)$. Now, by a generic choice of local coordinates $z' = (z_1, \ldots, z_p)$ and $z'' = (z_{p+1}, \ldots, z_n)$ on (X, x), the germ (A, x) is contained in a cone $|z''| \leqslant C|z'|$. If $B' \subset \mathbb{C}^p$ is a ball of center 0 and radius r' small, and $B'' \subset \mathbb{C}^{n-p}$ is the ball of center 0 and radius $r'' = Cr'$, the projection

$$\mathrm{pr} : A \cap (B' \times B'') \longrightarrow B'$$

is a ramified covering with finite sheet number m. When $z \in A$ tends to $x = 0$, the functions

$$\varphi(z) = \log|z| = \log(|z'|^2 + |z''|^2)^{1/2}, \qquad \psi(z) = \log|z'|.$$

satisfy $\lim_{z \to x} \psi(z)/\varphi(z) = 1$. Hence Theorem 2.15 implies

$$\nu([A], x) = \nu([A], \varphi) = \nu([A], \psi).$$

Now, Formula 2.13 with $\chi(t) = e^{2t}$ yields

$$\begin{aligned}
\nu([A], \psi, \log t) &= t^{-2p} \int_{\{\psi < \log t\}} [A] \wedge \Big(\frac{1}{2} dd^c e^{2\psi}\Big)^p \\
&= t^{-2p} \int_{A \cap \{|z'| < t\}} \Big(\frac{1}{2} \mathrm{pr}^* dd^c |z'|^2\Big)^p \\
&= m t^{-2p} \int_{\mathbb{C}^p \cap \{|z'| < t\}} \Big(\frac{1}{2} dd^c |z'|^2\Big)^p = m,
\end{aligned}$$

hence $\nu([A], \psi) = m$. Here, we have used the fact that pr is an étale covering with m sheets over the complement of the ramification locus $S \subset B'$, and the fact that S is of zero Lebesgue measure in B'.

(2.16) Proposition. *Under the assumptions of Proposition 2.3, we have*

$$\nu(dd^c u \wedge \Theta, x) \geqslant \nu(u, x)\, \nu(\Theta, x)$$

at every point $x \in X$.

Proof. Assume that $X = B(0, r)$ and $x = 0$. By definition

$$\nu(dd^c u \wedge \Theta, x) = \lim_{r \to 0} \int_{|z| \leqslant r} dd^c u \wedge \Theta \wedge (dd^c \log|z|)^{p-1}.$$

Set $\gamma = \nu(u, x)$ and

$$u_\nu(z) = \max\big(u(z), (\gamma - \varepsilon)\log|z| - \nu\big)$$

with $0 < \varepsilon < \gamma$ (if $\gamma = 0$, there is nothing to prove). Then u_ν decreases to u and

$$\int_{|z|\leqslant r} dd^c u \wedge \Theta \wedge (dd^c \log|z|)^{p-1} \geqslant \limsup_{\nu\to+\infty} \int_{|z|\leqslant r} dd^c u_\nu \wedge \Theta \wedge (dd^c \log|z|)^{p-1}$$

by the weak convergence of $dd^c u_\nu \wedge \Theta$; here $(dd^c \log|z|)^{p-1}$ is not smooth on $\overline{B}(0,r)$, but the integrals remain unchanged if we replace $\log|z|$ by $\chi(\log|z|/r)$ with a smooth convex function χ such that $\chi(t) = t$ for $t \geqslant -1$ and $\chi(t) = 0$ for $t \leqslant -2$. Now, we have $u(z) \leqslant \gamma\log|z| + C$ near 0, so $u_\nu(z)$ coincides with $(\gamma - \varepsilon)\log|z| - \nu$ on a small ball $B(0, r_\nu) \subset B(0, r)$ and we infer

$$\begin{aligned}\int_{|z|\leqslant r} dd^c u_\nu \wedge \Theta \wedge (dd^c \log|z|)^{p-1} &\geqslant (\gamma - \varepsilon)\int_{|z|\leqslant r_\nu} \Theta \wedge (dd^c \log|z|)^p \\ &\geqslant (\gamma - \varepsilon)\nu(\Theta, x).\end{aligned}$$

As $r \in\,]0, R[$ and $\varepsilon \in\,]0, \gamma[$ were arbitrary, the desired inequality follows. □

We will later need an important decomposition formula of [Siu74]. We start with the following lemma.

(2.17) Lemma. *If Θ is a closed positive current of bidimension (p,p) and Z is an irreducible analytic set in X, we set*

$$m_Z = \inf\{x \in Z\,;\ \nu(\Theta, x)\}.$$

(a) *There is a countable family of proper analytic subsets (Z'_j) of Z such that $\nu(\Theta, x) = m_Z$ for all $x \in Z \smallsetminus \bigcup Z'_j$. We say that m_Z is the generic Lelong number of Θ along Z.*

(b) *If $\dim Z = p$, then $\Theta \geqslant m_Z[Z]$ and $\mathbb{1}_Z\Theta = m_Z[Z]$.*

Proof. (a) By definition of m_Z and $E_c(\Theta)$, we have $\nu(\Theta, x) \geqslant m_Z$ for every $x \in Z$ and

$$\nu(\Theta, x) = m_Z \quad \text{on } Z \smallsetminus \bigcup_{c\in\mathbb{Q},\, c>m_Z} Z \cap E_c(\Theta).$$

However, for $c > m_Z$, the intersection $Z \cap E_c(\Theta)$ is a proper analytic subset of A.

(b) Left as an exercise to the reader. It is enough to prove that $\Theta \geqslant m_Z[Z_{\rm reg}]$ at regular points of Z, so one may assume that Z is a p-dimensional linear subspace in $\mathbb{C}^n$. Show that the measure $(\Theta - m_Z[Z]) \wedge (dd^c|z|^2)^p$ has nonnegative mass on every ball $|z - a| < r$ with center $a \in Z$. Conclude by using arbitrary affine changes of coordinates that $\Theta - m_Z[Z] \geqslant 0$. □

(2.18) Decomposition Formula ([Siu74]). *Let Θ be a closed positive current of bidimension (p,p). Then Θ can be written as a convergent series of closed positive currents*

$$\Theta = \sum_{k=1}^{+\infty} \lambda_k\,[Z_k] + R,$$

where $[Z_k]$ is a current of integration over an irreducible analytic set of dimension p, and R is a residual current with the property that $\dim E_c(R) < p$ *for every $c > 0$. This decomposition is locally and globally unique: the sets Z_k are precisely the p-dimensional components occurring in the upperlevel sets $E_c(\Theta)$, and $\lambda_k = \min_{x\in Z_k} \nu(\Theta, x)$ is the generic Lelong number of Θ along Z_k.*

Proof of uniqueness. If Θ has such a decomposition, the p-dimensional components of $E_c(\Theta)$ are $(Z_j)_{\lambda_j\geqslant c}$, for $\nu(\Theta,x) = \sum \lambda_j \nu([Z_j],x) + \nu(R,x)$ is non zero only on $\bigcup Z_j \cup \bigcup E_c(R)$, and is equal to λ_j generically on Z_j (more precisely, $\nu(\Theta,x) = \lambda_j$ at every regular point of Z_j which does not belong to any intersection $Z_j \cup Z_k$, $k \neq j$ or to $\bigcup E_c(R)$). In particular Z_j and λ_j are unique.

Proof of existence. Let $(Z_j)_{j\geqslant 1}$ be the countable collection of p-dimensional components occurring in one of the sets $E_c(\Theta)$, $c \in \mathbb{Q}_+^*$, and let $\lambda_j > 0$ be the generic Lelong number of Θ along Z_j. Then Lemma 2.17 shows by induction on N that $R_N = \Theta - \sum_{1\leqslant j\leqslant N} \lambda_j [Z_j]$ is positive. As R_N is a decreasing sequence, there must be a limit $R = \lim_{N\to+\infty} R_N$ in the weak topology. Thus we have the asserted decomposition. By construction, R has zero generic Lelong number along Z_j, so $\dim E_c(R) < p$ for every $c > 0$. □

It is very important to note that some components of lower dimension can actually occur in $E_c(R)$, but they cannot be subtracted because R has bidimension (p,p). A typical case is the case of a bidimension $(n-1,n-1)$ current $\Theta = dd^c u$ with $u = \log(|f_1|^{\gamma_1} + \cdots + |f_N|^{\gamma_N})$ and $f_j \in H^0(X, \mathcal{O}_X)$. In general $\bigcup E_c(\Theta) = \bigcap f_j^{-1}(0)$ has dimension $< n-1$.

(2.19) Corollary. *Let $\Theta_j = dd^c u_j$, $1 \leqslant j \leqslant p$, be closed positive $(1,1)$-currents on a complex manifold X. Suppose that there are analytic sets $A_2 \supset \cdots \supset A_p$ in X with* $\operatorname{codim} A_j \geqslant j$ *at every point such that each u_j, $j \geqslant 2$, is locally bounded on $X \smallsetminus A_j$. Let $\{A_{p,k}\}_{k\geqslant 1}$ be the irreducible components of A_p of codimension p exactly and let $\nu_{j,k} = \min_{x\in A_{p,k}} \nu(\Theta_j, x)$ be the generic Lelong number of Θ_j along $A_{p,k}$. Then $\Theta_1 \wedge \cdots \wedge \Theta_p$ is well-defined and*

$$\Theta_1 \wedge \cdots \wedge \Theta_p \geqslant \sum_{k=1}^{+\infty} \nu_{1,k} \cdot \cdots \cdot \nu_{p,k}\,[A_{p,k}].$$

Proof. By induction on p, Proposition 2.3 shows that $\Theta_1 \wedge \cdots \wedge \Theta_p$ is well defined. Moreover, Proposition 2.16 implies

$$\nu(\Theta_1 \wedge \cdots \wedge \Theta_p, x) \geqslant \nu(\Theta_1, x) \cdot \cdots \cdot \nu(\Theta_p, x) \geqslant \nu_{1,k} \cdot \cdots \cdot \nu_{p,k}$$

at every point $x \in A_{p,k}$. The desired inequality is then a consequence of Siu's decomposition theorem. □

Chapter 3

Hermitian Vector Bundles, Connections and Curvature

The goal of this chapter is to recall the most basic definitions of Hermitian differential geometry related to the concepts of connection, curvature and first Chern class of a line bundle.

Let F be a complex vector bundle of rank r over a smooth differentiable manifold M. A *connection* D on F is a linear differential operator of order 1:

$$D : \mathscr{C}^\infty(M, \Lambda^q T_M^* \otimes F) \to \mathscr{C}^\infty(M, \Lambda^{q+1} T_M^* \otimes F)$$

such that

$$D(f \wedge u) = df \wedge u + (-1)^{\deg f} f \wedge Du \tag{3.1}$$

for all forms $f \in \mathscr{C}^\infty(M, \Lambda^p T_M^*)$, $u \in \mathscr{C}^\infty(X, \Lambda^q T_M^* \otimes F)$. On an open set $\Omega \subset M$ where F admits a trivialization $\theta : F_{|\Omega} \xrightarrow{\simeq} \Omega \times \mathbb{C}^r$, a connection D can be written

$$Du \simeq_\theta du + \Gamma \wedge u$$

where $\Gamma \in \mathscr{C}^\infty(\Omega, \Lambda^1 T_M^* \otimes \mathrm{Hom}(\mathbb{C}^r, \mathbb{C}^r))$ is an arbitrary matrix of 1-forms and d acts componentwise (the coefficients of Γ are called the *Christoffel symbols* of the connection). It is then easy to check that

$$D^2 u \simeq_\theta (d\Gamma + \Gamma \wedge \Gamma) \wedge u \quad \text{on } \Omega.$$

Since D^2 is a globally defined operator, there is a global 2-form

$$\Theta_D \in \mathscr{C}^\infty(M, \Lambda^2 T_M^* \otimes \mathrm{Hom}(F, F)) \tag{3.2}$$

such that $D^2 u = \Theta_D \wedge u$ for every form u with values in F.

Assume now that F is endowed with a $\mathscr{C}^\infty$ Hermitian metric h along the fibers and that the isomorphism $F_{|\Omega} \simeq \Omega \times \mathbb{C}^r$ is given by a $\mathscr{C}^\infty$ frame (e_λ). We then have a canonical sesquilinear pairing

$$\begin{aligned} \mathscr{C}^\infty(M, \Lambda^p T_M^* \otimes F) \times \mathscr{C}^\infty(M, \Lambda^q T_M^* \otimes F) &\longrightarrow \mathscr{C}^\infty(M, \Lambda^{p+q} T_M^* \otimes \mathbb{C}) \\ (u, v) &\longmapsto \{u, v\}_h \end{aligned} \tag{3.3}$$

given by

$$\{u, v\}_h = \sum_{\lambda,\mu} u_\lambda \wedge \overline{v}_\mu \langle e_\lambda, e_\mu \rangle_h, \qquad u = \sum u_\lambda \otimes e_\lambda, \quad v = \sum v_\mu \otimes e_\mu.$$

The connection D is said to be *Hermitian* (with respect to h) if it satisfies the additional property

$$d\{u, v\}_h = \{Du, v\}_h + (-1)^{\deg u}\{u, Dv\}_h.$$

Assuming that (e_λ) is orthonormal, one easily checks that D is Hermitian if and only if $\Gamma^* = -\Gamma$. In this case $\Theta_D^* = -\Theta_D$, thus

$$\mathrm{i}\Theta_D \in \mathscr{C}^\infty(M, \Lambda^2 T_M^* \otimes \mathrm{Herm}(F, F)).$$

(3.4) Special Case. For a bundle F of rank 1, the connection form Γ of a Hermitian connection D can be seen as a 1-form with purely imaginary coefficients $\Gamma = \mathrm{i}A$ (A real). Then we have $\Theta_D = d\Gamma = \mathrm{i}dA$. In particular $\mathrm{i}\Theta_F$ is a closed 2-form. The *first Chern class* of F is defined to be the cohomology class

$$c_1(F)_{\mathbb{R}} = \Big\{\frac{\mathrm{i}}{2\pi}\Theta_F\Big\} \in H^2_{\mathrm{DR}}(M, \mathbb{R}).$$

The cohomology class is actually independent of the connection, since any other connection D_1 differs by a global 1-form, $D_1 u = Du + B \wedge u$, so that $\Theta_{D_1} = \Theta_D + dB$. It is well-known that $c_1(F)_{\mathbb{R}}$ is the image in $H^2(M, \mathbb{R})$ of an integral class $c_1(F) \in H^2(M, \mathbb{Z})$; by using the exponential exact sequence

$$0 \to \mathbb{Z} \to \mathscr{E} \to \mathscr{E}^* \to 0,$$

$c_1(F)$ can be defined in Čech cohomology theory as the image by the coboundary map $H^1(M, \mathscr{E}^*) \to H^2(M, \mathbb{Z})$ of the cocycle $\{g_{jk}\} \in H^1(M, \mathscr{E}^*)$ defining F; see e.g. [GrH78] for details. □

We now concentrate ourselves on the complex analytic case. If $M = X$ is a complex manifold X, every connection D on a complex $\mathscr{C}^\infty$ vector bundle F can be splitted in a unique way as a sum of a $(1,0)$ and of a $(0,1)$-connection, $D = D' + D''$. In a local trivialization θ given by a $\mathscr{C}^\infty$ frame, one can write

(3.5′) $$D'u \simeq_\theta d'u + \Gamma' \wedge u,$$

(3.5″) $$D''u \simeq_\theta d''u + \Gamma'' \wedge u,$$

with $\Gamma = \Gamma' + \Gamma''$. The connection is Hermitian if and only if $\Gamma' = -(\Gamma'')^*$ in any orthonormal frame. Thus there exists a unique Hermitian connection D corresponding to a prescribed $(0,1)$ part D''.

Assume now that the bundle F itself has a *holomorphic* structure, and is equipped with a Hermitian metric h. The unique Hermitian connection for which D'' is the d'' operator defined in Chapter 1 is called the *Chern connection* of F. In a local holomorphic frame (e_λ) of $E_{|\Omega}$, the metric is given by the Hermitian matrix $H = (h_{\lambda\mu})$, $h_{\lambda\mu} = \langle e_\lambda, e_\mu \rangle$. We have

$$\{u, v\}_h = \sum_{\lambda,\mu} h_{\lambda\mu} u_\lambda \wedge \overline{v}_\mu = u^\dagger \wedge H\overline{v},$$

where $u^\dagger$ is the transposed matrix of u, and easy computations yield

$$\begin{aligned} d\{u, v\}_h &= (du)^\dagger \wedge H\overline{v} + (-1)^{\deg u} u^\dagger \wedge (dH \wedge \overline{v} + H\overline{dv}) \\ &= \big(du + \overline{H}^{-1} d'\overline{H} \wedge u\big)^\dagger \wedge H\overline{v} + (-1)^{\deg u} u^\dagger \wedge \overline{(dv + \overline{H}^{-1} d'\overline{H} \wedge v)} \end{aligned}$$

using the fact that $dH = d'H + \overline{d'\overline{H}}$ and $\overline{H}^\dagger = H$. Therefore the Chern connection D coincides with the Hermitian connection defined by

$$\text{(3.6)} \qquad \begin{cases} Du \simeq_\theta du + \overline{H}^{-1} d'\overline{H} \wedge u, \\ D' \simeq_\theta d' + \overline{H}^{-1} d'\overline{H} \wedge \bullet = \overline{H}^{-1} d'(\overline{H} \bullet), \qquad D'' = d''. \end{cases}$$

It is clear from this relations that $D'^2 = D''^2 = 0$. Consequently D^2 is given by to $D^2 = D'D'' + D''D'$, and the curvature tensor Θ_D is of type $(1,1)$. Since $d'd'' + d''d' = 0$, we get

$$\begin{aligned} (D'D'' + D''D')u &\simeq_\theta \overline{H}^{-1} d'\overline{H} \wedge d''u + d''(\overline{H}^{-1} d'\overline{H} \wedge u) \\ &= d''(\overline{H}^{-1} d'\overline{H}) \wedge u. \end{aligned}$$

(3.7) Proposition. *The Chern curvature tensor $\Theta_{F,h} := \Theta_D$ of (F, h) is such that*

$$\mathrm{i}\,\Theta_{F,h} \in \mathscr{C}^\infty(X, \Lambda^{1,1} T_X^* \otimes \mathrm{Herm}(F, F)).$$

If $\theta : F_{\restriction\Omega} \to \Omega \times \mathbb{C}^r$ is a holomorphic trivialization and if H is the Hermitian matrix representing the metric along the fibers of $F_{\restriction\Omega}$, then

$$\mathrm{i}\,\Theta_{F,h} \simeq_\theta \mathrm{i}\, d''(\overline{H}^{-1} d'\overline{H}) \quad \text{on } \Omega.$$

In case there cannot be any confusion on which Hermitian metric h is used, we also sometimes simply write $\Theta_{F,h} = \Theta_F$.

Let $(z_1, \ldots, z_n)$ be holomorphic coordinates on X and let $(e_\lambda)_{1\leqslant\lambda\leqslant r}$ be an orthonormal frame of F. Writing

$$\mathrm{i}\Theta_{F,h} = \sum_{1\leqslant j,k\leqslant n,\ 1\leqslant\lambda,\mu\leqslant r} c_{jk\lambda\mu} dz_j \wedge d\overline{z}_k \otimes e_\lambda^* \otimes e_\mu,$$

we can identify the curvature tensor to a Hermitian form

$$\widetilde{\Theta}_{F,h}(\xi\otimes v)=\sum_{1\leqslant j,k\leqslant n,\,1\leqslant\lambda,\mu\leqslant r}c_{jk\lambda\mu}\xi_j\overline{\xi}_k v_\lambda\overline{v}_\mu \tag{3.8}$$

on $T_X\otimes F$. This leads in a natural way to positivity concepts, following definitions introduced by Kodaira [Kod53], Nakano [Nak55] and Griffiths [Gri69].

(3.9) Definition. *The Hermitian vector bundle (F,h) is said to be*

(a) *positive in the sense of Nakano if we have $\widetilde{\Theta}_{F,h}(\tau)>0$ for all non zero tensors $\tau=\sum\tau_{j\lambda}\,\partial/\partial z_j\otimes e_\lambda\in T_X\otimes F$.*

(b) *positive in the sense of Griffiths if $\widetilde{\Theta}_{F,h}(\xi\otimes v)>0$ for all non zero decomposable tensors $\xi\otimes v\in T_X\otimes F$,*

Corresponding semipositivity concepts are defined by relaxing the strict inequalities.

(3.10) Special Case of Rank 1 Bundles. Assume that F is a line bundle. The Hermitian matrix $H=(h_{11})$ associated to a trivialization $\theta : F_{\restriction\Omega}\simeq\Omega\times\mathbb{C}$ is simply a positive function. It is often convenient to denote it as an exponential, namely $e^{-2\varphi}$ (and also sometimes $e^{-\varphi}$ simply, if we do not want to stress that H is a quadratic form), with $\varphi\in\mathscr{C}^\infty(\Omega,\mathbb{R})$. In this case the curvature form $\Theta_{F,h}$ can be identified to the $(1,1)$-form $d'd''\varphi$, and

$$\frac{\mathrm{i}}{2\pi}\Theta_{F,h}=\frac{\mathrm{i}}{\pi}d'd''\varphi=dd^c\varphi$$

is a real $(1,1)$-form. Hence F is semi-positive (in either the Nakano or Griffiths sense) if and only if φ is psh, resp. positive if and only if φ is *strictly psh*. In this setting, the Lelong-Poincaré equation can be generalized as follows: let $\sigma\in H^0(X,F)$ be a non zero holomorphic section. Then

$$dd^c\log\|\sigma\|_h=[Z_\sigma]-\frac{\mathrm{i}}{2\pi}\Theta_{F,h}. \tag{3.11}$$

Formula (3.11) is immediate if we write $\|\sigma\|=|\theta(\sigma)|e^{-\varphi}$ and if we apply (1.20) to the holomorphic function $f=\theta(\sigma)$. As we shall see later, it is very important for the applications to consider also singular Hermitian metrics.

(3.12) Definition. *A singular (Hermitian) metric h on a line bundle F is a metric which is given in any trivialization $\theta : F_{\restriction\Omega}\xrightarrow{\simeq}\Omega\times\mathbb{C}$ by*

$$\|\xi\|_h=|\theta(\xi)|\,e^{-\varphi(x)},\qquad x\in\Omega,\ \xi\in F_x$$

where $\varphi\in L^1_{\mathrm{loc}}(\Omega)$ is an arbitrary function, called the weight of the metric with respect to the trivialization θ.

If $\theta' : F_{\restriction\Omega'}\longrightarrow\Omega'\times\mathbb{C}$ is another trivialization, φ' the associated weight and $g\in\mathscr{O}^*(\Omega\cap\Omega')$ the transition function, then $\theta'(\xi)=g(x)\,\theta(\xi)$ for $\xi\in F_x$, and so

$\varphi' = \varphi + \log|g|$ on $\Omega \cap \Omega'$. The curvature form of F is then given formally by the closed $(1,1)$-current $\frac{\mathrm{i}}{2\pi}\Theta_{F,h} = dd^c\varphi$ on Ω ; our assumption $\varphi \in L^1_{\mathrm{loc}}(\Omega)$ guarantees that $\Theta_{F,h}$ exists in the sense of distribution theory. As in the smooth case, $\frac{\mathrm{i}}{2\pi}\Theta_{F,h}$ is globally defined on X and independent of the choice of trivializations, and its De Rham cohomology class is the image of the first Chern class $c_1(F) \in H^2(X,\mathbb{Z})$ in $H^2_{DR}(X,\mathbb{R})$. Before going further, we discuss two basic examples.

(3.13) Example. Let $D = \sum \alpha_j D_j$ be a divisor with coefficients $\alpha_j \in \mathbb{Z}$ and let $F = \mathcal{O}(D)$ be the associated invertible sheaf of meromorphic functions u such that $\mathrm{div}(u) + D \geqslant 0$; the corresponding line bundle can be equipped with the singular metric defined by $\|u\| = |u|$. If g_j is a generator of the ideal of D_j on an open set $\Omega \subset X$ then $\theta(u) = u \prod g_j^{\alpha_j}$ defines a trivialization of $\mathcal{O}(D)$ over Ω, thus our singular metric is associated to the weight $\varphi = \sum \alpha_j \log|g_j|$. By the Lelong-Poincaré equation, we find

$$\frac{\mathrm{i}}{2\pi}\Theta_{\mathcal{O}(D)} = dd^c\varphi = [D],$$

where $[D] = \sum \alpha_j [D_j]$ denotes the current of integration over D. □

(3.14) Example. Assume that $\sigma_0, \sigma_1, \ldots, \sigma_N$ are non zero holomorphic sections of F. Then we can define a natural (possibly singular) Hermitian metric h^* on F^* by

$$\|\xi^*\|^2_{h^*} = \sum_{0\leqslant j\leqslant N} \big|\xi^* \cdot \sigma_j(x)\big|^2 \quad \text{for } \xi^* \in F^*_x.$$

The dual metric h on F is given by

$$\|\xi\|^2_h = \frac{|\theta(\xi)|^2}{|\theta(\sigma_0(x))|^2 + |\theta(\sigma_1(x))|^2 + \cdots + |\theta(\sigma_N(x))|^2}$$

with respect to any trivialization θ. The associated weight function is thus given by

$$\varphi(x) = \log\Big(\sum_{0\leqslant j\leqslant N} |\theta(\sigma_j(x))|^2\Big)^{1/2}.$$

In this case φ is a psh function, thus $\mathrm{i}\Theta_{F,h}$ is a closed positive current. Let us denote by Σ the linear system defined by $\sigma_0, \ldots, \sigma_N$ and by $B_\Sigma = \bigcap \sigma_j^{-1}(0)$ its base locus. We have a meromorphic map

$$\Phi_\Sigma : X \smallsetminus B_\Sigma \to \mathbb{P}^N, \qquad x \mapsto (\sigma_0(x) : \sigma_1(x) : \sigma_2(x) : \cdots : \sigma_N(x)).$$

Then $\frac{\mathrm{i}}{2\pi}\Theta_{F,h}$ is equal to the pull-back over $X \smallsetminus B_\Sigma$ of the Fubini-Study metric $\omega_{\mathrm{FS}} = \frac{\mathrm{i}}{2\pi}\log(|z_0|^2 + |z_1|^2 + \cdots + |z_N|^2)$ of $\mathbb{P}^N$ by Φ_Σ. □

(3.15) Ample and Very Ample Line Bundles. *A holomorphic line bundle F over a compact complex manifold X is said to be*

(a) *very ample if the map* $\Phi_{|F|} : X \to \mathbb{P}^N$ *associated to the complete linear system* $|F| = P(H^0(X, F))$ *is a regular embedding (by this we mean in particular that the base locus is empty, i.e.* $B_{|F|} = \emptyset$*).*

(b) *ample if some multiple* mF, $m > 0$, *is very ample.*

Here we use an additive notation for $\mathrm{Pic}(X) = H^1(X, \mathcal{O}^*)$, hence the symbol mF denotes the line bundle $F^{\otimes m}$. By Example 3.14, every ample line bundle F has a smooth Hermitian metric with positive definite curvature form; indeed, if the linear system $|mF|$ gives an embedding in projective space, then we get a smooth Hermitian metric on $F^{\otimes m}$, and the m-th root yields a metric h on F such that $\frac{i}{2\pi}\Theta_{F,h} = \frac{1}{m}\Phi^*_{|mF|}\omega_{\mathrm{FS}}$. Conversely, the Kodaira embedding theorem [Kod54] tells us that every positive line bundle F is ample (see Exercise 5.14 for a straightforward analytic proof of the Kodaira embedding theorem).

Chapter 4

Bochner Technique and Vanishing Theorems

4.A. Laplace-Beltrami Operators and Hodge Theory

We start by recalling briefly a few basic facts of Hodge theory. Assume for the moment that M is a differentiable manifold equipped with a Riemannian metric $g = \sum g_{ij} dx_i \otimes dx_j$ and that (F, h) is a Hermitian vector bundle over M. Given a q-form u on M with values in F, we consider the global L^2 norm

$$\|u\|^2 = \int_M |u(x)|^2 dV_g(x)$$

where $|u|$ is the pointwise Hermitian norm and dV_g is the Riemannian volume form (we omit the dependence on the metrics in the notation, but we should really put $|u(x)|_{g,h}$ and $\|u\|_{g,h}$ here). The Laplace-Beltrami operator associated to the connection D is by definition

$$\Delta = DD^* + D^*D$$

where

$$D^* : \mathscr{C}^\infty(M, \Lambda^q T_M^* \otimes F) \to \mathscr{C}^\infty(M, \Lambda^{q-1} T_M^* \otimes F)$$

is the (formal) adjoint of D with respect to the L^2 inner product. Assume that M is *compact*. Since

$$\Delta : \mathscr{C}^\infty(M, \Lambda^q T_M^* \otimes F) \to \mathscr{C}^\infty(M, \Lambda^q T_M^* \otimes F)$$

is a self-adjoint elliptic operator in each degree, standard results of PDE theory show that there is an orthogonal decomposition

$$\mathscr{C}^\infty(M, \Lambda^q T_M^* \otimes F) = \mathscr{H}^q(M, F) \oplus \operatorname{Im} \Delta$$

where $\mathscr{H}^q(M, F) = \operatorname{Ker} \Delta$ is the space of harmonic forms of degree q; $\mathscr{H}^q(M, F)$ is a finite dimensional space. Assume moreover that the connection D is *integrable*, i.e. that $D^2 = 0$. It is then easy to check that there is an orthogonal direct sum

$$\operatorname{Im} \Delta = \operatorname{Im} D \oplus \operatorname{Im} D^*,$$

indeed $\langle Du, D^*v\rangle = \langle D^2u, v\rangle = 0$ for all u, v. Hence we get an orthogonal decomposition

$$\mathscr{C}^\infty(M, \Lambda^q T^*_M \otimes F) = \mathscr{H}^q(M, F) \oplus \operatorname{Im} D \oplus \operatorname{Im} D^*,$$

and $\operatorname{Ker}\Delta$ is precisely equal to $\mathscr{H}^q(M, F) \oplus \operatorname{Im} D$. Especially, the q-th cohomology group $\operatorname{Ker}\Delta/\operatorname{Im}\Delta$ is isomorphic to $\mathscr{H}^q(M, F)$. All this can be applied for example in the case of the De Rham groups $H^q_{\mathrm{DR}}(M, \mathbb{C})$, taking F to be the trivial bundle $F = M \times \mathbb{C}$ (notice, however, that a nontrivial bundle F usually does not admit any integrable connection):

(4.1) Hodge Fundamental Theorem. *If M is a compact Riemannian manifold, there is an isomorphism*

$$H^q_{\mathrm{DR}}(M, \mathbb{C}) \simeq \mathscr{H}^q(M, \mathbb{C})$$

from De Rham cohomology groups onto spaces of harmonic forms.

A rather important consequence of the Hodge fundamental theorem is a proof of the *Poincaré duality theorem.* Assume that the Riemannian manifold (M, g) is oriented. Then there is a (conjugate linear) Hodge star operator

$$* : \Lambda^q T^*_M \otimes \mathbb{C} \to \Lambda^{m-q} T^*_M \otimes \mathbb{C}, \qquad m = \dim_{\mathbb{R}} M$$

defined by $u \wedge *v = \langle u, v\rangle dV_g$ for any two complex valued q-forms u, v. A standard computation shows that $*$ commutes with Δ, hence $*u$ is harmonic if and only if u is. This implies that the natural pairing

$$H^q_{\mathrm{DR}}(M, \mathbb{C}) \times H^{m-q}_{\mathrm{DR}}(M, \mathbb{C}), \qquad (\{u\}, \{v\}) \mapsto \int_M u \wedge v \tag{4.2}$$

is a nondegenerate duality, the dual of a class $\{u\}$ represented by a harmonic form being $\{*u\}$.

4.B. Serre Duality Theorem

Let us now suppose that X is a compact complex manifold equipped with a Hermitian metric $\omega = \sum \omega_{jk} dz_j \wedge d\overline{z}_k$. Let F be a holomorphic vector bundle on X equipped with a Hermitian metric, and let $D = D' + D''$ be its Chern curvature form. All that we said above for the Laplace-Beltrami operator Δ still applies to the complex Laplace operators

$$\Delta' = D'D'^* + D'^*D', \qquad \Delta'' = D''D''^* + D''^*D'',$$

with the great advantage that we always have $D'^2 = D''^2 = 0$. Especially, if X is a compact complex manifold, there are isomorphisms

$$H^{p,q}(X, F) \simeq \mathscr{H}^{p,q}(X, F) \tag{4.3}$$

between Dolbeault cohomology groups $H^{p,q}(X,F)$ and spaces $\mathcal{H}^{p,q}(X,F)$ of Δ''-harmonic forms of bidegree (p,q) with values in F. Now, there is a generalized Hodge star operator

$$* : \Lambda^{p,q}T_X^* \otimes F \to \Lambda^{n-p,n-q}T_X^* \otimes F^*, \qquad n = \dim_{\mathbb{C}} X,$$

such that $u \wedge *v = \langle u, v\rangle dV_g$, for any two F-valued (p,q)-forms, when the wedge product $u \wedge *v$ is combined with the pairing $F \times F^* \to \mathbb{C}$. This leads to the *Serre duality theorem* [Ser55]: the bilinear pairing

$$(4.4) \qquad H^{p,q}(X,F) \times H^{n-p,n-q}(X,F^*), \qquad (\{u\},\{v\}) \mapsto \int_X u \wedge v$$

is a nondegenerate duality. Combining this with the Dolbeault isomorphism, we may restate the result in the form of the duality formula

$$(4.4') \qquad H^q(X, \Omega_X^p \otimes \mathcal{O}(F))^* \simeq H^{n-q}(X, \Omega_X^{n-p} \otimes \mathcal{O}(F^*)).$$

4.C. Bochner-Kodaira-Nakano Identity on Kähler Manifolds

We now proceed to explain the basic ideas of the Bochner technique used to prove vanishing theorems. Great simplifications occur in the computations if the Hermitian metric on X is supposed to be *Kähler*, i.e. if the associated *fundamental* $(1,1)$-*form*

$$\omega = \mathrm{i} \sum \omega_{jk} dz_j \wedge d\overline{z}_k$$

satisfies $d\omega = 0$. It can be easily shown that ω is Kähler if and only if there are holomorphic coordinates $(z_1, \ldots, z_n)$ centered at any point $x_0 \in X$ such that the matrix of coefficients (ω_{jk}) is tangent to identity at order 2, i.e.

$$\omega_{jk}(z) = \delta_{jk} + O(|z|^2) \quad \text{at } x_0.$$

It follows that all order 1 operators D, D', D'' and their adjoints D^*, D'^*, D''^* admit at x_0 the same expansion as the analogous operators obtained when all Hermitian metrics on X or F are constant. From this, the basic commutation relations of Kähler geometry can be checked. If A, B are differential operators acting on the algebra $\mathcal{C}^\infty(X, \Lambda^{\bullet,\bullet}T_X^* \otimes F)$, their graded commutator (or graded Lie bracket) is defined by

$$[A,B] = AB - (-1)^{ab}BA$$

where a, b are the degrees of A and B respectively. If C is another endomorphism of degree c, the following purely formal *Jacobi identity* holds:

$$(-1)^{ca}\big[A,[B,C]\big] + (-1)^{ab}\big[B,[C,A]\big] + (-1)^{bc}\big[C,[A,B]\big] = 0.$$

(4.5) Basic Commutation Relations. *Let (X,ω) be a Kähler manifold and let L be the operators defined by $Lu = \omega \wedge u$ and $\Lambda = L^*$. Then*

$$[D''^*, L] = \mathrm{i}D', \qquad [D'^*, L] = -\mathrm{i}D'',$$
$$[\Lambda, D''] = -\mathrm{i}D'^*, \qquad [\Lambda, D'] = \mathrm{i}D''^*.$$

Proof (sketch). The first step is to check the identity $[d''^*, L] = \mathrm{i}d'$ for constant metrics on $X = \mathbb{C}^n$ and $F = X \times \mathbb{C}$, by a brute force calculation. All three other identities follow by taking conjugates or adjoints. The case of variable metrics follows by looking at Taylor expansions up to order 1. □

(4.6) Bochner-Kodaira-Nakano Identity. *If (X,ω) is Kähler, the complex Laplace operators Δ' and Δ'' acting on F-valued forms satisfy the identity*

$$\Delta'' = \Delta' + [\mathrm{i}\Theta_{F,h}, \Lambda].$$

Proof. The last equality in (4.5) yields $D''^* = -\mathrm{i}[\Lambda, D']$, hence

$$\Delta'' = [D'', \delta''] = -\mathrm{i}[D'', [\Lambda, D']].$$

By the Jacobi identity we get

$$[D'', [\Lambda, D']] = [\Lambda, [D', D'']] + [D', [D'', \Lambda]] = [\Lambda, \Theta_{F,h}] + \mathrm{i}[D', D'^*],$$

taking into account that $[D', D''] = D^2 = \Theta_{F,h}$. The formula follows. □

4.D. Vanishing Theorems

Assume that X is compact and that $u \in \mathscr{C}^\infty(X, \Lambda^{p,q}T^*X \otimes F)$ is an arbitrary (p,q)-form. An integration by parts yields

$$\langle \Delta' u, u\rangle = \|D'u\|^2 + \|D'^*u\|^2 \geqslant 0$$

and similarly for Δ'', hence we get the basic a priori inequality

$$\|D''u\|^2 + \|D''^*u\|^2 \geqslant \int_X \langle [\mathrm{i}\Theta_{F,h}, \Lambda]u, u\rangle dV_\omega. \tag{4.7}$$

This inequality is known as the *Bochner-Kodaira-Nakano* inequality (see [Boc48; Kod53; Nak55]). When u is Δ''-harmonic, we get

$$\int_X \big(\langle [\mathrm{i}\Theta_{F,h}, \Lambda]u, u\rangle + \langle T_\omega u, u\rangle\big) dV \leqslant 0.$$

If the Hermitian operator $[\mathrm{i}\Theta_{F,h}, \Lambda]$ acting on $\Lambda^{p,q}T^*_X \otimes F$ is positive on each fiber, we infer that u must be zero, hence

$$H^{p,q}(X,F) = \mathscr{H}^{p,q}(X,F) = 0$$

by Hodge theory. The main point is thus to compute the curvature form $\Theta_{F,h}$ and find sufficient conditions under which the operator $[\mathrm{i}\Theta_{F,h},\Lambda]$ is positive definite. Elementary (but somewhat tedious) calculations yield the following formulae: if the curvature of F is written as in (3.8) and $u=\sum u_{J,K,\lambda}dz_I\wedge d\overline{z}_J\otimes e_\lambda$, $|J|=p$, $|K|=q$, $1\leqslant\lambda\leqslant r$ is a (p,q)-form with values in F, then

$$(4.8)\qquad \langle[\mathrm{i}\Theta_{F,h},\Lambda]u,u\rangle=\sum_{j,k,\lambda,\mu,J,S} c_{jk\lambda\mu}\,u_{J,jS,\lambda}\,\overline{u_{J,kS,\mu}}+\sum_{j,k,\lambda,\mu,R,K} c_{jk\lambda\mu}\,u_{kR,K,\lambda}\,\overline{u_{jR,K,\mu}}-\sum_{j,\lambda,\mu,J,K} c_{jj\lambda\mu}\,u_{J,K,\lambda}\,\overline{u_{J,K,\mu}},$$

where the sum is extended to all indices $1\leqslant j,k\leqslant n$, $1\leqslant\lambda,\mu\leqslant r$ and multiindices $|R|=p-1$, $|S|=q-1$ (here the notation $u_{JK\lambda}$ is extended to non necessarily increasing multiindices by making it alternate with respect to permutations). It is usually hard to decide the sign of the curvature term (4.8), except in some special cases.

The easiest case is when $p=n$. Then all terms in the second summation of (4.8) must have $j=k$ and $R=\{1,\ldots,n\}\smallsetminus\{j\}$, therefore the second and third summations are equal. It follows that $[\mathrm{i}\Theta_{F,h},\Lambda]$ is positive on (n,q)-forms under the assumption that F is positive in the sense of Nakano. In this case X is automatically Kähler since

$$\omega=\mathrm{Tr}_F(\mathrm{i}\Theta_{F,h})=\mathrm{i}\sum_{j,k,\lambda}c_{jk\lambda\lambda}dz_j\wedge d\overline{z}_k=\mathrm{i}\Theta_{\det F,h}$$

is a Kähler metric.

(4.9) Nakano Vanishing Theorem ([Nak55]). *Let X be a compact complex manifold and let F be a Nakano positive vector bundle on X. Then*

$$H^{n,q}(X,F)=H^q(X,K_X\otimes F)=0\qquad \text{for every } q\geqslant 1.$$

Another tractable case is the case where F is a line bundle ($r=1$). Indeed, at each point $x\in X$, we may then choose a coordinate system which diagonalizes simultaneously the Hermitians forms $\omega(x)$ and $\mathrm{i}\Theta_{F,h}(x)$, in such a way that

$$\omega(x)=\mathrm{i}\sum_{1\leqslant j\leqslant n}dz_j\wedge d\overline{z}_j,\qquad \mathrm{i}\Theta_{F,h}(x)=\mathrm{i}\sum_{1\leqslant j\leqslant n}\gamma_j dz_j\wedge d\overline{z}_j$$

with $\gamma_1\leqslant\cdots\leqslant\gamma_n$. The curvature eigenvalues $\gamma_j=\gamma_j(x)$ are then uniquely defined and depend continuously on x. With our previous notation, we have

$\gamma_j = c_{jj11}$ and all other coefficients $c_{jk\lambda\mu}$ are zero. For any (p,q)-form $u = \sum u_{JK} dz_J \wedge d\overline{z}_K \otimes e_1$, this gives

$$(4.10)\quad \langle [\mathrm{i}\Theta_{F,h}, \Lambda]u, u\rangle = \sum_{|J|=p,\,|K|=q} \Big(\sum_{j\in J}\gamma_j + \sum_{j\in K}\gamma_j - \sum_{1\leqslant j\leqslant n}\gamma_j\Big)|u_{JK}|^2$$
$$\geqslant (\gamma_1 + \cdots + \gamma_q - \gamma_{n-p+1} - \cdots - \gamma_n)|u|^2.$$

Assume that $\mathrm{i}\Theta_{F,h}$ is positive. It is then natural to make the special choice $\omega = \mathrm{i}\Theta_{F,h}$ for the Kähler metric. Then $\gamma_j = 1$ for $j = 1, 2, \ldots, n$ and we obtain $\langle [\mathrm{i}\Theta_{F,h}, \Lambda]u, u\rangle = (p+q-n)|u|^2$. As a consequence:

(4.11) Akizuki-Kodaira-Nakano Vanishing Theorem ([AN54]). *If F is a positive line bundle on a compact complex manifold X, then*

$$H^{p,q}(X, F) = H^q(X, \Omega^p_X \otimes F) = 0 \qquad \textit{for}\ \ p+q \geqslant n+1.$$

More generally, if F is a Griffiths positive (or ample) vector bundle of rank $r \geqslant 1$, Le Potier [LP75] proved that $H^{p,q}(X, F) = 0$ for $p+q \geqslant n+r$. The proof is not a direct consequence of the Bochner technique. A rather easy proof has been found by M. Schneider [Sch74], using the Leray spectral sequence associated to the projectivized bundle projection $\mathbb{P}(F) \to X$, using the following more or less standard notation.

(4.12) Notation. If V is a complex vector space (resp. complex vector bundle), we let $P(V)$ be the projective space (resp. bundle) of *lines* of V, and $\mathbb{P}(V) = P(V^*)$ be the projective space (resp. bundle) of *hyperplanes* of V.

(4.13) Exercise. It is important for various applications to obtain vanishing theorems which are also valid in the case of semi-positive line bundles. The easiest case is the following result of Girbau [Gir76]: let (X, ω) be compact Kähler; assume that F is a line bundle and that $\mathrm{i}\Theta_{F,h} \geqslant 0$ has at least $n-k$ positive eigenvalues at each point, for some integer $k \geqslant 0$; show that $H^{p,q}(X, F) = 0$ for $p+q \geqslant n+k+1$.
Hint: use the Kähler metric $\omega_\varepsilon = \mathrm{i}\Theta_{F,h} + \varepsilon\omega$ with $\varepsilon > 0$ small.

A stronger and more natural "algebraic version" of this result has been obtained by Sommese [Som78]: define F to be k-ample if some multiple mF is such that the canonical map

$$\Phi_{|mF|} : X \smallsetminus B_{|mF|} \to \mathbb{P}^{N-1}$$

has at most k-dimensional fibers and $\dim B_{|mF|} \leqslant k$. If X is projective and F is k-ample, show that $H^{p,q}(X, F) = 0$ for $p+q \geqslant n+k+1$.
Hint: prove the dual result $H^{p,q}(X, F^{-1}) = 0$ for $p+q \leqslant n-k-1$ by induction on k. First show that F 0-ample $\Rightarrow$ F positive; then use hyperplane sections $Y \subset X$ to prove the induction step, thanks to the exact sequences

$$0 \longrightarrow \Omega^p_X \otimes F^{-1} \otimes \mathcal{O}(-Y) \longrightarrow \Omega^p_X \otimes F^{-1} \longrightarrow \big(\Omega^p_X \otimes F^{-1}\big)_{\restriction Y} \longrightarrow 0,$$

$$0 \longrightarrow \Omega^{p-1}_Y \otimes F^{-1}_{\restriction Y} \otimes \mathcal{O}(-Y)_{\restriction Y} \longrightarrow \big(\Omega^p_X \otimes F^{-1}\big)_{\restriction Y} \longrightarrow \Omega^p_Y \otimes F^{-1}_{\restriction Y} \longrightarrow 0. \qquad \square$$

Chapter 5

L^2 Estimates and Existence Theorems

5.A. Basic L^2 Existence Theorems

The starting point is the following L^2 existence theorem, which is essentially due to Hörmander [Hör65, 66], and Andreotti-Vesentini [AV65], following fundamental work by Kohn [Koh63, 64]. We will only present the strategy and the main ideas and tools, referring e.g. to [Dem82b] for a more detailed exposition of the technical situation considered here.

(5.1) Theorem. *Let (X, ω) be a Kähler manifold. Here X is not necessarily compact, but we assume that the geodesic distance δ_ω is complete on X. Let F be a Hermitian vector bundle of rank r over X, and assume that the curvature operator $A = A^{p,q}_{F,h,\omega} = [\mathrm{i}\Theta_{F,h}, \Lambda_\omega]$ is positive definite everywhere on $\Lambda^{p,q}T^*_X \otimes F$, $q \geqslant 1$. Then for any form $g \in L^2(X, \Lambda^{p,q}T^*_X \otimes F)$ satisfying $D''g = 0$ and $\int_X \langle A^{-1}g, g\rangle\, dV_\omega < +\infty$, there exists $f \in L^2(X, \Lambda^{p,q-1}T^*_X \otimes F)$ such that $D''f = g$ and*

$$\int_X |f|^2\, dV_\omega \leqslant \int_X \langle A^{-1}g, g\rangle\, dV_\omega.$$

Proof. The assumption that δ_ω is complete implies the existence of cut-off functions ψ_ν with arbitrarily large compact support such that $|d\psi_\nu| \leqslant 1$ (take ψ_ν to be a function of the distance $x \mapsto \delta_\omega(x_0, x)$, which is an almost everywhere differentiable 1-Lipschitz function, and regularize if necessary). From this, it follows that very form $u \in L^2(X, \Lambda^{p,q}T^*_X \otimes F)$ such that $D''u \in L^2$ and $D''^*u \in L^2$ in the sense of distribution theory is a limit of a sequence of smooth forms u_ν with compact support, in such a way that $u_\nu \to u$, $D''u_\nu \to D''u$ and $D''^*u_\nu \to D''^*u$ in L^2 (just take u_ν to be a regularization of $\psi_\nu u$). As a consequence, the basic a priori inequality (4.7) extends to arbitrary forms u such that $u,\ D''u, D''^*u \in L^2$. Now, consider the Hilbert space orthogonal decomposition:

$$L^2(X, \Lambda^{p,q}T^*_X \otimes F) = \mathrm{Ker}\, D'' \oplus (\mathrm{Ker}\, D'')^\perp,$$

observing that $\operatorname{Ker} D''$ is weakly (hence strongly) closed. Let $v = v_1 + v_2$ be the decomposition of a smooth form $v \in \mathcal{D}^{p,q}(X,F)$ with compact support according to this decomposition (v_1, v_2 do not have compact support in general!). Since $(\operatorname{Ker} D'')^\perp \subset \operatorname{Ker} D''^*$ by duality and $g, v_1 \in \operatorname{Ker} D''$ by hypothesis, we get $D''^* v_2 = 0$ and

$$|\langle g, v\rangle|^2 = |\langle g, v_1\rangle|^2 \leqslant \int_X \langle A^{-1}g, g\rangle \, dV_\omega \int_X \langle Av_1, v_1\rangle \, dV_\omega$$

thanks to the Cauchy-Schwarz inequality. The a priori inequality (4.7) applied to $u = v_1$ yields

$$\int_X \langle Av_1, v_1\rangle \, dV_\omega \leqslant \|D''v_1\|^2 + \|D''^* v_1\|^2 = \|D''^* v_1\|^2 = \|D''^* v\|^2.$$

Combining both inequalities, we find

$$|\langle g, v\rangle|^2 \leqslant \Big(\int_X \langle A^{-1}g, g\rangle \, dV_\omega\Big) \|D''^* v\|^2$$

for every smooth (p,q)-form v with compact support. This shows that we have a well defined linear form:

$$w = D''^* v \longmapsto \langle v, g\rangle, \qquad L^2(X, \Lambda^{p,q-1}T^*_X \otimes F) \supset D''^*(\mathcal{D}^{p,q}(F)) \longrightarrow \mathbb{C}$$

on the range of D''^*. This linear form is continuous in L^2 norm and has norm $\leqslant C$ with

$$C = \Big(\int_X \langle A^{-1}g, g\rangle \, dV_\omega\Big)^{1/2}.$$

By the Hahn-Banach theorem, there is an element $f \in L^2(X, \Lambda^{p,q-1}T^*_X \otimes F)$ with $||f|| \leqslant C$, such that $\langle v, g\rangle = \langle D''^* v, f\rangle$ for every v, hence $D''f = g$ in the sense of distributions. The inequality $||f|| \leqslant C$ is equivalent to the last estimate in the theorem. □

The above L^2 existence theorem can be applied in the fairly general context of *weakly pseudoconvex* manifolds. By this, we mean a complex manifold X such that there exists a smooth psh exhaustion function ψ on X (ψ is said to be an exhaustion if for every $c > 0$ the upperlevel set $X_c = \psi^{-1}(c)$ is relatively compact, i.e. $\psi(z)$ tends to $+\infty$ when z is taken outside larger and larger compact subsets of X). In particular, every compact complex manifold X is weakly pseudoconvex (take $\psi = 0$), as well as every Stein manifold, e.g. affine algebraic submanifolds of $\mathbb{C}^N$ (take $\psi(z) = |z|^2$), open balls $X = B(z_0, r)$ $\big($take $\psi(z) = 1/(r - |z - z_0|^2)\big)$, convex open subsets, etc. Now, a basic observation is that every weakly pseudoconvex Kähler manifold (X, ω) carries a *complete* Kähler metric: let $\psi \geqslant 0$ be a psh exhaustion function and set

$$\omega_\varepsilon = \omega + \varepsilon \, \mathrm{i} d'd''\psi^2 = \omega + 2\varepsilon(2\mathrm{i}\psi d'd''\psi + \mathrm{i} d'\psi \wedge d''\psi).$$

Then $|d\psi|_{\omega_\varepsilon} \leqslant 1/\varepsilon$ and $|\psi(x) - \psi(y)| \leqslant \varepsilon^{-1}\delta_{\omega_\varepsilon}(x, y)$. It follows easily from this estimate that the geodesic balls are relatively compact, hence $\delta_{\omega_\varepsilon}$ is complete for every $\varepsilon > 0$. Therefore, the L^2 existence theorem can be applied to each Kähler metric ω_ε, and by passing to the limit it can even be applied to the non necessarily complete metric ω. An important special case is the following

(5.2) Theorem. *Let (X, ω) be a Kähler manifold,* $\dim X = n$. *Assume that X is weakly pseudoconvex. Let F be a Hermitian line bundle and let*

$$\gamma_1(x) \leqslant \cdots \leqslant \gamma_n(x)$$

be the curvature eigenvalues (i.e. the eigenvalues of $\mathrm{i}\Theta_{F,h}$ *with respect to the metric ω) at every point. Assume that the curvature is positive, i.e. $\gamma_1 > 0$ everywhere. Then for any form $g \in L^2(X, \Lambda^{n,q}T_X^* \otimes F)$ satisfying $D''g = 0$ and $\int_X \langle(\gamma_1 + \cdots + \gamma_q)^{-1}|g|^2\, dV_\omega < +\infty$, there exists $f \in L^2(X, \Lambda^{p,q-1}T_X^* \otimes F)$ such that $D''f = g$ and*

$$\int_X |f|^2\, dV_\omega \leqslant \int_X (\gamma_1 + \cdots + \gamma_q)^{-1}|g|^2\, dV_\omega.$$

Proof. Indeed, for $p = n$, Formula (4.10) shows that

$$\langle Au, u\rangle \geqslant (\gamma_1 + \cdots + \gamma_q)|u|^2,$$

hence $\langle A^{-1}u, u\rangle \geqslant (\gamma_1 + \cdots + \gamma_q)^{-1}|u|^2$. □

An important observation is that the above theorem still applies when the Hermitian metric on F is a singular metric with positive curvature in the sense of currents. In fact, by standard regularization techniques (convolution of psh functions by smoothing kernels), the metric can be made smooth and the solutions obtained by Theorem 5.1 or Theorem 5.2 for the smooth metrics have limits satisfying the desired estimates. Especially, we get the following:

(5.3) Corollary. *Let (X, ω) be a Kähler manifold,* $\dim X = n$. *Assume that X is weakly pseudoconvex. Let F be a holomorphic line bundle equipped with a singular metric whose local weights are denoted $\varphi \in L^1_{\mathrm{loc}}$, i.e. $H = E^{-\varphi}$. Suppose that*

$$\mathrm{i}\Theta_{F,h} = \mathrm{i}d'd''\varphi \geqslant \varepsilon\omega$$

for some $\varepsilon > 0$. Then for any form $g \in L^2(X, \Lambda^{n,q}T_X^* \otimes F)$ satisfying $D''g = 0$, there exists $f \in L^2(X, \Lambda^{p,q-1}T_X^* \otimes F)$ such that $D''f = g$ and

$$\int_X |f|^2 e^{-\varphi}\, dV_\omega \leqslant \frac{1}{q\varepsilon}\int_X |g|^2 e^{-\varphi}\, dV_\omega.$$

Here we denoted somewhat incorrectly the metric by $|f|^2e^{-\varphi}$, as if the weight φ was globally defined on X (of course, this is so only if F is globally trivial). We will use this notation anyway, because it clearly describes the dependence of the L^2 norm on the psh weights.

5.B. Multiplier Ideal Sheaves and Nadel Vanishing Theorem

We now introduce the concept of *multiplier ideal sheaf*, following A. Nadel [Nad89]. The main idea actually goes back to the fundamental works of Bombieri [Bom70] and H. Skoda [Sko72a].

(5.4) Definition. *Let φ be a psh function on an open subset $\Omega \subset X$; to φ is associated the ideal subsheaf $\mathcal{I}(\varphi) \subset \mathcal{O}_\Omega$ of germs of holomorphic functions $f \in \mathcal{O}_{\Omega,x}$ such that $|f|^2 e^{-2\varphi}$ is integrable with respect to the Lebesgue measure in some local coordinates near x.*

The zero variety $V(\mathcal{I}(\varphi))$ is thus the set of points in a neighborhood of which $e^{-2\varphi}$ is non integrable. Of course, such points occur only if φ has logarithmic poles. This is made precise as follows.

(5.5) Definition. *A psh function φ is said to have a logarithmic pole of coefficient γ at a point $x \in X$ if the Lelong number*

$$\nu(\varphi, x) := \liminf_{z \to x} \frac{\varphi(z)}{\log|z-x|}$$

is non zero and if $\nu(\varphi, x) = \gamma$.

(5.6) Lemma (Skoda [Sko72a]). *Let φ be a psh function on an open set Ω and let $x \in \Omega$.*

(a) *If $\nu(\varphi, x) < 1$, then $e^{-2\varphi}$ is integrable in a neighborhood of x, in particular $\mathcal{I}(\varphi)_x = \mathcal{O}_{\Omega,x}$.*

(b) *If $\nu(\varphi, x) \geqslant n + s$ for some integer $s \geqslant 0$, then $e^{-2\varphi} \geqslant C|z-x|^{-2n-2s}$ in a neighborhood of x and $\mathcal{I}(\varphi)_x \subset \mathfrak{m}_{\Omega,x}^{s+1}$, where $\mathfrak{m}_{\Omega,x}$ is the maximal ideal of $\mathcal{O}_{\Omega,x}$.*

(c) *The zero variety $V(\mathcal{I}(\varphi))$ of $\mathcal{I}(\varphi)$ satisfies*

$$E_n(\varphi) \subset V(\mathcal{I}(\varphi)) \subset E_1(\varphi)$$

where $E_c(\varphi) = \{x \in X\,;\, \nu(\varphi, x) \geqslant c\}$ is the c-upperlevel set of Lelong numbers of φ.

Proof. (a) Set $\Theta = dd^c\varphi$ and $\gamma = \nu(\Theta, x) = \nu(\varphi, x)$. Let χ be a cut-off function with support in a small ball $B(x, r)$, equal to 1 in $B(x, r/2)$. As $(dd^c \log|z|)^n = \delta_0$, we get

$$\begin{aligned}\varphi(z) &= \int_{B(x,r)} \chi(\zeta)\varphi(\zeta)(dd^c \log|\zeta - z|)^n \\ &= \int_{B(x,r)} dd^c(\chi(\zeta)\varphi(\zeta)) \wedge \log|\zeta - z|(dd^c \log|\zeta - z|)^{n-1}\end{aligned}$$

for $z \in B(x, r/2)$. Expanding $dd^c(\chi\varphi)$ and observing that $d\chi = dd^c\chi = 0$ on $B(x, r/2)$, we find

$$\varphi(z) = \int_{B(x,r)} \chi(\zeta)\Theta(\zeta) \wedge \log|\zeta - z|(dd^c \log|\zeta - z|)^{n-1} + \text{smooth terms}$$

on $B(x, r/2)$. Fix r so small that

$$\int_{B(x,r)} \chi(\zeta)\Theta(\zeta) \wedge (dd^c \log|\zeta - x|)^{n-1} \leqslant \nu(\Theta, x, r) < 1.$$

By continuity, there exists $\delta, \varepsilon > 0$ such that

$$I(z) := \int_{B(x,r)} \chi(\zeta)\Theta(\zeta) \wedge (dd^c \log|\zeta - z|)^{n-1} \leqslant 1 - \delta$$

for all $z \in B(x, \varepsilon)$. Applying Jensen's convexity inequality to the probability measure

$$d\mu_z(\zeta) = I(z)^{-1}\chi(\zeta)\Theta(\zeta) \wedge (dd^c \log|\zeta - z|)^{n-1},$$

we find

$$-\varphi(z) = \int_{B(x,r)} I(z) \log|\zeta - z|^{-1}\, d\mu_z(\zeta) + O(1)$$

$$\Longrightarrow e^{-2\varphi(z)} \leqslant C \int_{B(x,r)} |\zeta - z|^{-2I(z)}\, d\mu_z(\zeta).$$

As

$$d\mu_z(\zeta) \leqslant C_1|\zeta - z|^{-(2n-2)}\Theta(\zeta) \wedge (dd^c|\zeta|^2)^{n-1} = C_2|\zeta - z|^{-(2n-2)} d\sigma_\Theta(\zeta),$$

we get

$$e^{-2\varphi(z)} \leqslant C_3 \int_{B(x,r)} |\zeta - z|^{-2(1-\delta)-(2n-2)} d\sigma_\Theta(\zeta),$$

and the Fubini theorem implies that $e^{-2\varphi(z)}$ is integrable on a neighborhood of x.

(b) If $\nu(\varphi, x) = \gamma$, the convexity properties of psh functions, namely, the convexity of $\log r \mapsto \sup_{|z-x|=r} \varphi(z)$ implies that

$$\varphi(z) \leqslant \gamma \log|z - x|/r_0 + M,$$

where M is the supremum on $B(x, r_0)$. Hence there exists a constant $C > 0$ such that $e^{-2\varphi(z)} \geqslant C|z - x|^{-2\gamma}$ in a neighborhood of x. The desired result follows from the estimate

$$\int_{B(0,r_0)} \frac{\left|\sum a_\alpha z^\alpha\right|^2}{|z|^{2\gamma}} dV(z) \sim \text{Const.} \int_0^{r_0} \left(\sum |a_\alpha|^2 r^{2|\alpha|}\right) r^{2n-1-2\gamma}\, dr,$$

which holds modulo multiplicative constants and is an easy consequence of Parseval's formula. Now, if γ has integral part $[\gamma] = n + s$, the integral converges if and only if $a_\alpha = 0$ for $|\alpha| \leqslant s$.

(c) is just a simple formal consequence of (a) and (b). □

(5.7) Proposition ([Nad89]). *For any psh function φ on $\Omega \subset X$, the sheaf $\mathcal{I}(\varphi)$ is a coherent sheaf of ideals over Ω. Moreover, if Ω is a bounded Stein open set, the sheaf $\mathcal{I}(\varphi)$ is generated by any Hilbert basis of the L^2 space $\mathcal{H}^2(\Omega, \varphi)$ of holomorphic functions f on Ω such that $\int_\Omega |f|^2 e^{-2\varphi} \, d\lambda < +\infty$.*

Proof. Since the result is local, we may assume that Ω is a bounded pseudoconvex open set in $\mathbb{C}^n$. By the strong noetherian property of coherent sheaves, the family of sheaves generated by finite subsets of $\mathcal{H}^2(\Omega, \varphi)$ has a maximal element on each compact subset of Ω, hence $\mathcal{H}^2(\Omega, \varphi)$ generates a coherent ideal sheaf $\mathcal{J} \subset \mathcal{O}_\Omega$. It is clear that $\mathcal{J} \subset \mathcal{I}(\varphi)$; in order to prove the equality, we need only check that $\mathcal{J}_x + \mathcal{I}(\varphi)_x \cap \mathfrak{m}_{\Omega,x}^{s+1} = \mathcal{I}(\varphi)_x$ for every integer s, in view of the Krull lemma. Let $f \in \mathcal{I}(\varphi)_x$ be defined in a neighborhood V of x and let θ be a cut-off function with support in V such that $\theta = 1$ in a neighborhood of x. We solve the equation $d''u = g := d''(\theta f)$ by means of Hörmander's L^2 estimates 5.3, where F is the trivial line bundle $\Omega \times \mathbb{C}$ equipped with the strictly psh weight

$$\widetilde{\varphi}(z) = \varphi(z) + (n + s) \log |z - x| + |z|^2.$$

We get a solution u such that $\int_\Omega |u|^2 e^{-2\varphi} |z - x|^{-2(n+s)} d\lambda < \infty$, thus $F = \theta f - u$ is holomorphic, $F \in \mathcal{H}^2(\Omega, \varphi)$ and $f_x - F_x = u_x \in \mathcal{I}(\varphi)_x \cap \mathfrak{m}_{\Omega,x}^{s+1}$. This proves the coherence. Now, $\mathcal{J}$ is generated by any Hilbert basis of $\mathcal{H}^2(\Omega, \varphi)$, because it is well-known that the space of sections of any coherent sheaf is a Fréchet space, therefore closed under local L^2 convergence. □

The multiplier ideal sheaves satisfy the following basic functoriality property with respect to direct images of sheaves by modifications.

(5.8) Proposition. *Let $\mu : X' \to X$ be a modification of non singular complex manifolds (i.e. a proper generically 1:1 holomorphic map), and let φ be a psh function on X. Then*

$$\mu_*\big(\mathcal{O}(K_{X'}) \otimes \mathcal{I}(\varphi \circ \mu)\big) = \mathcal{O}(K_X) \otimes \mathcal{I}(\varphi).$$

Proof. Let $n = \dim X = \dim X'$ and let $S \subset X$ be an analytic set such that $\mu : X' \smallsetminus S' \to X \smallsetminus S$ is a biholomorphism. By definition of multiplier ideal sheaves, $\mathcal{O}(K_X) \otimes \mathcal{I}(\varphi)$ is just the sheaf of holomorphic n-forms f on open sets $U \subset X$ such that $\mathrm{i}^{n^2} f \wedge \overline{f} \, e^{-2\varphi} \in L^1_{\mathrm{loc}}(U)$. Since φ is locally bounded from above, we may even consider forms f which are a priori defined only on $U \smallsetminus S$, because f will be in $L^2_{\mathrm{loc}}(U)$ and therefore will automatically extend through S. The change of variable formula yields

$$\int_U \mathrm{i}^{n^2} f \wedge \overline{f} \, e^{-2\varphi} = \int_{\mu^{-1}(U)} \mathrm{i}^{n^2} \mu^* f \wedge \overline{\mu^* f} \, e^{-2\varphi \circ \mu},$$

hence $f \in \Gamma(U, \mathcal{O}(K_X) \otimes \mathcal{I}(\varphi))$ iff $\mu^* f \in \Gamma(\mu^{-1}(U), \mathcal{O}(K_{X'}) \otimes \mathcal{I}(\varphi \circ \mu))$. Proposition 5.8 is proved. □

(5.9) Remark. If φ has analytic singularities (according to Definition 1.10), the computation of $\mathcal{I}(\varphi)$ can be reduced to a purely algebraic problem.

The first observation is that $\mathcal{I}(\varphi)$ can be computed easily if φ has the form $\varphi = \sum \alpha_j \log |g_j|$ where $D_j = g_j^{-1}(0)$ are nonsingular irreducible divisors with normal crossings. Then $\mathcal{I}(\varphi)$ is the sheaf of functions h on open sets $U \subset X$ such that

$$\int_U |h|^2 \prod |g_j|^{-2\alpha_j} dV < +\infty.$$

Since locally the g_j can be taken to be coordinate functions from a local coordinate system $(z_1, \ldots, z_n)$, the condition is that h is divisible by $\prod g_j^{m_j}$ where $m_j - \alpha_j > -1$ for each j, i.e. $m_j \geqslant \lfloor \alpha_j \rfloor$ (integer part). Hence

$$\mathcal{I}(\varphi) = \mathcal{O}(-\lfloor D \rfloor) = \mathcal{O}(-\sum \lfloor \alpha_j \rfloor D_j)$$

where $\lfloor D \rfloor$ denotes the integral part of the $\mathbb{Q}$-divisor $D = \sum \alpha_j D_j$.

Now, consider the general case of analytic singularities and suppose that

$$\varphi \sim \frac{\alpha}{2} \log \left(|f_1|^2 + \cdots + |f_N|^2\right)$$

near the poles. By the explanations given after Definition 1.10, we may assume that the $(f_j)'s$ are generators of the integrally closed ideal sheaf $\mathcal{J} = \mathcal{J}(\varphi/\alpha)$, defined as the sheaf of holomorphic functions h such that $|h| \leqslant C \exp(\varphi/\alpha)$. In this case, the computation is made as follows (see also L. Bonavero's work [Bon93], where similar ideas are used in connection with "singular" holomorphic Morse inequalities).

First, one computes a smooth modification $\mu : \widetilde{X} \to X$ of X such that $\mu^* \mathcal{J}$ is an invertible sheaf $\mathcal{O}(-D)$ associated with a normal crossing divisor $D = \sum \lambda_j D_j$, where (D_j) are the components of the exceptional divisor of $\widetilde{X}$ (take the blow-up X' of X with respect to the ideal $\mathcal{J}$ so that the pull-back of $\mathcal{J}$ to X' becomes an invertible sheaf $\mathcal{O}(-D')$, then blow up again by Hironaka [Hir64] to make X' smooth and D' have normal crossings). Now, we have $K_{\widetilde{X}} = \mu^* K_X + R$ where $R = \sum \rho_j D_j$ is the zero divisor of the Jacobi function J_μ of the blow-up map. By the direct image formula (5.8), we get

$$\mathcal{I}(\varphi) = \mu_*\big(\mathcal{O}(K_{\widetilde{X}} - \mu^* K_X) \otimes \mathcal{I}(\varphi \circ \mu)\big) = \mu_*\big(\mathcal{O}(R) \otimes \mathcal{I}(\varphi \circ \mu)\big).$$

Now, $\{f_j \circ \mu\}$ are generators of the ideal $\mathcal{O}(-D)$, hence

$$\varphi \circ \mu \sim \alpha \sum \lambda_j \log |g_j|$$

where g_j are local generators of $\mathcal{O}(-D_j)$. We are thus reduced to computing multiplier ideal sheaves in the case where the poles are given by a $\mathbb{Q}$-divisor with normal crossings $\sum \alpha\lambda_j D_j$. We obtain $\mathcal{I}(\varphi \circ \mu) = \mathcal{O}(-\sum \lfloor \alpha\lambda_j \rfloor D_j)$, hence

$$\mathcal{I}(\varphi) = \mu_* \mathcal{O}_{\widetilde{X}}\big(\sum (\rho_j - \lfloor \alpha\lambda_j \rfloor) D_j\big).$$ □

(5.10) Exercise. Compute the multiplier ideal sheaf $\mathcal{I}(\varphi)$ associated with the function $\varphi = \log(|z_1|^{\alpha_1} + \cdots + |z_p|^{\alpha_p})$ for arbitrary real numbers $\alpha_j > 0$.

Hint: using Parseval's formula and polar coordinates $z_j = r_j e^{i\theta_j}$, show that the problem is equivalent to determining for which p-tuples $(\beta_1, \ldots, \beta_p) \in \mathbb{N}^p$ the integral

$$\int_{[0,1]^p} \frac{r_1^{2\beta_1} \cdots r_p^{2\beta_p}\, r_1 dr_1 \cdots r_p dr_p}{r_1^{2\alpha_1} + \cdots + r_p^{2\alpha_p}} = \int_{[0,1]^p} \frac{t_1^{(\beta_1+1)/\alpha_1} \cdots t_p^{(\beta_p+1)/\alpha_p}}{t_1 + \cdots + t_p} \frac{dt_1}{t_1} \cdots \frac{dt_p}{t_p}$$

is convergent. Conclude from this that $\mathcal{I}(\varphi)$ is generated by the monomials $z_1^{\beta_1} \cdots z_p^{\beta_p}$ such that $\sum(\beta_p + 1)/\alpha_p > 1$. (This exercise shows that the analytic definition of $\mathcal{I}(\varphi)$ is sometimes also quite convenient for computations). □

Let F be a line bundle over X with a singular metric h of curvature current $\Theta_{F,h}$. If φ is the weight representing the metric in an open set $\Omega \subset X$, the ideal sheaf $\mathcal{I}(\varphi)$ is independent of the choice of the trivialization and so it is the restriction to Ω of a global coherent sheaf $\mathcal{I}(h)$ on X. We will sometimes still write $\mathcal{I}(h) = \mathcal{I}(\varphi)$ by abuse of notation. In this context, we have the following fundamental vanishing theorem, which is probably one of the most central results of analytic and algebraic geometry (as we will see later, it contains the Kawamata-Viehweg vanishing theorem as a special case).

(5.11) Nadel Vanishing Theorem ([Nad89; Dem93b]). *Let (X, ω) be a Kähler weakly pseudoconvex manifold, and let F be a holomorphic line bundle over X equipped with a singular Hermitian metric h of weight φ. Assume that $i\Theta_{F,h} \geqslant \varepsilon\omega$ for some continuous positive function ε on X. Then*

$$H^q\big(X, \mathcal{O}(K_X + F) \otimes \mathcal{I}(h)\big) = 0 \qquad \textit{for all } q \geqslant 1.$$

Proof. Let $\mathcal{L}^q$ be the sheaf of germs of (n, q)-forms u with values in F and with measurable coefficients, such that both $|u|^2 e^{-2\varphi}$ and $|d''u|^2 e^{-2\varphi}$ are locally integrable. The d'' operator defines a complex of sheaves $(\mathcal{L}^\bullet, d'')$ which is a resolution of the sheaf $\mathcal{O}(K_X + F) \otimes \mathcal{I}(\varphi)$: indeed, the kernel of d'' in degree 0 consists of all germs of holomorphic n-forms with values in F which satisfy the integrability condition; hence the coefficient function lies in $\mathcal{I}(\varphi)$; the exactness in degree $q \geqslant 1$ follows from Corollary 5.3 applied on arbitrary small balls. Each sheaf $\mathcal{L}^q$ is a $\mathcal{C}^\infty$-module, so $\mathcal{L}^\bullet$ is a resolution by acyclic sheaves. Let ψ be a smooth psh exhaustion function on X. Let us apply Corollary 5.3 globally on X, with the original metric of F multiplied by the factor $e^{-\chi\circ\psi}$, where χ is a convex increasing function of arbitrary fast growth at infinity. This factor can be used to ensure the convergence of integrals at infinity. By Corollary 5.3, we conclude that $H^q\big(\Gamma(X, \mathcal{L}^\bullet)\big) = 0$ for $q \geqslant 1$. The theorem follows. □

(5.12) Corollary. *Let (X, ω), F and φ be as in Theorem 5.11 and let $x_1, \ldots, x_N$ be isolated points in the zero variety $V(\mathcal{I}(\varphi))$. Then there is a surjective map*

$$H^0(X, K_X + F) \longrightarrow\!\!\!\!\!\rightarrow \bigoplus_{1 \leqslant j \leqslant N} \mathcal{O}(K_X + L)_{x_j} \otimes \big(\mathcal{O}_X/\mathcal{I}(\varphi)\big)_{x_j}.$$

Proof. Consider the long exact sequence of cohomology associated to the short exact sequence $0 \to \mathcal{I}(\varphi) \to \mathcal{O}_X \to \mathcal{O}_X/\mathcal{I}(\varphi) \to 0$ twisted by $\mathcal{O}(K_X + F)$, and apply Theorem 5.11 to obtain the vanishing of the first H^1 group. The asserted surjectivity property follows. □

(5.13) Corollary. *Let (X, ω), F and φ be as in Theorem 5.11 and suppose that the weight function φ is such that $\nu(\varphi, x) \geqslant n + s$ at some point $x \in X$ which is an isolated point of $E_1(\varphi)$. Then $H^0(X, K_X + F)$ generates all s-jets at x.*

Proof. The assumption is that $\nu(\varphi, y) < 1$ for y near x, $y \neq x$. By Skoda's Lemma 5.6 (b), we conclude that $e^{-2\varphi}$ is integrable at all such points y, hence $\mathcal{I}(\varphi)_y = \mathcal{O}_{X,y}$, whilst $\mathcal{I}(\varphi)_x \subset \mathfrak{m}_{X,x}^{s+1}$ by Lemma 5.6 (a). Corollary 5.13 is thus a special case of 5.12. □

The philosophy of these results (which can be seen as generalizations of the Hörmander-Bombieri-Skoda theorem [Bom70; Sko72a, 75]) is that the problem of constructing holomorphic sections of $K_X + F$ can be solved by constructing suitable Hermitian metrics on F such that the weight φ has isolated poles at given points x_j.

(5.14) Exercise. Assume that X is compact and that L is a positive line bundle on X. Let $\{x_1, \ldots, x_N\}$ be a finite set. Show that there are constants $a, b \geqslant 0$ depending only on L and N such that $H^0(X, mL)$ generates jets of any order s at all points x_j for $m \geqslant as + b$.
Hint: Apply Corollary 5.12 to $F = -K_X + mL$, with a singular metric on L of the form $h = h_0 e^{-\varepsilon\psi}$, where h_0 is smooth of positive curvature, $\varepsilon > 0$ small and $\psi(z) \sim \log|z - x_j|$ in a neighborhood of x_j. □

Derive the Kodaira embedding theorem from the above result:

(5.15) Theorem (Kodaira embedding theorem). *If L is a line bundle on a compact complex manifold, then L is ample if and only if L is positive.* □

(5.16) Exercise (solution of the Levi problem). Show that the following two properties are equivalent.

(a) X is strongly pseudoconvex, i.e. X admits a strongly psh exhaustion function.

(b) X is Stein, i.e. the global holomorphic functions $H^0(X, \mathcal{O}_X)$ separate points and yield local coordinates at any point, and X is holomorphically convex (this means that for any discrete sequence z_ν there is a function $f \in H^0(X, \mathcal{O}_X)$ such that $|f(z_\nu)| \to \infty$). □

(5.17) Remark. As long as forms of bidegree (n, q) are considered, the L^2 estimates can be extended to complex spaces with arbitrary singularities. In fact, if X is a complex space and φ is a psh weight function on X, we may still define a sheaf $K_X(\varphi)$ on X, such that the sections on an open set U are the holomorphic n-forms f on the regular part $U \cap X_{\rm reg}$, satisfying the integrability condition $\mathrm{i}^{n^2} f \wedge \overline{f}\, e^{-2\varphi} \in L^1_{\rm loc}(U)$. In this setting, the functoriality Property 5.8 becomes

$$\mu_*\big(K_{X'}(\varphi \circ \mu)\big) = K_X(\varphi)$$

for arbitrary complex spaces X, X' such that $\mu : X' \to X$ is a modification. If X is nonsingular we have $K_X(\varphi) = \mathcal{O}(K_X) \otimes \mathcal{I}(\varphi)$, however, if X is singular, the symbols K_X and $\mathcal{I}(\varphi)$ must not be dissociated. The statement of the Nadel vanishing theorem becomes $H^q(X, \mathcal{O}(F) \otimes K_X(\varphi)) = 0$ for $q \geqslant 1$, under the same assumptions (X Kähler and weakly pseudoconvex, curvature $\geqslant \varepsilon\omega$). The proof can be obtained by restricting everything to $X_{\rm reg}$. Although in general $X_{\rm reg}$ is not weakly pseudoconvex (e.g. in case $\operatorname{codim} X_{\rm sing} \geqslant 2$), $X_{\rm reg}$ is always Kähler complete (the complement of a proper analytic subset in a Kähler weakly pseudoconvex space is complete Kähler, see e.g. [Dem82b]). As a consequence, Nadel's vanishing theorem is essentially insensitive to the presence of singularities. □

Chapter 6

Numerically Effective and Pseudo-effective Line Bundles

6.A. Pseudo-effective Line Bundles and Metrics with Minimal Singularities

The concept of pseudo-effectivity is quite general and makes sense on an arbitrary compact complex manifold X (no projective or Kähler assumption is needed). In this general context, it is better to work with $\partial\overline{\partial}$-cohomology classes instead of De Rham cohomology classes: we define the *Bott-Chern cohomology* of X to be

$$H^{p,q}_{\mathrm{BC}}(X) = \{d\text{-closed } (p,q)\text{-forms}\}/\{\partial\overline{\partial}\text{-exact } (p,q)\text{-forms}\}. \tag{6.1}$$

By means of the Frölicher spectral sequence, it is easily shown that these cohomology groups are finite dimensional and can be computed either with spaces of smooth forms or with currents. In both cases, the quotient topology of $H^{p,q}_{\mathrm{BC}}(X)$ induced by the Fréchet topology of smooth forms or by the weak topology of currents is Hausdorff. Clearly $H^{\bullet}_{\mathrm{BC}}(X)$ is a bigraded algebra. This algebra can be shown to be isomorphic to the usual De Rham cohomology algebra $H^{\bullet}(X, \mathbb{C})$ if X is Kähler (or more generally if X is in the Fujiki class $\mathscr{C}$ of manifolds bimeromorphic to Kähler manifolds).

(6.2) Definition. *Let L be a holomorphic line bundle on a compact complex manifold X. be say that L pseudo-effective if $c_1(L) \in H^{1,1}_{\mathrm{BC}}(X)$ is the cohomology class of some closed positive current T, i.e. if L can be equipped with a singular Hermitian metric h with $T = \frac{\mathrm{i}}{2\pi}\Theta_{L,h} \geqslant 0$ as a current.*

The locus where h has singularities turns out to be extremely important. The following definition was introduced in [DPS00].

(6.3) Definition. *Let L be a pseudo-effective line bundle on a compact complex manifold X. Consider two Hermitian metrics h_1, h_2 on L with curvature $\mathrm{i}\Theta_{L,h_j} \geqslant 0$ in the sense of currents.*

(a) *We will write* $h_1 \preccurlyeq h_2$, *and say that* h_1 *is less singular than* h_2, *if there exists a constant* $C > 0$ *such that* $h_1 \leqslant Ch_2$.

(b) *We will write* $h_1 \sim h_2$, *and say that* h_1, h_2 *are equivalent with respect to singularities, if there exists a constant* $C > 0$ *such that* $C^{-1}h_2 \leqslant h_1 \leqslant Ch_2$.

Of course $h_1 \preccurlyeq h_2$ if and only if the associated weights in suitable trivializations locally satisfy $\varphi_2 \leqslant \varphi_1 + C$. This implies in particular $\nu(\varphi_1, x) \leqslant \nu(\varphi_2, x)$ at each point. The above definition is motivated by the following observation.

(6.4) Theorem. *For every pseudo-effective line bundle* L *over a compact complex manifold* X, *there exists up to equivalence of singularities a unique class of Hermitian metrics* h *with minimal singularities such that* $\mathrm{i}\Theta_{L,h} \geqslant 0$.

Proof. The proof is almost trivial. We fix once for all a smooth metric h_∞ (whose curvature is of random sign and signature), and we write singular metrics of L under the form $h = h_\infty e^{-\psi}$. The condition $\mathrm{i}\Theta_{L,h} \geqslant 0$ is equivalent to $\frac{\mathrm{i}}{2\pi}\partial\overline{\partial}\psi \geqslant -u$ where $u = \frac{\mathrm{i}}{2\pi}\Theta_{L,h_\infty}$. This condition implies that ψ is plurisubharmonic up to the addition of the weight φ_∞ of h_∞, and therefore locally bounded from above. Since we are concerned with metrics only up to equivalence of singularities, it is always possible to adjust ψ by a constant in such a way that $\sup_X \psi = 0$. We now set

$$h_{\min} = h_\infty e^{-\psi_{\min}}, \qquad \psi_{\min}(x) = \sup_\psi \psi(x)$$

where the supremum is extended to all functions ψ such that $\sup_X \psi = 0$ and $\frac{\mathrm{i}}{2\pi}\partial\overline{\partial}\psi \geqslant -u$. By standard results on plurisubharmonic functions (see Lelong [Lel69]), $\psi_{\min}$ still satisfies $\frac{\mathrm{i}}{2\pi}\partial\overline{\partial}\psi_{\min} \geqslant -u$ (i.e. the weight $\varphi_\infty + \psi_{\min}$ of $h_{\min}$ is plurisubharmonic), and $h_{\min}$ is obviously the metric with minimal singularities that we were looking for. [In principle one should take the upper semicontinuous regularization $\psi^*_{\min}$ of $\psi_{\min}$ to really get a plurisubharmonic weight, but since $\psi^*_{\min}$ also participates to the upper envelope, we obtain here $\psi_{\min} = \psi^*_{\min}$ automatically]. □

(6.5) Remark. In general, the supremum $\psi = \sup_{j \in I} \psi_j$ of a locally dominated family of plurisubharmonic functions ψ_j is not plurisubharmonic strictly speaking, but its "upper semi-continuous regularization" $\psi^*(z) = \limsup_{\zeta \to z} \psi(\zeta)$ is plurisubharmonic and coincides almost everywhere with ψ, with $\psi^* \geqslant \psi$. However, in the context of (6.5), ψ^* still satisfies $\psi^* \leqslant 0$ and $\frac{\mathrm{i}}{2\pi}\partial\overline{\partial}\psi \geqslant -u$, hence ψ^* participates to the upper envelope. As a consequence, we have $\psi^* \leqslant \psi$ and thus $\psi = \psi^*$ is indeed plurisubharmonic. Under a strict positivity assumption, namely if L is a big line bundle (i.e. the curvature can be taken to be strictly positive in the sense of currents, see Definition 6.12 and Theorem (6.17 b), then $h_{\min}$ can be shown to possess some regularity properties. The reader may consult [BmD09] for a rather general (but certainly non trivial) proof that $\psi_{\min}$ possesses locally bounded second derivatives $\partial^2\psi_{\min}/\partial z_j\partial\overline{z}_k$ outside an analytic set $Z \subset X$; in other words, $\mathrm{i}\Theta_{L,h_{\min}}$ has locally bounded coefficients on $X \smallsetminus Z$. See also (18.32) for further consequences. □

(6.6) Definition. *Let L be a pseudo-effective line bundle. If h is a singular Hermitian metric such that $\mathrm{i}\Theta_{L,h} \geqslant 0$ and*

$$H^0(X, mL \otimes \mathcal{I}(h^{\otimes m})) \simeq H^0(X, mL) \qquad \text{for all } m \geqslant 0,$$

we say that h is an analytic Zariski decomposition of L.

In other words, we require that h has singularities so mild that the vanishing conditions prescribed by the multiplier ideal sheaves $\mathcal{I}(h^{\otimes m})$ do not kill any sections of L and its multiples.

(6.7) Exercise. A special case is when there is an isomorphism $pL = A + E$ where A and E are effective divisors such that $H^0(X, mpL) = H^0(X, mA)$ for all m and $\mathcal{O}(A)$ is generated by sections. Then A possesses a smooth Hermitian metric h_A, and this metric defines a singular Hermitian metric h on L with poles $\frac{1}{p}E$ and curvature $\frac{1}{p}\Theta_{A,h_A} + \frac{1}{p}[E]$. Show that this metric h is an analytic Zariski decomposition.
Note: when X projective and there is a decomposition $pL = A + E$ with A nef (see (6.9)), E effective and $H^0(X, mpL) = H^0(X, mA)$ for all m, one says that the $\mathbb{Q}$-divisor equality $L = \frac{1}{p}A + \frac{1}{p}E$ is an *algebraic Zariski decomposition* of L. It can be shown that Zariski decompositions exist in dimension 2, but in higher dimension it can be seen that they do not exist in general. □

(6.8) Theorem. *The metric $h_{\min}$ with minimal singularities provides an analytic Zariski decomposition.*

It follows that an analytic Zariski decomposition always exists (while algebraic decompositions do not exist in general, especially in dimension 3 and more).

Proof. Let $\sigma \in H^0(X, mL)$ be any section. Then we get a singular metric h on L by putting $|\xi|_h = |\xi/\sigma(x)^{1/m}|$ for $\xi \in L_x$, and it is clear that $|\sigma|_{h^m} = 1$ for this metric. Hence $\sigma \in H^0(X, mL \otimes \mathcal{I}(h^{\otimes m}))$, and a fortiori $\sigma \in H^0(X, mL \otimes \mathcal{I}(h_{\min}^{\otimes m}))$ since $h_{\min}$ is less singular than h. □

6.B. Nef Line Bundles

Many problems of algebraic geometry (e.g. problems of classification of algebraic surfaces or higher dimensional varieties) lead in a natural way to the study of line bundles satisfying semipositivity conditions. It turns out that semipositivity in the sense of curvature (at least, as far as smooth metrics are considered) is not a very satisfactory notion. A more flexible notion perfectly suitable for algebraic purposes is the notion of *numerical effectivity*. The goal of this section is to give a few fundamental algebraic definitions and to discuss their differential geometric counterparts. We first suppose that X is a projective algebraic manifold, $\dim X = n$.

(6.9) Definition. *A holomorphic line bundle L over a projective manifold X is said to be numerically effective, nef for short, if $L \cdot C = \int_C c_1(L) \geqslant 0$ for every curve $C \subset X$.*

If L is nef, it can be shown that $L^p \cdot Y = \int_Y c_1(L)^p \geqslant 0$ for any p-dimensional subvariety $Y \subset X$ (see e.g. [Har70]). In relation to this, let us recall the Nakai-Moishezon ampleness criterion: a line bundle L is ample if and only if $L^p \cdot Y > 0$ for every p-dimensional subvariety Y (related stronger statements will be proved in Chapter 17). From this, we easily infer

(6.10) Proposition. *Let L be a line bundle on a projective algebraic manifold X, on which an ample line bundle A and a Hermitian metric ω are given. The following properties are equivalent:*

(a) *L is nef;*

(b) *for any integer $k \geqslant 1$, the line bundle $kL + A$ is ample;*

(c) *for every $\varepsilon > 0$, there is a smooth metric h_ε on L such that $\mathrm{i}\Theta_{L,h_\varepsilon} \geqslant -\varepsilon\omega$.*

Proof. (a) $\Rightarrow$ (b). If L is nef and A is ample then clearly $kL + A$ satisfies the Nakai-Moishezon criterion, hence $kL + A$ is ample.

(b) $\Rightarrow$ (c). Condition (c) is independent of the choice of the Hermitian metric, so we may select a metric h_A on A with positive curvature and set $\omega = \mathrm{i}\Theta_{A,h_A}$. If $kL + A$ is ample, this bundle has a metric h_{kL+A} of positive curvature. Then the metric $h_L = (h_{kL+A} \otimes h_A^{-1})^{1/k}$ has curvature

$$\mathrm{i}\Theta_{L,h_L} = \frac{1}{k}\big(\mathrm{i}\Theta_{kL+A} - \mathrm{i}\Theta_A\big) \geqslant -\frac{1}{k}\mathrm{i}\Theta_{A,h_A}\,;$$

in this way the negative part can be made smaller than $\varepsilon\,\omega$ by taking k large enough.

(c) $\Rightarrow$ (a). Under hypothesis (c), we get $L \cdot C = \int_C \frac{\mathrm{i}}{2\pi}\Theta_{L,h_\varepsilon} \geqslant -\frac{\varepsilon}{2\pi}\int_C \omega$ for every curve C and every $\varepsilon > 0$, hence $L \cdot C \geqslant 0$ and L is nef. □

Let now X be an arbitrary compact complex manifold. Since there need not exist any curve in X, Property 6.10 (c) is simply taken as a definition of nefness ([DPS94]):

(6.11) Definition. *A line bundle L on a compact complex manifold X is said to be nef if for every $\varepsilon > 0$, there is a smooth Hermitian metric h_ε on L such that $\mathrm{i}\Theta_{L,h_\varepsilon} \geqslant -\varepsilon\omega$.*

In general, it is not possible to extract a smooth limit h_0 such that $\mathrm{i}\Theta_{L,h_0} \geqslant 0$. The following simple example is given in [DPS94] (Example 1.7). Let E be a non trivial extension $0 \to \mathcal{O} \to E \to \mathcal{O} \to 0$ over an elliptic curve C and let $X = \mathbb{P}(E)$ (with notation as in (4.12)) be the corresponding ruled surface over C. Then $L = \mathcal{O}_{\mathbb{P}(E)}(1)$ is nef but does not admit any smooth metric of nonnegative curvature. In fact one can show that up to a constant factor there is only one singular Hermitian metric with semi-positive curvature current, associated with the section of L defined by the inclusion $\mathcal{O} \to E$; its curvature current is the current of integration $[C]$ on a curve $C \subset X$ which is a section of $X \to C$. This example answers negatively a question raised by Fujita [Fuj83].

Let us now introduce the important concept of *Kodaira-Iitaka dimension* of a line bundle.

(6.12) Definition. *If L is a line bundle, the Kodaira-Iitaka dimension $\kappa(L)$ is the supremum of the rank of the canonical maps:*

$$\Phi_m : X \smallsetminus B_m \longrightarrow \mathbb{P}(V_m), \qquad x \longmapsto H_x = \{\sigma \in V_m\,;\, \sigma(x) = 0\}, \qquad m \geqslant 1$$

with $V_m = H^0(X, mL)$ and $B_m = \bigcap_{\sigma \in V_m} \sigma^{-1}(0) =$ base locus of V_m. In case $V_m = \{0\}$ for all $m \geqslant 1$, we set $\kappa(L) = -\infty$. A line bundle is said to be big *if $\kappa(L) = \dim X$.*

The following lemma is well-known (the proof is a rather elementary consequence of the Schwarz lemma).

(6.13) Serre-Siegel Lemma ([Ser54; Sie55]). *Let L be a holomorphic line bundle on a compact complex manifold. Then we have :*

(a) *$h^0(X, mL) \leqslant O(m^{\kappa(L)})$ for $m \geqslant 1$;*

(b) *$\kappa(L)$ is the smallest constant for which this estimate holds;*

(c) *the volume of L defined as*

$$\mathrm{Vol}(L) = \limsup_{k \to +\infty} \frac{n!}{k^n} h^0(X, kL)$$

is finite, and L is big if and only if $\mathrm{Vol}(L) > 0$.

Notice that if L is ample, we have $h^q(X, kL) = 0$ for $q \geqslant 1$ and $k \gg 1$ by the Kodaira-Serre vanishing theorem, hence

$$h^0(X, kL) \sim \chi(X, kL) \sim \frac{L^n}{n!} k^n$$

by the Riemann-Roch formula. Thus $\mathrm{Vol}(L) = L^n \;(= c_1(L)^n)$ if L is ample. This is still true if X is Kähler and L is nef. In fact, in that case, we will show later (see Corollary 8.3) that $h^q(X, kL) = o(k^n)$ for $q \geqslant 1$ (in the projective algebraic case, one can even show that $h^q(X, kL) = O(k^{n-q})$, see Lemma 6.18).

6.C. Description of the Positive Cones

Let us recall that an integral cohomology class in $H^2(X,\mathbb{Z})$ is the first Chern class of a holomorphic (or algebraic) line bundle if and only if it lies in the *Neron-Severi* group

$$\mathrm{NS}(X) = \mathrm{Ker}\left(H^2(X,\mathbb{Z}) \to H^2(X,\mathcal{O}_X)\right) \tag{6.14}$$

(this fact is just an elementary consequence of the exponential exact sequence $0 \to \mathbb{Z} \to \mathcal{O} \to \mathcal{O}^* \to 0$). If X is compact Kähler, as we will suppose from now on in this section, this is the same as saying that the class is of type $(1,1)$ with respect to Hodge decomposition.

Let us consider the real vector space $\mathrm{NS}_{\mathbb{R}}(X) = \mathrm{NS}(X) \otimes_{\mathbb{Z}} \mathbb{R}$, which can be viewed as a subspace of the space $H^{1,1}(X,\mathbb{R})$ of real $(1,1)$ cohomology classes. Its dimension is by definition the Picard number

$$\rho(X) = \mathrm{rank}_{\mathbb{Z}}\,\mathrm{NS}(X) = \dim_{\mathbb{R}} \mathrm{NS}_{\mathbb{R}}(X). \tag{6.15}$$

We thus have $0 \leqslant \rho(X) \leqslant h^{1,1}(X)$, and the example of complex tori shows that all intermediate values can occur when $n = \dim X \geqslant 2$.

The positivity concepts for line bundles considered in Sections 6.A and 6.B possess in fact natural generalizations to $(1,1)$ classes which are not necessarily integral or rational — and this works at least in the category of compact Kähler manifolds (in fact, by using Bott-Chern cohomology, one could even extend these concepts to arbitrary compact complex manifolds).

(6.16) Definition. *Let (X,ω) be a compact Kähler manifold.*

(a) *The Kähler cone is the set $\mathcal{K} \subset H^{1,1}(X,\mathbb{R})$ of cohomology classes $\{\omega\}$ of Kähler forms. This is an* open convex cone.

(b) *The closure $\overline{\mathcal{K}}$ of the Kähler cone consists of classes $\{\alpha\} \in H^{1,1}(X,\mathbb{R})$ such that for every $\varepsilon > 0$ the sum $\{\alpha + \varepsilon\omega\}$ is Kähler, or equivalently, for every $\varepsilon > 0$, there exists a smooth function φ_ε on X such that $\alpha + i\partial\overline{\partial}\varphi_\varepsilon \geqslant -\varepsilon\omega$. We say that $\overline{\mathcal{K}}$ is the cone of* nef *$(1,1)$-classes.*

(c) *The pseudo-effective cone is the set $\mathcal{E} \subset H^{1,1}(X,\mathbb{R})$ of cohomology classes $\{T\}$ of closed positive currents of type $(1,1)$. This is a* closed convex cone.

(d) *The interior $\mathcal{E}^\circ$ of $\mathcal{E}$ consists of classes which still contain a closed positive current after one subtracts $\varepsilon\{\omega\}$ for $\varepsilon > 0$ small, in other words, they are classes of closed $(1,1)$-currents T such that $T \geqslant \varepsilon\omega$. Such a current will be called a* Kähler current, *and we say that $\{T\} \in H^{1,1}(X,\mathbb{R})$ is a* big $(1,1)$-class.

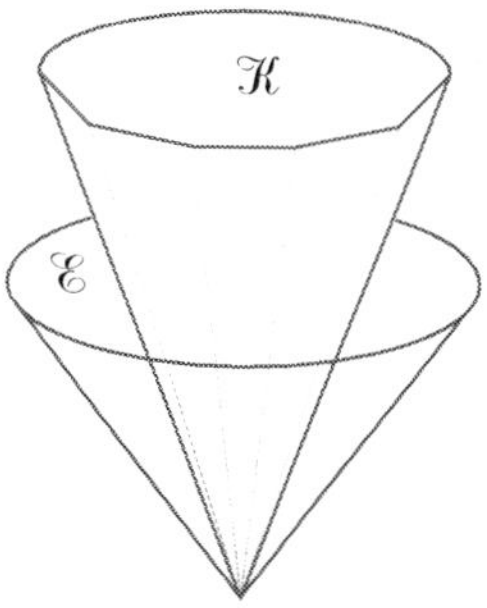

$\mathcal{K}$ = Kähler cone in $H^{1,1}(X,\mathbb{R})$ [open]

$\overline{\mathcal{K}}$ = nef cone in $H^{1,1}(X,\mathbb{R})$ [closure of $\mathcal{K}$]

$\mathcal{E}$ = pseudo-effective cone in $H^{1,1}(X,\mathbb{R})$ [closed]

$\mathcal{E}^\circ$ = big cone in $H^{1,1}(X,\mathbb{R})$ [interior of $\mathcal{E}$]

The openness of $\mathcal{K}$ is clear by definition, and the closedness of $\mathcal{E}$ is a consequence of the fact that bounded sets of currents are weakly compact (as follows from the similar weak compactness property for bounded sets of positive measures). It is then clear that $\overline{\mathcal{K}} \subset \mathcal{E}$.

In spite of the fact that cohomology groups can be defined either in terms of forms or currents, it turns out that the cones $\overline{\mathcal{K}}$ and $\mathcal{E}$ are in general different. To see this, it is enough to observe that a Kähler class $\{\alpha\}$ satisfies $\int_Y \alpha^p > 0$ for every p-dimensional analytic set. On the other hand, if X is the surface obtained by blowing-up $\mathbb{P}^2$ in one point, then the exceptional divisor $E \simeq \mathbb{P}^1$ has a cohomology class $\{\alpha\}$ such that $\int_E \alpha = E^2 = -1$, hence $\{\alpha\} \notin \overline{\mathcal{K}}$, although $\{\alpha\} = \{[E]\} \in \mathcal{E}$.

In case X is projective, all Chern classes $c_1(L)$ of line bundles lie by definition in $\mathrm{NS}(X)$, and likewise, all classes of real divisors $D = \sum c_j D_j$, $c_j \in \mathbb{R}$, lie in $\mathrm{NS}_\mathbb{R}(X)$. In order to deal with such *algebraic classes*, we therefore introduce the intersections

$$\mathcal{K}_{\mathrm{NS}} = \mathcal{K} \cap \mathrm{NS}_\mathbb{R}(X), \qquad \mathcal{E}_{\mathrm{NS}} = \mathcal{E} \cap \mathrm{NS}_\mathbb{R}(X),$$

and refer to classes of $H^{1,1}(X,\mathbb{R})$ not contained in $\mathrm{NS}_\mathbb{R}(X)$ as *transcendental classes.*

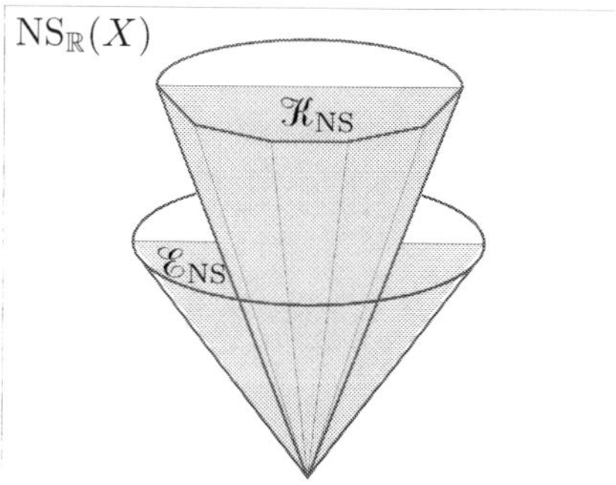

A very important fact is that all four cones $\mathcal{K}_{\mathrm{NS}}$, $\mathcal{E}_{\mathrm{NS}}$, $\overline{\mathcal{K}}_{\mathrm{NS}}$, $\mathcal{E}^\circ_{\mathrm{NS}}$ have simple algebraic interpretations.

(6.17) Theorem. *Let X be a projective manifold. Then*

(a) *$\mathcal{K}_{\mathrm{NS}}$ is equal to the open cone* $\mathrm{Amp}(X)$ *generated by classes of* ample (*or very ample*) *divisors A (Recall that a divisor A is said to be very ample if the linear system $H^0(X,\mathcal{O}(A))$ provides an embedding of X in projective space).*

(b) *The interior $\mathcal{E}^{\circ}_{\mathrm{NS}}$ is the cone* $\mathrm{Big}(X)$ *generated by classes of* big *divisors, namely divisors D such that $h^0(X,\mathcal{O}(kD)) \geqslant c\,k^{\dim X}$ for k large.*

(c) *$\mathcal{E}_{\mathrm{NS}}$ is the closure* $\overline{\mathrm{Eff}(X)}$ *of the cone generated by classes of* effective *divisors, i.e. divisors $D = \sum c_j D_j$, $c_j \in \mathbb{R}_+$.*

(d) *The closed cone $\overline{\mathcal{K}}_{\mathrm{NS}}$ consists of the closure* $\overline{\mathrm{Nef}(X)}$ *of the cone generated by* nef *divisors D (or nef line bundles L), namely effective integral divisors D such that $D \cdot C \geqslant 0$ for every curve C, also equal to* $\overline{\mathrm{Amp}(X)}$.

In other words, the terminology "nef", "big", "pseudo-effective" used for classes of the full transcendental cones appear to be a natural extrapolation of the algebraic case.

Proof. First notice that since all of our cones $\mathcal{C}$ have non empty interior in $\mathrm{NS}_{\mathbb{R}}(X)$ (which is a rational vector space in terms of a basis of elements in $H^2(X,\mathbb{Q})$), the rational points $\mathcal{C}_{\mathbb{Q}} := \mathcal{C} \cap \mathrm{NS}_{\mathbb{Q}}(X)$, $\mathrm{NS}_{\mathbb{Q}}(X) = \mathrm{NS}(X) \otimes_{\mathbb{Z}} \mathbb{Q}$, are dense in each of them.

(a) is therefore just Kodaira's embedding theorem when we look at rational points, and properties (c) and (d) are obtained easily by passing to the closure of the open cones. We will now give details of the proof only for (b) which is possibly slightly more involved.

By looking at points of $\mathcal{E}^{\circ}_{\mathbb{Q}} = \mathcal{E}^{\circ} \cap \mathrm{NS}_{\mathbb{Q}}(X)$ and multiplying by a denominator, it is enough to check that a line bundle L such that $c_1(L) \in \mathcal{E}^{\circ}$ is big. However, this means that L possesses a singular Hermitian metric h_L such that $\Theta_{L,h_L} \geqslant \varepsilon\omega$ for some Kähler metric ω. For some integer $p_0 > 0$, we can then produce a singular Hermitian metric with positive curvature and with a given logarithmic pole $h_L^{p_0} e^{-\theta(z)\log|z-x_0|^2}$ in a neighborhood of every point $x_0 \in X$ (here θ is a smooth cut-off function supported on a neighborhood of x_0). Then Hörmander's L^2 existence theorem [Hör65] can be used to produce sections of L^k which generate all jets of order $(k/p_0) - n$ at points x_0, so that L is big.

Conversely, if L is big and A is a (smooth) very ample divisor, the exact sequence $0 \to \mathcal{O}_X(kL - A) \to \mathcal{O}_X(kL) \to \mathcal{O}_A(kL_{\restriction A}) \to 0$ and the estimates $h^0(X,\mathcal{O}_X(kL)) \geqslant ck^n$, $h^0(A,\mathcal{O}_A(kL_{\restriction A})) = O(k^{n-1})$ imply that $\mathcal{O}_X(kL - A)$ has a section for k large, thus $kL - A \equiv D$ for some effective divisor D. This means that there exists a singular metric h_L on L such that

$$\frac{\mathrm{i}}{2\pi}\Theta_{L,h_L} = \frac{1}{k}\Big(\frac{\mathrm{i}}{2\pi}\Theta_{A,h_A} + [D]\Big) \geqslant \frac{1}{k}\omega$$

where $\omega = \frac{\mathrm{i}}{2\pi}\Theta_{A,h_A}$, hence $c_1(L) \in \mathcal{E}^{\circ}$. □

Before going further, we need a lemma.

(6.18) Lemma. *Let X be a compact Kähler n-dimensional manifold, let L be a nef line bundle on X, and let E be an arbitrary holomorphic vector bundle. Then $h^q(X, \mathcal{O}(E) \otimes \mathcal{O}(kL)) = o(k^n)$ as $k \to +\infty$, for every $q \geqslant 1$. If X is projective algebraic, the following more precise bound holds:*

$$h^q(X, \mathcal{O}(E) \otimes \mathcal{O}(kL)) = O(k^{n-q}), \qquad \forall q \geqslant 0.$$

Proof. The Kähler case will be proved in Chapter 8, as a consequence of the holomorphic Morse inequalities. In the projective algebraic case, we proceed by induction on $n = \dim X$. If $n = 1$ the result is clear, as well as if $q = 0$. Now let A be a nonsingular ample divisor such that $E \otimes \mathcal{O}(A - K_X)$ is Nakano positive. Then the Nakano vanishing theorem applied to the vector bundle $F = E \otimes \mathcal{O}(kL + A - K_X)$ shows that $H^q(X, \mathcal{O}(E) \otimes \mathcal{O}(kL + A)) = 0$ for all $q \geqslant 1$. The exact sequence

$$0 \to \mathcal{O}(kL) \to \mathcal{O}(kL + A) \to \mathcal{O}(kL + A)_{\restriction A} \to 0$$

twisted by E implies

$$H^q(X, \mathcal{O}(E) \otimes \mathcal{O}(kL)) \simeq H^{q-1}(A, \mathcal{O}(E_{\restriction A} \otimes \mathcal{O}(kL + A)_{\restriction A}),$$

and we easily conclude by induction since $\dim A = n - 1$. Observe that the argument does not work any more if X is not algebraic. It seems to be unknown whether the $O(k^{n-q})$ bound still holds in that case. □

(6.19) Corollary. *If L is nef, then L is big (i.e. $\kappa(L) = n$) if and only if $L^n > 0$. Moreover, if L is nef and big, then for every $\delta > 0$, L has a singular metric $h = e^{-2\varphi}$ such that $\max_{x \in X} \nu(\varphi, x) \leqslant \delta$ and $\mathrm{i}\Theta_{L,h} \geqslant \varepsilon\omega$ for some $\varepsilon > 0$. The metric h can be chosen to be smooth on the complement of a fixed divisor D, with logarithmic poles along D.*

Proof. By Lemma 6.18 and the Riemann-Roch formula, we have $h^0(X, kL) = \chi(X, kL) + o(k^n) = k^n L^n / n! + o(k^n)$, whence the first statement. By the proof of Theorem 6.17 (b), there exists a singular metric h_1 on L such that

$$\frac{\mathrm{i}}{2\pi}\Theta_{L,h_1} = \frac{1}{k}\Big(\frac{\mathrm{i}}{2\pi}\Theta_{A,h_A} + [D]\Big) \geqslant \frac{1}{k}\omega, \qquad \omega = \frac{\mathrm{i}}{2\pi}\Theta_{A,h_A}.$$

Now, for every $\varepsilon > 0$, there is a smooth metric h_ε on L such that $\frac{\mathrm{i}}{2\pi}\Theta_{L,h_\varepsilon} \geqslant -\varepsilon\omega$. The convex combination of metrics $h'_\varepsilon = h_1^{k\varepsilon} h_\varepsilon^{1-k\varepsilon}$ is a singular metric with poles along D which satisfies

$$\frac{\mathrm{i}}{2\pi}\Theta_{L,h'_\varepsilon} \geqslant \varepsilon(\omega + [D]) - (1 - k\varepsilon)\varepsilon\omega \geqslant k\varepsilon^2\omega.$$

Its Lelong numbers are $\varepsilon\nu(D, x)$ and they can be made smaller than δ by choosing $\varepsilon > 0$ small. □

We still need a few elementary facts about the numerical dimension of nef line bundles.

(6.20) Definition. *Let L be a nef line bundle on a compact Kähler manifold X. One defines the numerical dimension of L to be*

$$\mathrm{nd}(L) = \max\big\{k = 0, \ldots, n\,;\ c_1(L)^k \neq 0 \text{ in } H^{2k}(X, \mathbb{R})\big\}.$$

By Corollary 6.19, we have $\kappa(L) = n$ if and only if $\mathrm{nd}(L) = n$. In general, we merely have an inequality.

(6.21) Proposition. *If L is a nef line bundle on a compact Kähler manifold, then $\kappa(L) \leqslant \mathrm{nd}(L)$.*

Proof. By induction on $n = \dim X$. If $\mathrm{nd}(L) = n$ or $\kappa(L) = n$ the result is true, so we may assume $r := \kappa(L) \leqslant n - 1$ and $k := \mathrm{nd}(L) \leqslant n - 1$. Fix $m > 0$ so that $\Phi = \Phi_{|mL|}$ has generic rank r. Select a nonsingular ample divisor A in X such that the restriction of $\Phi_{|mL|}$ to A still has rank r (for this, just take A passing through a point $x \notin B_{|mL|}$ at which $\mathrm{rank}(d\Phi_x) = r < n$, in such a way that the tangent linear map $d\Phi_{x\restriction T_{A,x}}$ still has rank r). Then $\kappa(L_{\restriction A}) \geqslant r = \kappa(L)$ (we just have an equality because there might exist sections in $H^0(A, mL_{\restriction A})$ which do not extend to X). On the other hand, we claim that $\mathrm{nd}(L_{\restriction A}) = k = \mathrm{nd}(L)$. The inequality $\mathrm{nd}(L_{\restriction A}) \geqslant \mathrm{nd}(L)$ is clear. Conversely, if we set $\omega = \frac{\mathrm{i}}{2\pi}\Theta_{A,h_A} > 0$, the cohomology class $c_1(L)^k$ can be represented by a closed positive current of bidegree (k, k)

$$T = \lim_{\varepsilon \to 0}\Big(\frac{\mathrm{i}}{2\pi}\Theta_{L,h_\varepsilon} + \varepsilon\omega\Big)^k$$

after passing to some subsequence (there is a uniform bound for the mass thanks to the Kähler assumption, taking wedge products with ω^{n-k}). The current T must be non zero since $c_1(L)^k \neq 0$ by definition of $k = \mathrm{nd}(L)$. Then $\{[A]\} = \{\omega\}$ as cohomology classes, and

$$\int_A c_1(L_{\restriction A})^k \wedge \omega^{n-1-k} = \int_X c_1(L)^k \wedge [A] \wedge \omega^{n-1-k} = \int_X T \wedge \omega^{n-k} > 0.$$

This implies $\mathrm{nd}(L_{\restriction A}) \geqslant k$, as desired. The induction hypothesis with X replaced by A yields

$$\kappa(L) \leqslant \kappa(L_{\restriction A}) \leqslant \mathrm{nd}(L_{\restriction A}) \leqslant \mathrm{nd}(L).$$

□

(6.22) Remark. It may happen that $\kappa(L) < \mathrm{nd}(L)$: take e.g.

$$L \to X = X_1 \times X_2$$

equals to the total tensor product of an ample line bundle L_1 on a projective manifold X_1 and of a unitary flat line bundle L_2 on an elliptic curve X_2 given by a representation $\pi_1(X_2) \to U(1)$ such that no multiple kL_2 with $k \neq 0$ is trivial. Then $H^0(X,kL) = H^0(X_1,kL_1)\otimes H^0(X_2,kL_2) = 0$ for $k>0$, and thus $\kappa(L)=-\infty$. However $c_1(L)=\mathrm{pr}_1^*c_1(L_1)$ has numerical dimension equal to $\dim X_1$. The same example shows that the Kodaira dimension may increase by restriction to a subvariety (if $Y = X_1\times\{\text{point}\}$, then $\kappa(L_{\restriction Y})=\dim Y$). □

6.D. The Kawamata-Viehweg Vanishing Theorem

We derive here an algebraic version of the Nadel vanishing theorem in the context of nef line bundles. This algebraic vanishing theorem has been obtained independently by Kawamata [Kaw82] and Viehweg [Vie82], who both reduced it to the Akizuki-Kodaira-Nakano vanishing theorem [AN54] by cyclic covering constructions. Since then, a number of other proofs have been given, one based on connections with logarithmic singularities [EV86], another on Hodge theory for twisted coefficient systems [Kol85], a third one on the Bochner technique [Dem89] (see also [EV92] for a general survey). Since the result is best expressed in terms of multiplier ideal sheaves (avoiding then any unnecessary desingularization in the statement), we feel that the direct approach via Nadel's vanishing theorem is extremely natural.

If $D=\sum\alpha_jD_j\geqslant 0$ is an effective $\mathbb{Q}$-divisor, we define the *multiplier ideal sheaf* $\mathcal{I}(D)$ to be equal to $\mathcal{I}(\varphi)$ where $\varphi=\sum\alpha_j|g_j|$ is the corresponding psh function defined by generators g_j of $\mathcal{O}(-D_j)$. If D is a divisor with normal crossings,we know that

$$\text{(6.23)}\qquad \mathcal{I}(D)=\mathcal{O}(-\lfloor D\rfloor),\qquad\text{where}\quad \lfloor D\rfloor=\sum\lfloor\alpha_j\rfloor D_j$$

is the integer part of D. In general, the computation of $\mathcal{I}(D)$ can be made algebraically by using a desingularization $\mu:\widetilde{X}\to X$ such that μ^*D becomes a divisor with normal crossings (Hironaka [Hir64]), and the direct image formula proved in (5.8):

$$\text{(6.24)}\qquad \mathcal{O}_X(K_X)\otimes\mathcal{I}(\varphi)=\mu_*\big(\mathcal{O}_{\widetilde{X}}(K_{\widetilde{X}})\otimes\mathcal{I}(\varphi\circ\mu)\big),$$

$$\text{(6.24}'\text{)}\qquad \mathcal{I}(\varphi)=\mu_*\big(\mathcal{O}_{\widetilde{X}}(K_{\widetilde{X}/X})\otimes\mathcal{I}(\varphi\circ\mu)\big)$$

in terms of the relative canonical sheaf $K_{\widetilde{X}/X}=K_{\widetilde{X}}\otimes\mu^*(K_X^{-1})$.

(6.25) Kawamata-Viehweg Vanishing Theorem. *Let X be a projective algebraic manifold and let F be a line bundle over X such that some positive multiple mF can be written $mF=L+D$ where L is a nef line bundle and D an effective divisor. Then*

$$H^q\big(X,\mathcal{O}(K_X+F)\otimes\mathcal{I}(m^{-1}D)\big)=0\qquad\textit{for}\quad q>n-\mathrm{nd}(L).$$

(6.26) Special Case. *If F is a nef line bundle, then*

$$H^q\big(X,\mathcal{O}(K_X+F)\big)=0\qquad\textit{for}\quad q>n-\mathrm{nd}(F).$$

Proof of Theorem 6.25. First suppose that $\mathrm{nd}(L) = n$, i.e. that L is big. By the proof of Theorem 6.17 (b), there is a singular Hermitian metric h_0 on L such that the corresponding weight φ_0 has algebraic singularities and

$$\mathrm{i}\Theta_{L,h_0} = 2\mathrm{i}d'd''\varphi_0 \geqslant \varepsilon_0\omega$$

for some $\varepsilon_0 > 0$. On the other hand, since L is nef, there are metrics given by weights φ_ε such that $\frac{\mathrm{i}}{2\pi}\Theta_{L,h_\varepsilon} \geqslant -\varepsilon\omega$ for every $\varepsilon > 0$, ω being a Kähler metric. Let $\varphi_D = \sum \alpha_j \log |g_j|$ be the weight of the singular metric on $\mathcal{O}(D)$. We define a singular metric on F by

$$\varphi_F = \frac{1}{m}\big((1-\delta)\varphi_{L,\varepsilon} + \delta\varphi_{L,0} + \varphi_D\big)$$

with $\varepsilon \ll \delta \ll 1$, δ rational. Then φ_F has algebraic singularities, and by taking δ small enough we find $\mathcal{I}(\varphi_F) = \mathcal{I}(\frac{1}{m}\varphi_D) = \mathcal{I}(\frac{1}{m}D)$. In fact, $\mathcal{I}(\varphi_F)$ can be computed by taking integer parts of certain $\mathbb{Q}$-divisors, and adding $\delta\varphi_{L,0}$ does not change the integer part of the rational numbers involved when δ is small. Now

$$\begin{aligned} dd^c\varphi_F &= \frac{1}{m}\big((1-\delta)dd^c\varphi_{L,\varepsilon} + \delta dd^c\varphi_{L,0} + dd^c\varphi_D\big) \\ &\geqslant \frac{1}{m}\big(-(1-\delta)\varepsilon\omega + \delta\varepsilon_0\omega + [D]\big) \geqslant \frac{\delta\varepsilon}{m}\omega, \end{aligned}$$

if we choose $\varepsilon \leqslant \delta\varepsilon_0$. Nadel's theorem 5.11 thus implies the desired vanishing result for all $q \geqslant 1$.

Now, if $\mathrm{nd}(L) < n$, we use hyperplane sections and argue by induction on $n = \dim X$. Since the sheaf $\mathcal{O}(K_X) \otimes \mathcal{I}(m^{-1}D)$ behaves functorially with respect to modifications (and since the L^2 cohomology complex is "the same" upstairs and downstairs), we may assume after blowing-up that D is a divisor with normal crossings. Then the multiplier ideal sheaf $\mathcal{I}(m^{-1}D) = \mathcal{O}(-\lfloor m^{-1}D\rfloor)$ is locally free. By Serre duality, the expected vanishing is equivalent to

$$H^q(X, \mathcal{O}(-F) \otimes \mathcal{O}(\lfloor m^{-1}D\rfloor)) = 0 \qquad \text{for } q < \mathrm{nd}(L).$$

Select a nonsingular ample divisor A such that A meets all components D_j transversally, and take A positive enough so that $\mathcal{O}(A + F - \lfloor m^{-1}D\rfloor)$ is ample. Then $H^q(X, \mathcal{O}(-A-F)\otimes\mathcal{O}(\lfloor m^{-1}D\rfloor)) = 0$ for $q < n$ by Kodaira vanishing, and the exact sequence $0 \to \mathcal{O}_X(-A) \to \mathcal{O}_X \to (i_A)_*\mathcal{O}_A \to 0$ twisted by $\mathcal{O}(-F) \otimes \mathcal{O}(\lfloor m^{-1}D\rfloor)$ yields an isomorphism

$$H^q(X, \mathcal{O}(-F) \otimes \mathcal{O}(\lfloor m^{-1}D\rfloor)) \simeq H^q(A, \mathcal{O}(-F_{\restriction A}) \otimes \mathcal{O}(\lfloor m^{-1}D_{\restriction A}\rfloor).$$

The proof of Proposition 6.21 showed that $\mathrm{nd}(L_{\restriction A}) = \mathrm{nd}(L)$, hence the induction hypothesis implies that the cohomology group on A on the right hand side is zero for $q < \mathrm{nd}(L)$. □

6.E. A Uniform Global Generation Property due to Y.T. Siu

Let X be a projective manifold, and (L,h) a pseudo-effective line bundle. The "uniform global generation property" states in some sense that the tensor product sheaf $L\otimes\mathcal{I}(h)$ has a uniform positivity property, for any singular Hermitian metric h with nonnegative curvature on L.

(6.27) Theorem (Y.T. Siu, [Siu98]). *Let X be a projective manifold. There exists an ample line bundle G on X such that for every pseudo-effective line bundle (L,h), the sheaf $\mathcal{O}(G+L)\otimes\mathcal{I}(h)$ is generated by its global sections. In fact, G can be chosen as follows: pick any very ample line bundle A, and take G such that $G-(K_X+nA)$ is ample, e.g. $G=K_X+(n+1)A$.*

Proof. Let φ be the weight of the metric h on a small neighborhood of a point $z_0\in X$. Assume that we have a local section u of $\mathcal{O}(G+L)\otimes\mathcal{I}(h)$ on a coordinate open ball $B=B(z_0,\delta)$, such that

$$\int_B |u(z)|^2 e^{-2\varphi(z)}|z-z_0|^{-2(n+\varepsilon)}dV(z)<+\infty.$$

Then Skoda's division theorem [Sko72b] (see also Corollary 11.13) implies $u(z)=\sum(z_j-z_{j,0})v_j(z)$ with

$$\int_B |v_j(z)|^2 e^{-2\varphi(z)}|z-z_0|^{-2(n-1+\varepsilon)}dV(z)<+\infty,$$

in particular $u_{z_0}\in\mathcal{O}(G+L)\otimes\mathcal{I}(h)\otimes\mathfrak{m}_{X,z_0}$. Select a very ample line bundle A on X. We take a basis $\sigma=(\sigma_j)$ of sections of $H^0(X,G\otimes\mathfrak{m}_{X,z_0})$ and multiply the metric h of G by the factor $|\sigma|^{-2(n+\varepsilon)}$. The weight of the above metric has singularity $(n+\varepsilon)\log|z-z_0|^2$ at z_0, and its curvature is

$$\mathrm{i}\Theta_G+(n+\varepsilon)\mathrm{i}\partial\overline{\partial}\log|\sigma|^2\geqslant\mathrm{i}\Theta_G-(n+\varepsilon)\Theta_A. \tag{6.28}$$

Now, let f be a local section in $H^0(B,\mathcal{O}(G+L)\otimes\mathcal{I}(h))$ on $B=B(z_0,\delta)$, δ small. We solve the global $\overline{\partial}$ equation

$$\overline{\partial}u=\overline{\partial}(\theta f)\qquad\text{on } X$$

with a cut-off function θ supported near z_0 and with the weight associated with our above choice of metric on $G+L$. Thanks to Nadel's theorem, the solution exists if the metric of $G+L-K_X$ has positive curvature. As $\mathrm{i}\Theta_{L,h}\geqslant 0$ in the sense of currents, (6.28) shows that a sufficient condition is $G-K_X-nA>0$ (provided that ε is small enough). We then find a smooth solution u such that $u_{z_0}\in\mathcal{O}(G+L)\otimes\mathcal{I}(h)\otimes\mathfrak{m}_{X,z_0}$, hence

$$F:=\theta f-u\in H^0(X,\mathcal{O}(G+L)\otimes\mathcal{I}(h))$$

is a global section differing from f by a germ in $\mathcal{O}(G+L)\otimes\mathcal{I}(h)\otimes\mathfrak{m}_{X,z_0}$. Nakayama's lemma implies that $H^0(X,\mathcal{O}(G+L)\otimes\mathcal{I}(h))$ generates the stalks of $\mathcal{O}(G+L)\otimes\mathcal{I}(h)$.

Chapter 7

A Simple Algebraic Approach to Fujita's Conjecture

This chapter is devoted to a proof of various results related to the Fujita conjecture. The main ideas occurring here are inspired by a recent work of Y.T. Siu [Siu96]. His method, which is algebraic in nature and quite elementary, consists in a combination of the Riemann-Roch formula together with Nadel's vanishing theorem (in fact, only the algebraic case is needed, thus the original Kawamata-Viehweg vanishing theorem would be sufficient). Slightly later, Angehrn and Siu [AS95; Siu95] introduced other closely related methods, producing better bounds for the global generation question; since their method is rather delicate, we can only refer the reader to the above references. In the sequel, X denotes a projective algebraic n-dimensional manifold. The first observation is the following well-known consequence of the Riemann-Roch formula.

(7.1) Special Case of Riemann-Roch. *Let $\mathcal{J} \subset \mathcal{O}_X$ be a coherent ideal sheaf on X such that the subscheme $Y = V(\mathcal{J})$ has dimension d (with possibly some lower dimensional components). Let $[Y] = \sum \lambda_j [Y_j]$ be the effective algebraic cycle of dimension d associated to the d dimensional components of Y (taking into account multiplicities λ_j given by the ideal $\mathcal{J}$). Then for any line bundle F, the Euler characteristic*

$$\chi(Y, \mathcal{O}(F + mL)_{\restriction Y}) = \chi(X, \mathcal{O}(F + mL) \otimes \mathcal{O}_X/\mathcal{J})$$

is a polynomial $P(m)$ of degree d and leading coefficient $L^d \cdot [Y]/d!$

The second fact is an elementary lemma about numerical polynomials (polynomials with rational coefficients, mapping $\mathbb{Z}$ into $\mathbb{Z}$).

(7.2) Lemma. *Let $P(m)$ be a numerical polynomial of degree $d > 0$ and leading coefficient $a_d/d!$, $a_d \in \mathbb{Z}$, $a_d > 0$. Suppose that $P(m) \geqslant 0$ for $m \geqslant m_0$. Then*

(a) *For every integer $N \geqslant 0$, there exists $m \in [m_0, m_0 + Nd]$ such that $P(m) \geqslant N$.*

(b) *For every $k \in \mathbb{N}$, there exists $m \in [m_0, m_0 + kd]$ such that $P(m) \geqslant a_d k^d / 2^{d-1}$.*

(c) *For every integer $N \geqslant 2d^2$, there exists $m \in [m_0, m_0 + N]$ such that $P(m) \geqslant N$.*

Proof. (a) Each of the N equations $P(m) = 0$, $P(m) = 1$, ..., $P(m) = N - 1$ has at most d roots, so there must be an integer $m \in [m_0, m_0 + dN]$ which is not a root of these.

(b) By Newton's formula for iterated differences $\Delta P(m) = P(m+1) - P(m)$, we get

$$\Delta^d P(m) = \sum_{1 \leqslant j \leqslant d} (-1)^j \binom{d}{j} P(m + d - j) = a_d, \qquad \forall m \in \mathbb{Z}.$$

Hence if $j \in \{0, 2, 4, \ldots, 2\lfloor d/2 \rfloor\} \subset [0, d]$ is the even integer achieving the maximum of $P(m_0 + d - j)$ over this finite set, we find

$$2^{d-1} P(m_0 + d - j) = \left(\binom{d}{0} + \binom{d}{2} + \cdots \right) P(m_0 + d - j) \geqslant a_d,$$

whence the existence of an integer $m \in [m_0, m_0 + d]$ with $P(m) \geqslant a_d / 2^{d-1}$. The case $k = 1$ is thus proved. In general, we apply the above case to the polynomial $Q(m) = P(km - (k-1)m_0)$, which has leading coefficient $a_d k^d / d!$

(c) If $d = 1$, part (a) already yields the result. If $d = 2$, a look at the parabola shows that

$$\max_{m \in [m_0, m_0 + N]} P(m) \geqslant \begin{cases} a_2 N^2 / 8, & \text{if } N \text{ is even,} \\ a_2 (N^2 - 1)/8, & \text{if } N \text{ is odd;} \end{cases}$$

thus $\max_{m \in [m_0, m_0 + N]} P(m) \geqslant N$ whenever $N \geqslant 8$. If $d \geqslant 3$, we apply (b) with k equal to the smallest integer such that $k^d / 2^{d-1} \geqslant N$, i.e. $k = \lceil 2(N/2)^{1/d} \rceil$, where $\lceil x \rceil \in \mathbb{Z}$ denotes the round-up of $x \in \mathbb{R}$. Then $kd \leqslant (2(N/2)^{1/d} + 1)d \leqslant N$ whenever $N \geqslant 2d^2$, as a short computation shows. □

We now apply Nadel's vanishing theorem pretty much in the same way as Siu [Siu96], but with substantial simplifications in the technique and improvements in the bounds. Our method yields simultaneously a simple proof of the following basic result.

(7.3) Theorem. *If L is an ample line bundle over a projective n-fold X, then the adjoint line bundle $K_X + (n+1)L$ is nef.*

By using Mori theory and the base point free theorem ([Mor82; Kaw84]), one can even show that $K_X + (n+1)L$ is semiample, i.e., there exists a positive integer m such that $m(K_X + (n+1)L)$ is generated by sections (see [Kaw85; Fuj87]). The proof rests on the observation that $n+1$ is the maximal length of extremal rays of smooth projective n-folds. Our proof of (7.3) is different and will be given simultaneously with the proof of Theorem 7.4 below.

(7.4) Theorem. *Let L be an ample line bundle and let G be a nef line bundle on a projective n-fold X. Then the following properties hold.*

(a) $2K_X+mL+G$ *generates simultaneous jets of order* $s_1, \dots, s_p \in \mathbb{N}$ *at arbitrary points* $x_1, \dots, x_p \in X$, *i.e., there is a surjective map*

$$H^0(X, 2K_X+mL+G) \longrightarrow\!\!\!\!\!\rightarrow \bigoplus_{1\leqslant j\leqslant p} \mathcal{O}(2K_X+mL+G)\otimes \mathcal{O}_{X,x_j}/\mathfrak{m}_{X,x_j}^{s_j+1},$$

provided that $m \geqslant 2+\sum\limits_{1\leqslant j\leqslant p}\binom{3n+2s_j-1}{n}$. *In particular* $2K_X+mL+G$ *is very ample for* $m\geqslant 2+\binom{3n+1}{n}$.

(b) $2K_X+(n+1)L+G$ *generates simultaneous jets of order* $s_1, \dots, s_p$ *at arbitrary points* $x_1, \dots, x_p \in X$ *provided that the intersection numbers* $L^d\cdot Y$ *of* L *over all d-dimensional algebraic subsets* Y *of* X *satisfy*

$$L^d\cdot Y > \frac{2^{d-1}}{\lfloor n/d\rfloor^d}\sum_{1\leqslant j\leqslant p}\binom{3n+2s_j-1}{n}.$$

Proof. The proofs of Theorem 7.3 and Theorem 7.4 (a), (b) go along the same lines, so we deal with them simultaneously (in the case of (7.3), we simply agree that $\{x_1, \dots, x_p\} = \emptyset$). The idea is to find an integer (or rational number) m_0 and a singular Hermitian metric h_0 on K_X+m_0L with strictly positive curvature current $\mathrm{i}\Theta_{h_0}\geqslant \varepsilon\omega$, such that $V(\mathcal{J}(h_0))$ is 0-dimensional and the weight φ_0 of h_0 satisfies $\nu(\varphi_0, x_j)\geqslant n+s_j$ for all j. As L and G are nef, $(m-m_0)L+G$ has for all $m\geqslant m_0$ a metric h' whose curvature $\mathrm{i}\Theta_{h'}$ has arbitrary small negative part (see [Dem90]), e.g., $\mathrm{i}\Theta_{h'}\geqslant -\frac{\varepsilon}{2}\omega$. Then $\mathrm{i}\Theta_{h_0}+\mathrm{i}\Theta_{h'}\geqslant\frac{\varepsilon}{2}\omega$ is again positive definite. An application of Corollary 6.12 to $F=K_X+mL+G=(K_X+m_0L)+((m-m_0)L+G)$ equipped with the metric $h_0\otimes h'$ implies the existence of the desired sections in $K_X+F=2K_X+mL+G$ for $m\geqslant m_0$.

Fix an embedding $\Phi_{|\mu L|}: X\to\mathbb{P}^N$, $\mu\gg 0$, given by sections $\lambda_0, \dots, \lambda_N$ of $H^0(X,\mu L)$, and let h_L be the associated metric on L of positive definite curvature form $\omega=\frac{\mathrm{i}}{2\pi}\Theta_{L,h_L}$. In order to obtain the desired metric h_0 on K_X+m_0L, we fix $a\in\mathbb{N}^*$ and use a double induction process to construct singular metrics $(h_{k,\nu})_{\nu\geqslant 1}$ on aK_X+b_kL for a non increasing sequence of positive integers $b_1\geqslant b_2\geqslant\cdots\geqslant b_k\geqslant\cdots$. Such a sequence much be stationary and m_0 will just be the stationary limit $m_0=\lim b_k/a$. The metrics $h_{k,\nu}$ are taken to satisfy the following properties:

(a) $h_{k,\nu}$ is an algebraic metric of the form

$$\|\xi\|^2_{h_{k,\nu}} = \frac{|\tau_k(\xi)|^2}{\left(\sum_{1\leqslant i\leqslant\nu,\,0\leqslant j\leqslant N}\left|\tau_k^{(a+1)\mu}(\sigma_i^{a\mu}\cdot\lambda_j^{(a+1)b_k-am_i})\right|^2\right)^{1/(a+1)\mu}},$$

defined by sections $\sigma_i\in H^0(X,(a+1)K_X+m_iL)$, $m_i<\frac{a+1}{a}b_k$, $1\leqslant i\leqslant\nu$, where $\xi\mapsto\tau_k(\xi)$ is an arbitrary local trivialization of aK_X+b_kL; note that $\sigma_i^{a\mu}\cdot\lambda_j^{(a+1)b_k-am_i}$ is a section of

$$a\mu((a+1)K_X+m_iL)+((a+1)b_k-am_i)\mu L=(a+1)\mu(aK_X+b_kL).$$

(b) $\mathrm{ord}_{x_j}(\sigma_i) \geqslant (a+1)(n+s_j)$ for all i, j;

(c) $\mathcal{I}(h_{k,\nu+1}) \supset \mathcal{I}(h_{k,\nu})$ and $\mathcal{I}(h_{k,\nu+1}) \neq \mathcal{I}(h_{k,\nu})$ whenever the zero variety $V(\mathcal{I}(h_{k,\nu}))$ has positive dimension.

The weight $\varphi_{k,\nu} = \frac{1}{2(a+1)\mu} \log \sum \left|\tau_k^{(a+1)\mu}(\sigma_i^{a\mu} \cdot \lambda_j^{(a+1)b_k - am_i})\right|^2$ of $h_{k,\nu}$ is plurisubharmonic and the condition $m_i < \frac{a+1}{a} b_k$ implies $(a+1)b_k - am_i \geqslant 1$, thus the difference $\varphi_{k,\nu} - \frac{1}{2(a+1)\mu} \log \sum |\tau(\lambda_j)|^2$ is also plurisubharmonic. Hence we get $\frac{\mathrm{i}}{2\pi}\Theta_{h_{k,\nu}}(aK_X + b_k L) = \frac{\mathrm{i}}{\pi} d'd''\varphi_{k,\nu} \geqslant \frac{1}{(a+1)}\omega$. Moreover, condition b) clearly implies $\nu(\varphi_{k,\nu}, x_j) \geqslant a(n+s_j)$. Finally, condition c) combined with the strong Noetherian property of coherent sheaves ensures that the sequence $(h_{k,\nu})_{\nu \geqslant 1}$ will finally produce a zero dimensional subscheme $V(\mathcal{I}(h_{k,\nu}))$. We agree that the sequence $(h_{k,\nu})_{\nu \geqslant 1}$ stops at this point, and we denote by $h_k = h_{k,\nu}$ the final metric, such that $\dim V(\mathcal{I}(h_k)) = 0$.

For $k = 1$, it is clear that the desired metrics $(h_{1,\nu})_{\nu \geqslant 1}$ exist if b_1 is taken large enough (so large, say, that $(a+1)K_X + (b_1 - 1)L$ generates jets of order $(a+1)(n + \max s_j)$ at every point; then the sections $\sigma_1, \ldots, \sigma_\nu$ can be chosen with $m_1 = \cdots = m_\nu = b_1 - 1$). Suppose that the metrics $(h_{k,\nu})_{\nu \geqslant 1}$ and h_k have been constructed and let us proceed with the construction of $(h_{k+1,\nu})_{\nu \geqslant 1}$. We do this again by induction on ν, assuming that $h_{k+1,\nu}$ is already constructed and that $\dim V(\mathcal{I}(h_{k+1,\nu})) > 0$. We start in fact the induction with $\nu = 0$, and agree in this case that $\mathcal{I}(h_{k+1,0}) = 0$ (this would correspond to an infinite metric of weight identically equal to $-\infty$). By Nadel's vanishing theorem applied to

$$F_m = aK_X + mL = (aK_X + b_k L) + (m - b_k)L$$

with the metric $h_k \otimes (h_L)^{\otimes m - b_k}$, we get

$$H^q(X, \mathcal{O}((a+1)K_X + mL) \otimes \mathcal{I}(h_k)) = 0 \qquad \text{for } q \geqslant 1,\ m \geqslant b_k.$$

As $V(\mathcal{I}(h_k))$ is 0-dimensional, the sheaf $\mathcal{O}_X/\mathcal{I}(h_k)$ is a skyscraper sheaf, and the exact sequence $0 \to \mathcal{I}(h_k) \to \mathcal{O}_X \to \mathcal{O}_X/\mathcal{I}(h_k) \to 0$ twisted with the invertible sheaf $\mathcal{O}((a+1)K_X + mL)$ shows that

$$H^q(X, \mathcal{O}((a+1)K_X + mL)) = 0 \qquad \text{for } q \geqslant 1,\ m \geqslant b_k.$$

Similarly, we find

$$H^q(X, \mathcal{O}((a+1)K_X + mL) \otimes \mathcal{I}(h_{k+1,\nu})) = 0 \qquad \text{for } q \geqslant 1,\ m \geqslant b_{k+1}$$

(also true for $\nu = 0$, since $\mathcal{I}(h_{k+1,0}) = 0$), and when $m \geqslant \max(b_k, b_{k+1}) = b_k$, the exact sequence $0 \to \mathcal{I}(h_{k+1,\nu}) \to \mathcal{O}_X \to \mathcal{O}_X/\mathcal{I}(h_{k+1}, \nu) \to 0$ implies

$$H^q(X, \mathcal{O}((a+1)K_X + mL) \otimes \mathcal{O}_X/\mathcal{I}(h_{k+1,\nu})) = 0 \qquad \text{for } q \geqslant 1,\ m \geqslant b_k.$$

In particular, since the H^1 group vanishes, every section u' of $(a+1)K_X + mL$ on the subscheme $V(\mathcal{I}(h_{k+1,\nu}))$ has an extension u to X. Fix a basis $u'_1, \ldots, u'_N$

of the sections on $V(\mathcal{I}(h_{k+1,\nu}))$ and take arbitrary extensions $u_1, \ldots, u_N$ to X. Look at the linear map assigning the collection of jets of order $(a+1)(n+s_j)-1$ at all points x_j

$$u = \sum_{1 \leqslant j \leqslant N} a_j u_j \longmapsto \bigoplus J_{x_j}^{(a+1)(n+s_j)-1}(u).$$

Since the rank of the bundle of s-jets is $\binom{n+s}{n}$, the target space has dimension

$$\delta = \sum_{1 \leqslant j \leqslant p} \binom{n+(a+1)(n+s_j)-1}{n}.$$

In order to get a section $\sigma_{\nu+1} = u$ satisfying condition b) with non trivial restriction $\sigma'_{\nu+1}$ to $V(\mathcal{I}(h_{k+1,\nu}))$, we need at least $N = \delta + 1$ independent sections $u'_1, \ldots, u'_N$. This condition is achieved by applying Lemma 7.2 to the numerical polynomial

$$\begin{aligned} P(m) &= \chi(X, \mathcal{O}((a+1)K_X + mL) \otimes \mathcal{O}_X/\mathcal{I}(h_{k+1,\nu})) \\ &= h^0(X, \mathcal{O}((a+1)K_X + mL) \otimes \mathcal{O}_X/\mathcal{I}(h_{k+1,\nu})) \geqslant 0, \qquad m \geqslant b_k. \end{aligned}$$

The polynomial P has degree $d = \dim V(\mathcal{I}(h_{k+1,\nu})) > 0$. We get the existence of an integer $m \in [b_k, b_k + \eta]$ such that $N = P(m) \geqslant \delta + 1$ with some explicit integer $\eta \in \mathbb{N}$ (for instance $\eta = n(\delta+1)$ always works by Lemma 7.2 (a), but we will also use other possibilities to find an optimal choice in each case). Then we find a section $\sigma_{\nu+1} \in H^0(X, (a+1)K_X + mL)$ with non trivial restriction $\sigma'_{\nu+1}$ to $V(\mathcal{I}(h_{k+1,\nu}))$, vanishing at order $\geqslant (a+1)(n+s_j)$ at each point x_j. We just set $m_{\nu+1} = m$, and the condition $m_{\nu+1} < \frac{a+1}{a} b_{k+1}$ is satisfied if $b_k + \eta < \frac{a+1}{a} b_{k+1}$. This shows that we can take inductively

$$b_{k+1} = \left\lfloor \frac{a}{a+1}(b_k + \eta) \right\rfloor + 1.$$

By definition, $h_{k+1,\nu+1} \leqslant h_{k+1,\nu}$, hence $\mathcal{I}(h_{k+1,\nu+1}) \supset \mathcal{I}(h_{k+1,\nu})$. We necessarily have $\mathcal{I}(h_{k+1,\nu+1}) \neq \mathcal{I}(h_{k+1,\nu})$, for $\mathcal{I}(h_{k+1,\nu+1})$ contains the ideal sheaf associated with the zero divisor of $\sigma_{\nu+1}$, whilst $\sigma_{\nu+1}$ does not vanish identically on $V(\mathcal{I}(h_{k+1,\nu}))$. Now, an easy computation shows that the iteration of $b_{k+1} = \lfloor \frac{a}{a+1}(b_k + \eta) \rfloor + 1$ stops at $b_k = a(\eta+1)+1$ for any large initial value b_1. In this way, we obtain a metric h_∞ of positive definite curvature on $aK_X + (a(\eta+1)+1)L$, with $\dim V(\mathcal{I}(h_\infty)) = 0$ and $\nu(\varphi_\infty, x_j) \geqslant a(n+s_j)$ at each point x_j.

Proof of Theorem 7.3. In this case, the set $\{x_j\}$ is taken to be empty, thus $\delta = 0$. By Theorem 7.2 (a), the condition $P(m) \geqslant 1$ is achieved for some $m \in [b_k, b_k + n]$ and we can take $\eta = n$. As μL is very ample, there exists on μL a metric with an isolated logarithmic pole of Lelong number 1 at any given point x_0 (e.g., the algebraic metric defined with all sections of μL vanishing at x_0). Hence

$$F'_a = aK_X + (a(n+1)+1)L + n\mu L$$

has a metric h'_a such that $V(\mathcal{I}(h'_a))$ is zero dimensional and contains $\{x_0\}$. By Corollary 6.12, we conclude that

$$K_X + F'_a = (a+1)K_X + (a(n+1)+1+n\mu)L$$

is generated by sections, in particular $K_X + \frac{a(n+1)+1+n\mu}{a+1}L$ is nef. As a tends to $+\infty$, we infer that $K_X + (n+1)L$ is nef. □

Proof of Theorem 7.4 (a). Here, the choice $a = 1$ is sufficient for our purposes. Then

$$\delta = \sum_{1 \leqslant j \leqslant p} \binom{3n + 2s_j - 1}{n}.$$

If $\{x_j\} \neq \emptyset$, we have $\delta + 1 \geqslant \binom{3n-1}{n} + 1 \geqslant 2n^2$ for $n \geqslant 2$. Lemma 7.2 (c) shows that $P(m) \geqslant \delta + 1$ for some $m \in [b_k, b_k + \eta]$ with $\eta = \delta + 1$. We can start in fact the induction procedure $k \mapsto k+1$ with $b_1 = \eta + 1 = \delta + 2$, because the only property needed for the induction step is the vanishing property

$$H^0(X, 2K_X + mL) = 0 \qquad \text{for } q \geqslant 1,\ m \geqslant b_1,$$

which is true by the Kodaira vanishing theorem and the ampleness of $K_X + b_1 L$ (here we use Fujita's result 7.3, observing that $b_1 > n+1$). Then the recursion formula $b_{k+1} = \lfloor \frac{1}{2}(b_k + \eta) \rfloor + 1$ yields $b_k = \eta + 1 = \delta + 2$ for all k, and Theorem 7.4 (a) follows. □

Proof of Theorem 7.4 (b). Quite similar to Theorem 7.4 (a), except that we take $\eta = n$, $a = 1$ and $b_k = n+1$ for all k. By Lemma 7.2 (b), we have $P(m) \geqslant a_d k^d / 2^{d-1}$ for some integer $m \in [m_0, m_0 + kd]$, where $a_d > 0$ is the coefficient of highest degree in P. By Lemma 7.2 we have $a_d \geqslant \inf_{\dim Y = d} L^d \cdot Y$. We take $k = \lfloor n/d \rfloor$. The condition $P(m) \geqslant \delta + 1$ can thus be realized for some $m \in [m_0, m_0 + kd] \subset [m_0, m_0 + n]$ as soon as

$$\inf_{\dim Y = d} L^d \cdot Y \ \lfloor n/d \rfloor^d / 2^{d-1} > \delta,$$

which is equivalent to the condition given in Theorem 7.4 (b). □

(7.5) Corollary. *Let X be a smooth projective n-fold, let L be an ample line bundle and G a nef line bundle over X. Then $m(K_X + (n+2)L) + G$ is very ample for $m \geqslant \binom{3n+1}{n} - 2n$.*

Proof. Apply Theorem 7.4 (a) with $G' = a(K_X + (n+1)L) + G$, so that

$$2K_X + mL + G' = (a+2)(K_X + (n+2)L) + (m - 2n - 4 - a)L + G,$$

and take $m = a + 2n + 4 \geqslant 2 + \binom{3n+1}{n}$. □

The main drawback of the above technique is that multiples of L at least equal to $(n+1)L$ are required to avoid zeroes of the Hilbert polynomial. In particular, it is not possible to obtain directly a very ampleness criterion for $2K_X + L$ in the statement of Theorem 7.4 (b). Nevertheless, using different ideas from Angehrn-Siu [AS95; Siu96] has obtained such a criterion. We derive here a slightly weaker version, thanks to the following elementary lemma.

(7.6) Lemma. *Assume that for some integer $\mu \in \mathbb{N}^*$ the line bundle μF generates simultaneously all jets of order $\mu(n+s_j)+1$ at any point x_j in a subset $\{x_1, \ldots, x_p\}$ of X. Then $K_X + F$ generates simultaneously all jets of order s_j at x_j.*

Proof. Take the algebraic metric on F defined by a basis of sections $\sigma_1, \ldots, \sigma_N$ of μF which vanish at order $\mu(n+s_j)+1$ at all points x_j. Since we are still free to choose the homogeneous term of degree $\mu(n+s_j)+1$ in the Taylor expansion at x_j, we find that $x_1, \ldots, x_p$ are isolated zeroes of $\bigcap \sigma_j^{-1}(0)$. If φ is the weight of the metric of F near x_j, we thus have $\varphi(z) \sim (n+s_j+\frac{1}{\mu})\log|z-x_j|$ in suitable coordinates. We replace φ in a neighborhood of x_j by

$$\varphi'(z) = \max\big(\varphi(z),\, |z|^2 - C + (n+s_j)\log|z-x_j|\big)$$

and leave φ elsewhere unchanged (this is possible by taking $C > 0$ very large). Then $\varphi'(z) = |z|^2 - C + (n+s_j)\log|z-x_j|$ near x_j, in particular φ' is strictly plurisubharmonic near x_j. In this way, we get a metric h' on F with semipositive curvature everywhere on X, and with positive definite curvature on a neighborhood of $\{x_1, \ldots, x_p\}$. The conclusion then follows directly from Hörmander's L^2 estimates (5.1) and (5.2). □

(7.7) Theorem. *Let X be a smooth projective n-fold, and let L be an ample line bundle over X. Then $2K_X + L$ generates simultaneous jets of order $s_1, \ldots, s_p$ at arbitrary points $x_1, \ldots, x_p \in X$ provided that the intersection numbers $L^d \cdot Y$ of L over all d-dimensional algebraic subsets Y of X satisfy*

$$L^d \cdot Y > \frac{2^{d-1}}{\lfloor n/d \rfloor^d} \sum_{1 \leqslant j \leqslant p} \binom{(n+1)(4n+2s_j+1)-2}{n}, \qquad 1 \leqslant d \leqslant n.$$

Proof. By Lemma 7.6 applied with $F = K_X + L$ and $\mu = n+1$, the desired jet generation of $2K_X + L$ occurs if $(n+1)(K_X+L)$ generates jets of order $(n+1)(n+s_j)+1$ at x_j. By Lemma 7.6 again with $F = aK_X + (n+1)L$ and $\mu = 1$, we see by backward induction on a that we need the simultaneous generation of jets of order

$$(n+1)(n+s_j)+1+(n+1-a)(n+1)$$

at x_j. In particular, for $2K_X + (n+1)L$ we need the generation of jets of order $(n+1)(2n+s_j-1)+1$. Theorem 7.4 (b) yields the desired condition. □

We now list a few immediate consequences of Theorem 7.4, in connection with some classical questions of algebraic geometry.

(7.8) Corollary. *Let X be a projective n-fold of general type with K_X ample. Then mK_X is very ample for $m \geqslant m_0 = \binom{3n+1}{n} + 4$.*

(7.9) Corollary. *Let X be a Fano n-fold, that is, a n-fold such that $-K_X$ is ample. Then $-mK_X$ is very ample for $m \geqslant m_0 = \binom{3n+1}{n}$.*

Proof. Corollaries 7.8, 7.9 follow easily from Theorem 7.4 (a) applied to $L = \pm K_X$. Hence we get pluricanonical embeddings $\Phi : X \to \mathbb{P}^N$ such that $\Phi^*\mathcal{O}(1) = \pm m_0 K_X$. The image $Y = \Phi(X)$ has degree

$$\deg(Y) = \int_Y c_1\big(\mathcal{O}(1)\big)^n = \int_X c_1\big(\pm m_0 K_X\big)^n = m_0^n |K_X^n|.$$

It can be easily reproved from this that there are only finitely many deformation types of Fano n-folds, as well as of n-folds of general type with K_X ample, corresponding to a given discriminant $|K_X^n|$ (from a theoretical viewpoint, this result is a consequence of Matsusaka's big theorem [Mat72; KoM83], but the bounds which can be obtained from it are probably extremely huge). In the Fano case, a fundamental result obtained independently by Kollár-Miyaoka-Mori [KoMM92] and Campana [Cam92] shows that the discriminant K_X^n is in fact bounded by a constant C_n depending only on n. Therefore, one can find an explicit bound C'_n for the degree of the embedding Φ, and it follows that there are only finitely many families of Fano manifolds in each dimension. □

In the case of surfaces, much more is known. We will content ourselves with a brief account of recent results. If X is a surface, the failure of an adjoint bundle $K_X + L$ to be globally generated or very ample is described in a very precise way by the following result of I. Reider [Rei88].

(7.10) Reider's Theorem. *Let X be a smooth projective surface and let L be a nef line bundle on X.*

(a) *Assume that $L^2 \geqslant 5$ and let $x \in X$ be a given point. Then $K_X + L$ has a section which does not vanish at x, unless there is an effective divisor $D \subset X$ passing through x such that either*

$$L \cdot D = 0 \quad \textit{and} \quad D^2 = -1; \qquad \textit{or}$$
$$L \cdot D = 1 \quad \textit{and} \quad D^2 = 0.$$

(b) *Assume that $L^2 \geqslant 10$. Then any two points $x, y \in X$ (possibly infinitely near) are separated by sections of $K_X + L$, unless there is an effective divisor $D \subset X$ passing through x and y such that either*

$$L \cdot D = 0 \quad \textit{and} \quad D^2 = -1 \ \textit{or} \ -2; \qquad \textit{or}$$
$$L \cdot D = 1 \quad \textit{and} \quad D^2 = 0 \quad \textit{or} \ -1; \qquad \textit{or}$$
$$L \cdot D = 2 \quad \textit{and} \quad D^2 = 0.$$

(7.11) Corollary. *Let L be an ample line bundle on a smooth projective surface X. Then $K_X + 3L$ is globally generated and $K_X + 4L$ is very ample. If $L^2 \geqslant 2$ then $K_X + 2L$ is globally generated and $K_X + 3L$ is very ample.*

The case of higher order jets can be treated similarly. The most general result in this direction has been obtained by Beltrametti and Sommese [BeS93].

(7.12) Theorem ([BeS93]). *Let X be a smooth projective surface and let L be a nef line bundle on X. Let p be a positive integer such that $L^2 > 4p$. Then for every 0-dimensional subscheme $Z \subset X$ of length $h^0(Z, \mathcal{O}_Z) \leqslant p$ the restriction*

$$\rho_Z : H^0(X, \mathcal{O}_X(K_X + L)) \longrightarrow H^0(Z, \mathcal{O}_Z(K_X + L))$$

is surjective, unless there is an effective divisor $D \subset X$ intersecting the support $|Z|$ such that

$$L \cdot D - p \leqslant D^2 < \frac{1}{2} L \cdot D. \qquad \square$$

Proof (*Sketch*). The proof the above theorems rests in an essential way on the construction of rank 2 vector bundles sitting in an exact sequence

$$0 \to \mathcal{O}_X \to E \to L \otimes \mathcal{I}_Z \to 0.$$

Arguing by induction on the length of Z, we may assume that Z is a 0-dimensional subscheme such that ρ_Z is not surjective, but such that $\rho_{Z'}$ is surjective for every proper subscheme $Z' \subset Z$. The existence of E is obtained by a classical construction of Serre (unfortunately, this construction only works in dimension 2). The numerical condition on L^2 in the hypotheses ensures that $c_1(E)^2 - 4\,c_2(E) > 0$, hence E is unstable in the sense of Bogomolov. The existence of the effective divisor D asserted in Theorems 7.10 or 7.12 follows. We refer to [Rei88], [BeS93] and [Laz97] for details. The reader will find in [FdB95] a proof of the Bogomolov inequality depending only on the Kawamata-Viehweg vanishing theorem. $\square$

(7.13) Exercise. Prove the Fujita conjecture in the case of dimension 1, according to the following steps.

(a) By using Hodge theory, show that for every smooth function f on a compact Kähler manifold (X, ω), the equation $\Delta u = f$ has a solution if and only if $\int_X f \, dV_\omega = 0$.

(b) Derive from (a), by using the local solvability of elliptic operators, that one has a similar result when f is a distribution.

(c) If $X = C$ is a compact complex curve and L a positive line bundle, for every positive measure μ on X such that $\int_C \mu = \deg(L) = \int_C c_1(L)$, there exists a singular Hermitian metric h on L such that $\frac{i}{2\pi}\Theta_h(L) = \mu$ (with the obvious identification of measures and currents of bidegree $(1,1)$).

(d) Given a finite collection of points $x_j \in C$ and integers $s_j > 0$, then $K_C + L$ generates jets of order s_j at all points x_j as soon as $\deg(L) > \sum_j (s_j + 1)$.

(e) If L is positive on C, then $K_C + 2L$ is globally generated and $K_C + 3L$ is very ample.

(7.14) Exercise. The goal of the exercise is to prove the following weaker form of Theorems 7.10 and 7.12, by a simple direct method based on Nadel's vanishing theorem:

Let L be a nef line bundle on a smooth projective surface X. Fix points $x_1, \ldots, x_N$ and corresponding multiplicities $s_1, \ldots, s_N$, and set $p = \sum (2+s_j)^2$. Then $H^0(X, K_X + L)$ generates simultaneously jets of order s_j at all points x_j provided that $L^2 > p$ and $L \cdot C > p$ for all curves C passing through one of the points x_j.

(a) Using the Riemann-Roch formula, show that the condition $L^2 > p$ implies the existence of a section of a large multiple mL vanishing at order $> m(2 + s_j)$ at each of the points.

(b) Construct a sequence of singular Hermitian metrics on L with positive definite curvature, such that the weights φ_ν have algebraic singularities, $\nu(\varphi_\nu, x_j) \geqslant 2 + s_j$ at each point, and such that for some integer $m_1 > 0$ the multiplier ideal sheaves satisfy $\mathcal{I}(m_1\varphi_{\nu+1}) \supsetneq \mathcal{I}(m_1\varphi_\nu)$ if $V(\mathcal{I}(\varphi_\nu))$ is not 0-dimensional near some x_j.

Hint: (a) starts the procedure. Fix $m_0 > 0$ such that $m_0L - K_X$ is ample. Use Nadel's vanishing theorem to show that

$$H^q(X, \mathcal{O}((m + m_0)L) \otimes \mathcal{I}(\lambda m\varphi_\nu)) = 0 \qquad \text{for all } q \geqslant 1,\ m \geqslant 0,\ \lambda \in [0, 1].$$

Let D_ν be the effective $\mathbb{Q}$-divisor describing the 1-dimensional singularities of φ_ν. Then $\mathcal{I}(\lambda m\varphi_\nu) \subset \mathcal{O}(-\lfloor \lambda m D_\nu \rfloor)$ and the quotient has 0-dimensional support, hence

$$H^q(X, \mathcal{O}((m + m_0)L) \otimes \mathcal{O}(-\lfloor \lambda m D_\nu \rfloor)) = 0 \qquad \text{for all } q \geqslant 1,\ m \geqslant 0,\ \lambda \in [0, 1].$$

By the Riemann-Roch formule again prove that

$$(*)\quad h^0(X, \mathcal{O}((m + m_0)L) \otimes \mathcal{O}/\mathcal{O}(-\lfloor \lambda m D_\nu \rfloor)) = \frac{m^2}{2}(2\lambda\, L \cdot D_\nu - \lambda^2 D_\nu^2) + O(m).$$

As the left hand side of $(*)$ is increasing with λ, one must have $D_\nu^2 \leqslant L \cdot D_\nu$. If $V(\mathcal{I}(\varphi_\nu))$ is not 0-dimensional at x_j, then the coefficient of some component of D_ν passing through x_j is at least 1, hence

$$2\, L \cdot D_\nu - D_\nu^2 \geqslant L \cdot D_\nu \geqslant p + 1.$$

Show the existence of an integer $m_1 > 0$ independent of ν such that

$$h^0(X, \mathcal{O}((m + m_0)L) \otimes \mathcal{O}/\mathcal{O}(-\lfloor m D_\nu \rfloor)) > \sum_{1 \leqslant j \leqslant N} \binom{(m + m_0)(2 + s_j) + 2}{2}$$

for $m \geqslant m_1$, and infer the existence of a suitable section of $(m_1 + m_0)L$ which is not in $H^0(X, \mathcal{O}((m_1 + m_0)L - \lfloor m_1 D_\nu \rfloor))$. Use this section to construct $\varphi_{\nu+1}$ such that $\mathcal{I}(m_1\varphi_{\nu+1}) \supsetneq \mathcal{I}(m_1\varphi_\nu)$.

Chapter 8

Holomorphic Morse Inequalities

Holomorphic Morse inequalities were first introduced in [Dem85] to improve Siu's solution of the Grauert-Riemenschneider conjecture. They express asymptotic bounds on the cohomology of tensor bundles of holomorphic line bundles, and appear to be a useful complement to the Riemann-Roch formula. We present here the main results and several important applications. The reader is referred to [Dem85b, 91] for the required analytic details in the spectral theory of operators.

8.A. General Analytic Statement on Compact Complex Manifolds

Let X be a compact complex manifold, E a holomorphic vector bundle of rank r and L a line bundle over X. If L is equipped with a smooth metric h of curvature form $\Theta_{L,h}$, we define the q-index set of L to be the open subset

$$X(L,h,q) = \left\{ x \in X \,;\, \mathrm{i}\Theta_{L,h}(x) \text{ has } \begin{matrix} q & \text{negative eigenvalues} \\ n-q & \text{positive eigenvalues} \end{matrix} \right\} \tag{8.1}$$

for $0 \leqslant q \leqslant n$. Hence X admits a partition $X = \Delta \cup \bigcup_q X(L,h,q)$ where

$$\Delta = \{x \in X \,;\, \det(\Theta_{L,h}(x)) = 0\}$$

is the degeneracy set. We also introduce

$$X(L,h,\leqslant q) = \bigcup_{0\leqslant j\leqslant q} X(L,h,j). \tag{8.1$'$}$$

(8.2) Morse inequalities ([Dem85b]). *For any Hermitian holomorphic line bundle (L,h) and any holomorphic vector bundle E over a compact complex manifold X, the cohomology groups $H^q(X, E\otimes\mathcal{O}(kL))$ satisfy the following asymptotic inequalities as $k \to +\infty$:*

(a) *Weak Morse inequalities*

$$h^q\big(X, \mathcal{O}(E)\otimes\mathcal{O}(kL)\big) \leqslant r\frac{k^n}{n!}\int_{X(L,h,q)} (-1)^q \Big(\frac{\mathrm{i}}{2\pi}\Theta_{L,h}\Big)^n + o(k^n).$$

(b) *Strong Morse inequalities*

$$\sum_{0\leqslant j\leqslant q}(-1)^{q-j}h^j(X,\mathcal{O}(E)\otimes\mathcal{O}(kL))\leqslant r\frac{k^n}{n!}\int_{X(L,h,\leqslant q)}(-1)^q\Big(\frac{\mathrm{i}}{2\pi}\Theta_{L,h}\Big)^n+o(k^n).$$

The proof is based on the spectral theory of the complex Laplace operator, using either a localization procedure or, alternatively, a heat kernel technique. These inequalities are a useful complement to the Riemann-Roch formula when information is needed about individual cohomology groups, and not just about the Euler-Poincaré characteristic. One of the typical consequences is a solution of the Grauert-Riemenschneider conjecture, which was first announced by [Siu84] in the case of a semi-positive line bundle L, and by [Dem85b] in general.

(8.3) Corollary (solution of the Grauert-Riemenschneider conjecture, [Siu84; Dem85b]). *Let X be a compact complex manifold carrying a holomorphic Hermitian line bundle (L,h) such that*

$$\int_{X(L,h,\leqslant 1)}\Big(\frac{\mathrm{i}}{2\pi}\Theta_{L,h}\Big)^n>0.$$

Then L is a big line bundle, and as a consequence, X is a Moishezon manifold, i.e. is bimeromorphic to a projective manifold.

Proof. In the case $q=1$, the strong Morse inequalities yield

$$h^0(X,\mathcal{O}(kL))-h^1(X,\mathcal{O}(kL))\geqslant\frac{k^n}{n!}\int_{X(L,h,\leqslant 1)}\Big(\frac{\mathrm{i}}{2\pi}\Theta_{L,h}\Big)^n-o(k^n)\geqslant ck^n,\quad c>0,$$

hence L is big. □

(8.4) Corollary. *If X is compact Kähler and L is nef, then*

$$h^q(X,\mathcal{O}(E)\otimes\mathcal{O}(kL))=o(k^n)\qquad\textit{for all } q\geqslant 1.$$

Proof. Let ω be a Kähler metric. The nefness of L implies that there exists a smooth Hermitian metric h_ε on L such that $\frac{\mathrm{i}}{2\pi}\Theta_{L,h_\varepsilon}\geqslant-\varepsilon\omega$. On $X(L,h_\varepsilon,1)$ we have exactly 1 negative eigenvalue λ_1 which is belongs to $[-\varepsilon,0[$ and the other ones λ_j $(j\geqslant 2)$ are positive. The product $\lambda_1\cdots\lambda_n$ satisfies $|\lambda_1\cdots\lambda_n|\leqslant\varepsilon\prod_{j\geqslant 2}(\varepsilon+\lambda_j)$, hence

$$\frac{1}{n!}\Big|\Big(\frac{\mathrm{i}}{2\pi}\Theta_{L,h_\varepsilon}\Big)^n\Big|\leqslant\frac{1}{(n-1)!}\varepsilon\omega\wedge\Big(\varepsilon\omega+\frac{\mathrm{i}}{2\pi}\Theta_{L,h_\varepsilon}\Big)^{n-1}\qquad\text{on } X(L,h_\varepsilon,1).$$

By integrating, we find

$$\int_{X(L,h_\varepsilon,1)}\Big(\frac{\mathrm{i}}{2\pi}\Theta_{L,h_\varepsilon}\Big)^n\leqslant n\varepsilon\int_X\omega\wedge(c_1(L)+\omega)^{n-1}$$

and the result follows. (Note: when X is non Kähler, D. Popovici [Pop08] has announced bounds for the Monge-Ampère masses of Θ_{L,h_ε} which still imply the result, but the proof is much harder in that case.) □

8.B. Algebraic Counterparts of the Holomorphic Morse Inequalities

One difficulty in the application of the analytic form of the inequalities is that the curvature integral is in general quite uneasy to compute, since it is neither a topological nor an algebraic invariant. However, the Morse inequalities can be reformulated in a more algebraic setting in which only algebraic invariants are involved. We give here two such reformulations.

(8.5) Theorem. *Let $L = F - G$ be a holomorphic line bundle over a compact Kähler manifold X, where F and G are numerically effective line bundles. Then for every $q = 0, 1, \ldots, n = \dim X$, there is an asymptotic strong Morse inequality*

$$\sum_{0\leqslant j\leqslant q} (-1)^{q-j} h^j(X, kL) \leqslant \frac{k^n}{n!} \sum_{0\leqslant j\leqslant q} (-1)^{q-j} \binom{n}{j} F^{n-j} \cdot G^j + o(k^n).$$

Proof. By adding ε times a Kähler metric ω to the curvature forms of F and G, $\varepsilon > 0$ one can write $\frac{\mathrm{i}}{2\pi}\Theta_L = \theta_{F,\varepsilon} - \theta_{G,\varepsilon}$ where $\theta_{F,\varepsilon} = \frac{\mathrm{i}}{2\pi}\Theta_F + \varepsilon\omega$ and $\theta_{G,\varepsilon} = \frac{\mathrm{i}}{2\pi}\Theta_G + \varepsilon\omega$ are positive definite. Let $\lambda_1 \geqslant \cdots \geqslant \lambda_n > 0$ be the eigenvalues of $\theta_{G,\varepsilon}$ with respect to $\theta_{F,\varepsilon}$. Then the eigenvalues of $\frac{\mathrm{i}}{2\pi}\Theta_L$ with respect to $\theta_{F,\varepsilon}$ are the real numbers $1 - \lambda_j$ and the set $X(L, h, \leqslant q)$ is the set $\{\lambda_{q+1} < 1\}$ of points $x \in X$ such that $\lambda_{q+1}(x) < 1$. The strong Morse inequalities yield

$$\sum_{0\leqslant j\leqslant q} (-1)^{q-j} h^j(X, kL) \leqslant \frac{k^n}{n!} \int_{\{\lambda_{q+1}<1\}} (-1)^q \prod_{1\leqslant j\leqslant n} (1 - \lambda_j)\theta_{F,\varepsilon}^n + o(k^n).$$

On the other hand we have

$$\binom{n}{j} \theta_{F,\varepsilon}^{n-j} \wedge \theta_{G,\varepsilon}^j = \sigma_n^j(\lambda)\, \theta_{F,\varepsilon}^n,$$

where $\sigma_n^j(\lambda)$ is the j-th elementary symmetric function in $\lambda_1, \ldots, \lambda_n$, hence

$$\sum_{0\leqslant j\leqslant q} (-1)^{q-j} \binom{n}{j} F^{n-j} \cdot G^j = \lim_{\varepsilon\to 0} \int_X \sum_{0\leqslant j\leqslant q} (-1)^{q-j} \sigma_n^j(\lambda)\, \theta_{F,\varepsilon}^n.$$

Thus, to prove the lemma, we only have to check that

$$\sum_{0\leqslant j\leqslant n} (-1)^{q-j} \sigma_n^j(\lambda) - 1\!\!1_{\{\lambda_{q+1}<1\}} (-1)^q \prod_{1\leqslant j\leqslant n} (1 - \lambda_j) \geqslant 0$$

for all $\lambda_1 \geqslant \cdots \geqslant \lambda_n \geqslant 0$, where $1\!\!1_{\{\ldots\}}$ denotes the characteristic function of a set. This is easily done by induction on n (just split apart the parameter λ_n and write $\sigma_n^j(\lambda) = \sigma_{n-1}^j(\lambda) + \sigma_{n-1}^{j-1}(\lambda)\, \lambda_n$). □

In the case $q = 1$, we get an especially interesting lower bound (this bound has been observed and used by S. Trapani [Tra95] in a similar context).

(8.6) Consequence. $h^0(X,kL) - h^1(X,kL) \geqslant \frac{k^n}{n!}(F^n - nF^{n-1}\cdot G) - o(k^n)$. *Therefore some multiple kL has a section as soon as $F^n - nF^{n-1}\cdot G > 0$.*

(8.7) Remark. The weaker inequality

$$h^0(X,kL) \geqslant \frac{k^n}{n!}(F^n - nF^{n-1}\cdot G) - o(k^n)$$

is easy to prove if X is projective algebraic. Indeed, by adding a small ample $\mathbb{Q}$-divisor to F and G, we may assume that F, G are ample. Let m_0G be very ample and let k' be the smallest integer $\geqslant k/m_0$. Then $h^0(X,kL) \geqslant h^0(X,kF - k'm_0G)$. We select k' smooth members G_j, $1 \leqslant j \leqslant k'$ in the linear system $|m_0G|$ and use the exact sequence

$$0 \to H^0(X, kF - \sum G_j) \to H^0(X,kF) \to \bigoplus H^0(G_j, kF_{|G_j}).$$

Kodaira's vanishing theorem yields $H^q(X,kF) = 0$ and $H^q(G_j, kF_{|G_j}) = 0$ for $q \geqslant 1$ and $k \geqslant k_0$. By the exact sequence combined with Riemann-Roch, we get

$$\begin{aligned} h^0(X,kL) &\geqslant h^0(X, kF - \sum G_j) \\ &\geqslant \frac{k^n}{n!}F^n - O(k^{n-1}) - \sum\Big(\frac{k^{n-1}}{(n-1)!}F^{n-1}\cdot G_j - O(k^{n-2})\Big) \\ &\geqslant \frac{k^n}{n!}\Big(F^n - n\frac{k'm_0}{k}F^{n-1}\cdot G\Big) - O(k^{n-1}) \\ &\geqslant \frac{k^n}{n!}\Big(F^n - n\,F^{n-1}\cdot G\Big) - O(k^{n-1}). \end{aligned}$$

(This simple proof is due to F. Catanese.) □

(8.8) Corollary. *Suppose that F and G are nef and that F is big. Some multiple of $mF - G$ has a section as soon as*

$$m > n\,\frac{F^{n-1}\cdot G}{F^n}.$$

In the last condition, the factor n is sharp: this is easily seen by taking $X = \mathbb{P}_1^n$ and $F = \mathcal{O}(a,\ldots,a)$ and $G = \mathcal{O}(b_1,\ldots,b_n)$ over $\mathbb{P}_1^n$; the condition of the corollary is then $m > \sum b_j/a$, whereas $k(mF - G)$ has a section if and only if $m \geqslant \sup b_j/a$; this shows that we cannot replace n by $n(1-\varepsilon)$.

8.C. Asymptotic Cohomology Groups

In order to estimate the growth of individual cohomology groups, it is interesting to consider appropriate "asymptotic cohomology functions". We mostly follow here notation and concepts introduced by A. Küronya [Kur06; FKL07].

(8.9) Definition. *Let X be a compact complex manifold. One defines the asymptotic (analytic) q-cohomology function on* $\mathrm{NS}_{\mathbb{R}}(X)$ *to be*

$$\begin{aligned}\widehat{h}^q(X,\alpha) &= \limsup_{k\to+\infty,\,\frac{1}{k}c_1(L)\to\alpha} \frac{n!}{k^n} h^q(X,L)\\ &= \inf_{\varepsilon>0,\,k_0>0} \sup_{k\geqslant k_0,\|\frac{1}{k}c_1(L)-\alpha\|\leqslant\varepsilon} \frac{n!}{k^n} h^q(X,L).\end{aligned}$$

where the pair (k,L) runs over $\mathbb{N}^*\times\mathrm{Pic}(X)$.

From the very definition, $\widehat{h}^q$ is an upper semi-continuous function on $NS_{\mathbb{R}}(X)$ and it is positively homogeneous of degree n, namely

$$\widehat{h}^q(X,\lambda\alpha) = \lambda^n \widehat{h}^q(X,\alpha) \tag{8.10}$$

for all $\alpha\in\mathrm{NS}_{\mathbb{R}}(X)$ and all $\lambda\geqslant 0$. In the general case of compact complex manifolds, the fact that $\widehat{h}^q(X,\alpha)$ is finite follows from spectral theory estimates for the complex Laplace-Beltrami operators (this will become quite clear from the discussions below).

For a line bundle L, we simply denote $\widehat{h}^q(X,L)=\widehat{h}^q(X,c_1(L))$. In this case, things become a little bit simpler, and especially, for $q=0$, one recovers the usual concept of volume of a line bundle.

(8.11) Proposition. *If X is projective algebraic or $q=0$, then*

$$\widehat{h}^q(X,L) = \limsup_{k\to+\infty} \frac{n!}{k^n} h^q(X,L^{\otimes k}) = \lim_{k\to+\infty} \frac{n!}{k^n} h^q(X,L^{\otimes k}).$$

Moreover, in these cases, the map $\alpha\mapsto\widehat{h}^q(X,\alpha)$ is (locally) Lipschitz continuous on $\mathrm{NS}_{\mathbb{R}}(X)$.

The proof is derived from arguments quite similar to those already developed in [Kur05]. Actually, let us introduce $\mathrm{DNS}_{\mathbb{R}}(X)\subset\mathrm{NS}_{\mathbb{R}}(X)$ to be the subspace generated by classes of integral divisors D on X ("divisorial Neron-Severi group"). If X is projective algebraic then $\mathrm{DNS}_{\mathbb{R}}(X)=\mathrm{NS}_{\mathbb{R}}(X)$, but the inclusion can be strict in general (e.g. on complex 2-tori which only have indefinite integral $(1,1)$-classes, cf. [BL04]). If $D=\sum p_jD_j$ is an integral divisor, we define its norm to be $\|D\|=\sum|p_j|\,\mathrm{Vol}_\omega(D_j)$, where the volume of an irreducible divisor is computed by means of a given Hermitian metric ω on X; in other words, this is precisely the mass of the current of integration $[D]$ with respect to ω. Clearly, since X is compact, we get equivalent norms for all choices of Hermitian metrics ω on X. We can also use ω to fix a normalized metric on $H^{1,1}_{\mathrm{BC}}(X,\mathbb{R})$. Elementary properties of potential theory show that $\|c_1(\mathcal{O}(D))\|\leqslant C\|D\|$ for some constant $C>0$ (but the converse inequality is of course wrong in most cases). Proposition 8.11 is a simple consequence of the more precise cohomology estimates (8.15) which will be obtained below.

(8.12) Lemma. *Let X be a compact complex n-fold. Then for every coherent sheaf $\mathcal{F}$ on X, there is a constant $C_{\mathcal{F}} > 0$ such that for every holomorphic line bundle L on X we have*

$$h^q(X, \mathcal{F} \otimes \mathcal{O}_X(L)) \leqslant C_{\mathcal{F}}(\|c_1(L)\| + 1)^p$$

where $p = \dim \operatorname{Supp} \mathcal{F}$.

Proof. We prove the result by induction on p; it is indeed clear for $p = 0$ since we then have cohomology only in degree 0 and the dimension of $H^0(X, \mathcal{F} \otimes \mathcal{O}_X(L))$ does not depend on L when $\mathcal{F}$ has finite support. Let us consider the support Y of $\mathcal{F}$ and a resolution of singularity $\mu : \widehat{Y} \to Y$ of the corresponding (reduced) analytic space. Then $\mathcal{F}$ is an $\mathcal{O}_Y$-module for some non necessarily reduced complex structure $\mathcal{O}_Y = \mathcal{O}_X/\mathcal{J}$ on J. We can look at the reduced structure $\mathcal{O}_{Y,\mathrm{red}} = \mathcal{O}_X/\mathcal{I}$, $\mathcal{I} = \sqrt{\mathcal{J}}$, and filter $\mathcal{F}$ by $\mathcal{I}^k\mathcal{F}$, $k \geqslant 0$. Since $\mathcal{I}^k\mathcal{F}/\mathcal{I}^{k+1}\mathcal{F}$ is a coherent $\mathcal{O}_{Y,\mathrm{red}}$-module, we can easily reduce the situation to the case where Y is reduced and $\mathcal{F}$ is an $\mathcal{O}_Y$-module. In that case the cohomology

$$H^q(X, \mathcal{F} \otimes \mathcal{O}_X(L)) = H^q(Y, \mathcal{F} \otimes \mathcal{O}_Y(L_{|Y}))$$

just lives on the reduced space Y.

Now, we have an injective sheaf morphism $\mathcal{F} \to \mu_*\mu^*\mathcal{F}$ whose cokernel $\mathcal{G}$ has support in dimension $< p$. By induction on p, we conclude from the exact sequence that

$$\left|h^q(X, \mathcal{F} \otimes \mathcal{O}_X(L)) - h^q(X, \mu_*\mu^*\mathcal{F} \otimes \mathcal{O}_X(L))\right| \leqslant C_1(\|c_1(L)\| + 1)^{p-1}.$$

The functorial morphisms

$$\mu^* : H^q(Y, \mathcal{F} \otimes \mathcal{O}_Y(L_{|Y})) \to H^q(\widehat{Y}, \mu^*\mathcal{F} \otimes \mathcal{O}_{\widehat{Y}}(\mu^*L)_{|Y}),$$
$$\mu_* : H^q(\widehat{Y}, \mu^*\mathcal{F} \otimes \mathcal{O}_{\widehat{Y}}(\mu^*L)_{|Y}) \to H^q(Y, \mu_*\mu^*\mathcal{F} \otimes \mathcal{O}_Y(L_{|Y}))$$

yield a composition

$$\mu_* \circ \mu^* : H^q(Y, \mathcal{F} \otimes \mathcal{O}_Y(L_{|Y})) \to H^q(Y, \mu_*\mu^*\mathcal{F} \otimes \mathcal{O}_Y(L_{|Y}))$$

induced by the natural injection $\mathcal{F} \to \mu_*\mu^*\mathcal{F}$. This implies

$$h^q(Y, \mathcal{F} \otimes \mathcal{O}_Y(L_{|Y})) \leqslant h^q(\widehat{Y}, \mu^*\mathcal{F} \otimes \mathcal{O}_{\widehat{Y}}(\mu^*L_{|Y})) + C_1(\|c_1(L)\| + 1)^{p-1}.$$

By taking a suitable modification $\mu' : Y' \to Y$ of the desingularization $\widehat{Y}$, we can assume that $(\mu')^*\mathcal{F}$ is locally free modulo torsion. Then we are reduced to the case where $\mathcal{F}' = (\mu')^*\mathcal{F}$ is a locally free sheaf on a smooth manifold Y', and $L' = (\mu')^*L_{|Y}$. In this case, we apply Morse inequalities to conclude that

$h^q(Y', \mathcal{F}' \otimes \mathcal{O}_{Y'}(L')) \leqslant C_2(\|c_1(L')\| + 1)^p$. Since $\|c_1(L')\| \leqslant C_3\|c_1(L)\|$ by pulling-back, the statement follows easily. □

(8.13) Corollary. *For every irreducible divisor D on X, there exists a constant C_D such that*

$$h^q(D, \mathcal{O}_D(L_{|D})) \leqslant C_D(\|c_1(L)\| + 1)^{n-1}$$

Proof. It is enough to apply Lemma 8.12 with $\mathcal{F} = (i_D)_*\mathcal{O}_D$ where $i_D : D \to X$ is the injection. □

(8.14) Remark. It is very likely that one can get an "elementary" proof of Lemma 8.12 without invoking resolutions of singularities, e.g. by combining the Cartan-Serre finiteness argument along with the standard Serre-Siegel proof based ultimately on the Schwarz lemma. In this context, one would invoke L^2 estimates to get explicit bounds for the homotopy operators between Čech complexes relative to two coverings $\mathcal{U} = (B(x_j, r_j))$, $\mathcal{U}' = (B(x_j, r_j/2))$ of X by concentric balls. By exercising enough care in the estimates, it is likely that one could reach an explicit dependence $C_D \leqslant C'\|D\|$ for the constant C_D of Corollary 8.13. The proof would of course become much more technical than the rather naive brute force approach we have used.

(8.15) Theorem. *Let X be a compact complex manifold. Fix a finitely generated subgroup Γ of the group of $\mathbb{Z}$-divisors on X. Then there are constants C, C' depending only on X, its Hermitian metric ω and the subgroup Γ, satisfying the following properties.*

(a) *Let L and $L' = L \otimes \mathcal{O}(D)$ be holomorphic line bundles on X, where $D \in \Gamma$ is an integral divisor. Then*

$$\big|h^q(X, L') - h^q(X, L)\big| \leqslant C(\|c_1(L)\| + \|D\|)^{n-1}\|D\|.$$

(b) *On the subspace $\mathrm{DNS}_{\mathbb{R}}(X)$, the asymptotic q-cohomology function $\widehat{h}^q$ satisfies a global estimate*

$$\big|\widehat{h}^q(X, \beta) - \widehat{h}^q(X, \alpha)\big| \leqslant C'(\|\alpha\| + \|\beta\|)^{n-1}\|\beta - \alpha\|.$$

In particular (without any further assumption on X), $\widehat{h}^q$ is locally Lipschitz continuous on $\mathrm{DNS}_{\mathbb{R}}(X)$.

Proof. (a) We want to compare the cohomology of L and $L' = L \otimes \mathcal{O}(D)$ on X. For this we write $D = D_+ - D_-$, and compare the cohomology of the pairs L and $L_1 = L \otimes \mathcal{O}(-D_-)$ one one hand, and of L' and $L_1 = L' \otimes \mathcal{O}(-D_+)$ on the other hand. Since $\|c_1(\mathcal{O}(D))\| \leqslant C\|D\|$ by elementary potential theory, we see that is is enough to consider the case of a negative divisor, i.e. $L' = L \otimes \mathcal{O}(-D)$, $D \geqslant 0$. If D is an irreducible divisor, we use the exact sequence

$$0 \to L \otimes \mathcal{O}(-D) \to L \to \mathcal{O}_D \otimes L_{|D} \to 0$$

and conclude by Corollary 8.13 that

$$\begin{aligned}\left|h^q(X, L\otimes\mathcal{O}(-D)) - h^q(X,L)\right| &\leqslant h^q(D,\mathcal{O}_D\otimes L_{|D}) + h^{q-1}(D,\mathcal{O}_D\otimes L_{|D})\\ &\leqslant 2C_D(\|c_1(L)\|+1)^{n-1}.\end{aligned}$$

For $D=\sum p_jD_j \geqslant 0$, we easily get by induction

$$\left|h^q(X, L\otimes\mathcal{O}(-D)) - h^q(X,L)\right| \leqslant 2\sum_j p_jC_{D_j}\Big(\|c_1(L)\| + \sum_k p_k\|\nabla_k\|+1\Big)^{n-1}.$$

If we knew that $C_D \leqslant C'\|D\|$ as expected in Remark 8.14, then the argument would be complete without any restriction on D. The trouble disappears if we fix D in a finitely generated subgroup Γ of divisors, because only finitely many irreducible components appear in that case, and so we have to deal with only finitely many constants C_{D_j}. Property 8.15 (a) is proved.

(b) Fix once for all a finite set of divisors $(\Delta_j)_{1\leqslant j\leqslant t}$ providing a basis of $\mathrm{DNS}_{\mathbb{R}}(X) \subset H^{1,1}_{\mathrm{BC}}(X,\mathbb{R})$. Take two elements α and β in $\mathrm{DNS}_{\mathbb{R}}(X)$, and fix $\varepsilon>0$. Then $\beta-\alpha$ can be ε-approximated by a $\mathbb{Q}$-divisor $\sum\lambda_jD_j$, $\lambda_j\in\mathbb{Q}$, and we can find a pair (k,L) with k arbitrary large such that $\frac{1}{k}c_1(L)$ is ε-close to α and $n!/k^nh^q(X,L)$ approaches $\widehat{h}^q(X,\alpha)$ by ε. Then $\frac{1}{k}L+\sum\lambda_j\Delta_j$ approaches β as closely as we want. When approximating $\beta-\alpha$, we can arrange that $k\lambda_j$ is an integer by taking k large enough. Then β is approximated by $\frac{1}{k}c_1(L')$ with $L' = L\otimes\mathcal{O}(\sum k\lambda_j\Delta_j)$. Property (a) implies

$$\begin{aligned}h^q(X,L') - h^q(X,L) &\geqslant -C\Big(\|c_1(L)\| + \Big\|\sum k\lambda_j\Delta_j\Big\|\Big)^{n-1}\Big\|\sum k\lambda_j\Delta_j\Big\|\\ &\geqslant -Ck^n\big(\|\alpha\|+\varepsilon+\|\beta-\alpha\|+\varepsilon\big)^{n-1}(\|\beta-\alpha\|+\varepsilon).\end{aligned}$$

We multiply the previous inequality by $n!/k^n$ and get in this way

$$\frac{n!}{k^n}h^q(X,L') \geqslant \widehat{h}^q(X,\alpha) - \varepsilon - C'\big(\|\alpha\|+\|\beta\|+\varepsilon\big)^{n-1}(\|\beta-\alpha\|+\varepsilon).$$

By taking the limsup and letting $\varepsilon\to0$, we finally obtain

$$\widehat{h}^q(X,\beta) - \widehat{h}^q(X,\alpha) \geqslant -C'\big(\|\alpha\|+\|\beta\|\big)^{n-1}\|\beta-\alpha\|.$$

Property 8.15 (b) follows by exchanging the roles of α and β. □

8.D. Transcendental Asymptotic Cohomology Functions

Our ambition is to extend the function $\widehat{h}^q$ in a natural way to the full cohomology group $H^{1,1}_{\mathrm{BC}}(X,\mathbb{R})$. The main trouble, already when X is projective algebraic, is that the Picard number $\rho(X)=\dim_{\mathbb{R}}\mathrm{NS}_{\mathbb{R}}(X)$ may be much smaller than $\dim_{\mathbb{R}} H^{1,1}_{\mathrm{BC}}(X,\mathbb{R})$, namely, there can be rather few integral classes of type $(1,1)$

on X. It is well known for instance that $\rho(X) = 0$ for a generic complex torus a dimension $n \geqslant 2$, while $\dim_{\mathbb{R}} H^{1,1}_{\mathrm{BC}}(X,\mathbb{R}) = n^2$. However, if we look at the natural morphism

$$H^{1,1}_{\mathrm{BC}}(X,\mathbb{R}) \to H^2_{\mathrm{DR}}(X,\mathbb{R}) \simeq H^2(X,\mathbb{R})$$

to de Rham cohomology, then $H^2(X,\mathbb{Q})$ is dense in $H^2(X,\mathbb{R})$. Therefore, given a class $\alpha \in H^{1,1}_{\mathrm{BC}}(X,\mathbb{R})$ and a smooth d-closed $(1,1)$-form u in α, we can find an infinite sequence $\frac{1}{k}L_k$ ($k \in S \subset \mathbb{N}$) of topological $\mathbb{Q}$-line bundles, equipped with Hermitian metrics h_k and compatible connections ∇_k such that the curvature forms $\frac{1}{k}\Theta_{\nabla_k}$ converge to u. By using Kronecker's approximation with respect to the integral lattice $H^2(X,\mathbb{Z})/\mathrm{torsion} \subset H^2(X,\mathbb{R})$, we can even achieve a fast diophantine approximation

$$\|\Theta_{\nabla_k} - ku\| \leqslant Ck^{-1/b_2} \tag{8.16}$$

for a suitable infinite subset $k \in S \subset \mathbb{N}$ of multipliers. Then in particular

$$\|\Theta^{0,2}_{\nabla_k}\| = \|\Theta^{0,2}_{\nabla_k} - u^{0,2}\| \leqslant Ck^{-1/b_2},$$

and we see that (L_k, h_k, ∇_k) is a C^∞ Hermitian line bundle which is extremely close to being holomorphic, since $(\nabla^{0,1}_k)^2 = \Theta^{0,2}_{\nabla_k}$ is very small. We introduce the complex Laplace-Beltrami operator

$$\overline{\square}_k = (\nabla^{0,1}_k)(\nabla^{0,1}_k)^* + (\nabla^{0,1}_k)^*(\nabla^{0,1}_k)$$

and look at its eigenspaces in $L^2(X, \Lambda^{0,q}T^*X \otimes L_k)$ with the metric induced by ω on X and h_k on L_k. In the holomorphic case, Hodge theory tells us that the 0-eigenspace is isomorphic to $H^q(X, \mathcal{O}(L_k))$, but in the "almost holomorphic case" the 0-eigenvalues deviate from 0, essentially by a shift of the order of magnitude of $\|\Theta^{0,2}_{\nabla_k}\| \sim k^{-1/b_2}$ (see [Lae02], Chapter 4). It is thus natural to introduce in this case

(8.17) Definition. *Let X be a compact complex manifold and $\alpha \in H^{1,1}_{\mathrm{BC}}(X,\mathbb{R})$ an arbitrary Bott-Chern $(1,1)$-class. We define the "transcendental" asymptotic q-cohomology function to be*

$$\widehat{h}^q_{\mathrm{tr}}(X,\alpha) = \inf_{u\in\alpha} \limsup_{\varepsilon\to 0,\, k\to+\infty,\, L_k,\, h_k,\, \nabla_k,\, \frac{1}{k}\Theta_{\nabla_k}\to u} \frac{n!}{k^n} N(\overline{\square}_k, k\varepsilon)$$

where the lim sup *runs over all 5-tuples $(\varepsilon, k, L_k, h_k, \nabla_k)$, and where $N(\overline{\square}_k, k\varepsilon)$ denotes the sum of dimensions of all eigenspaces of eigenvalues at most equal to $k\varepsilon$ for the Laplace-Beltrami operator $\overline{\square}_k$ associated with (L_k, h_k, ∇_k) and the base Hermitian metric ω.*

The word "transcendental" refers here to the fact that we deal with classes α of type $(1,1)$ which are not algebraic or even analytic. Of course, in the definition, we could have restricted the limsup to families satisfying a better approximation property $\|\frac{1}{k}\Theta_{\nabla_k} - u\| \leqslant Ck^{-1-1/b_2}$ for some large constant C (this would lead a priori to a smaller limsup, but there is enough stability in the parameter dependence of the spectrum for making such a change irrelevant). The minimax principle easily shows that Definition 8.16 does not depend on ω, as the eigenvalues are at most multiplied or divided by constants under a change of base metric. When $\alpha \in \mathrm{NS}_{\mathbb{R}}(X)$, by restricting our families $\{(\varepsilon, k, L_k, h_k, \nabla_k)\}$ to the case of holomorphic line bundles only, we get the obvious inequality

$$\widehat{h}^q(X,\alpha) \leqslant \widehat{h}^q_{\mathrm{tr}}(X,\alpha), \qquad \forall \alpha \in \mathrm{NS}_{\mathbb{R}}(X). \tag{8.18}$$

It is natural to raise the question whether this is always an equality. Hopefully, the calculation of the quantities $\limsup \frac{n!}{k^n} N(\overline{\square}_k, k\varepsilon)$ is a problem of spectral theory which is completely understood since a long time. In fact, as a consequence of the techniques of [Dem85b; Dem91; Lae02], one gets

(8.19) Theorem. *With the above notations and assumptions, one has*

$$\limsup_{\varepsilon\to 0,\, k\to+\infty,\, L_k,\, h_k,\, \nabla_k,\, \frac{1}{k}\Theta_{\nabla_k}\to u} \frac{n!}{k^n} N(\overline{\square}_k, k\varepsilon) = \int_{X(u,q)} (-1)^q u^n,$$

where $X(u,q)$ is the open set of points $x \in X$ where $u(x)$ has signature $(n-q,q)$. Therefore

$$\widehat{h}^q_{\mathrm{tr}}(X,\alpha) = \inf_{u\in\alpha} \int_{X(u,q)} (-1)^q u^n \qquad (u\ \textit{smooth}).$$

The first equality follows mainly from Theorems 2.16 and 3.14 of [Dem85b], which even yield explicitly the limit for any given ε outside a countable set (the limit as $\varepsilon \to 0$ is then obtained from the calculations of page 224 after Cor. 4.3). One has to observe, in the case of sequences of "almost holomorphic line bundles" considered here, that the perturbation indeed goes to 0, and also that all constants involved in the calculations of [Dem85b] are uniformly bounded; see [Dem91] and [Lae02] for more details on this. Therefore, we can reformulate more explicitly our previous question in the following terms.

(8.20) Question. *For every $\alpha \in \mathrm{NS}_{\mathbb{R}}(X)$, is it true that*

$$\widehat{h}^q(X,\alpha) = \inf_{u\in\alpha} \int_{X(u,q)} (-1)^q u^n \qquad (u\ \textit{smooth})\ ?$$

(Note: it follows from the holomorphic Morse inequalities that the inequality $\leqslant$ always holds true).

In general, equality (8.20) seems rather hard to prove. In some sense, this would be an asymptotic converse of the Andreotti-Grauert theorem [AG62] : under a suitable q-convexity assumption, the latter asserts the vanishing of related cohomology groups in degree q; here, conversely, assuming a known growth of these groups in degree q, we expect to be able to say something about the q-index sets of suitable Hermitian metrics on the line bundles under consideration. The only case where we have a positive answer to Question 8.20 is when X is projective of dimension at most 2 (see 18.31). In the general setting of compact complex manifolds, we also hope for the following "transcendental" case of holomorphic Morse inequalities.

(8.21) Conjecture. *Let X be a compact complex n-fold and α an arbitrary cohomology class in $H^{1,1}_{\mathrm{BC}}(X,\mathbb{R})$. Then the volume, defined as the supremum*

$$\mathrm{Vol}(\alpha) := \sup_{0<T\in\alpha} \int_{X\smallsetminus \mathrm{Sing}(T)} T^n, \tag{8.22}$$

extended to all Kähler currents $T \in \alpha$ with analytic singularities(see Definition 13.15 in Chapter 13), satisfies

$$\mathrm{Vol}(\alpha) \geqslant \sup_{u\in\alpha} \int_{X(u,0)\cup X(u,1)} u^n \tag{8.23}$$

where u runs over all smooth closed $(1,1)$ forms. In particular, if the right hand side is positive, then α contains a Kähler current.

By the holomorphic Morse inequalities, Conjecture 8.21 holds true in case α is an integral class. Our hope is that the general case can be attained by the diophantine approximation technique described earlier; there are however major hurdles, see [Lae02] for a few hints on these issues.

Chapter 9

Effective Version of Matsusaka's Big Theorem

An important problem of algebraic geometry is to find effective bounds m_0 such that multiples mL of an ample line bundle become very ample for $m \geqslant m_0$. From a theoretical point of view, this problem has been solved by Matsusaka [Mat72] and Kollár-Matsusaka [KoM83]. Their result is that there is a bound $m_0 = m_0(n, L^n, L^{n-1} \cdot K_X)$ depending only on the dimension and on the first two coefficients L^n and $L^{n-1} \cdot K_X$ in the Hilbert polynomial of L. Unfortunately, the original proof does not tell much on the actual dependence of m_0 in terms of these coefficients. The goal of this chapter is to find effective bounds for such an integer m_0, along the lines of [Siu93]. However, one of the technical lemmas used in [Siu93] to deal with dualizing sheaves can be sharpened. Using this sharpening of the lemma, Siu's bound will be here substantially improved. We first start with the simpler problem of obtaining merely a nontrivial section in mL. The idea, more generally, is to obtain a criterion for the ampleness of $mL - B$ when B is nef. In this way, one is able to subtract from mL any multiple of K_X which happens to get added by the application of Nadel's vanishing theorem (for this, replace B by B plus a multiple of $K_X + (n+1)L$).

(9.1) Proposition. *Let L be an ample line bundle over a projective n-fold X and let B be a nef line bundle over X. Then $K_X + mL - B$ has a nonzero section for some integer m such that*

$$m \leqslant n \frac{L^{n-1} \cdot B}{L^n} + n + 1.$$

Proof. Let m_0 be the smallest integer $> n \frac{L^{n-1} \cdot B}{L^n}$. Then $m_0 L - B$ can be equipped with a singular Hermitian metric of positive definite curvature. Let φ be the weight of this metric. By Nadel's vanishing theorem, we have

$$H^q(X, \mathcal{O}(K_X + mL - B) \otimes \mathcal{I}(\varphi)) = 0 \qquad \text{for } q \geqslant 1,$$

thus $P(m) = h^0(X, \mathcal{O}(K_X + mL - B) \otimes \mathcal{I}(\varphi))$ is a polynomial for $m \geqslant m_0$. Since P is a polynomial of degree n and is not identically zero, there must be an integer $m \in [m_0, m_0 + n]$ which is not a root. Hence there is a nontrivial section in

$$H^0(X, \mathcal{O}(K_X + mL - B)) \supset H^0(X, \mathcal{O}(K_X + mL - B) \otimes \mathcal{I}(\varphi))$$

for some $m \in [m_0, m_0 + n]$, as desired. $\square$

(9.2) Corollary. *If L is ample and B is nef, then $mL - B$ has a nonzero section for some integer*

$$m \leqslant n\Big(\frac{L^{n-1} \cdot B + L^{n-1} \cdot K_X}{L^n} + n + 1\Big).$$

Proof. By *Fujita's result* 9.3 (a), $K_X + (n+1)L$ is nef. We can thus replace B by $B + K_X + (n+1)L$ in the result of Proposition 9.1. Corollary 9.2 holds. $\square$

(9.3) Remark. We do not know if the above Corollary is sharp, but it is certainly not far from being so. Indeed, for $B = 0$, the initial constant n cannot be replaced by anything smaller than $n/2$: take X to be a product of curves C_j of large genus g_j and $B = 0$; our bound for $L = \mathcal{O}(a_1[p_1]) \otimes \cdots \otimes \mathcal{O}(a_n[p_n])$ to have $|mL| \neq \emptyset$ becomes $m \leqslant \sum(2g_j - 2)/a_j + n(n+1)$, which fails to be sharp only by a factor 2 when $a_1 = \cdots = a_n = 1$ and $g_1 \gg g_2 \gg \cdots \gg g_n \to +\infty$. On the other hand, the additive constant $n+1$ is already best possible when $B = 0$ and $X = \mathbb{P}^n$. $\square$

So far, the method is not really sensitive to singularities (the Morse inequalities are indeed still true in the singular case as is easily seen by using desingularizations of the ambient variety). The same is true with Nadel's vanishing theorem, provided that K_X is replaced by the L^2 dualizing sheaf ω_X (according to the notation introduced in Remark 6.22, $\omega_X = K_X(0)$ is the sheaf of holomorphic n-forms u on X_{reg} such that $i^{n^2} u \wedge \overline{u}$ is integrable in a neighborhood of the singular set). Then Proposition 9.1 can be generalized as

(9.4) Proposition. *Let L be an ample line bundle over a projective n-fold X and let B be a nef line bundle over X. For every p-dimensional (reduced) algebraic subvariety Y of X, there is an integer*

$$m \leqslant p\frac{L^{p-1} \cdot B \cdot Y}{L^p \cdot Y} + p + 1$$

such that the sheaf $\omega_Y \otimes \mathcal{O}_Y(mL - B)$ has a nonzero section.

To proceed further, we need the following useful "upper estimate" about L^2 dualizing sheaves (this is one of the crucial steps in Siu's approach; unfortunately, it has the effect of producing rather large final bounds when the dimension increases).

(9.5) Proposition. *Let H be a very ample line bundle on a projective algebraic manifold X, and let $Y \subset X$ be a p-dimensional irreducible algebraic subvariety. If $\delta = H^p \cdot Y$ is the degree of Y with respect to H, the sheaf*

$$\mathrm{Hom}\big(\omega_Y, \mathcal{O}_Y((\delta - p - 2)H)\big)$$

has a nontrivial section.

Observe that if Y is a smooth hypersurface of degree δ in $(X,H) = (\mathbb{P}^{p+1},\mathcal{O}(1))$, then $\omega_Y = \mathcal{O}_Y(\delta - p - 2)$ and the estimate is optimal. On the other hand, if Y is a smooth complete intersection of multidegree $(\delta_1, \ldots, \delta_r)$ in $\mathbb{P}^{p+r}$, then $\delta = \delta_1 \cdots \delta_r$ whilst $\omega_Y = \mathcal{O}_Y(\delta_1 + \cdots + \delta_r - p - r - 1)$; in this case, Proposition 9.5 is thus very far from being sharp.

Proof. Let $X \subset \mathbb{P}^N$ be the embedding given by H, so that $H = \mathcal{O}_X(1)$. There is a linear projection $\mathbb{P}^n \dashrightarrow \mathbb{P}^{p+1}$ whose restriction $\pi : Y \to \mathbb{P}^{p+1}$ to Y is a finite and regular birational map of Y onto an algebraic hypersurface Y' of degree δ in $\mathbb{P}^{p+1}$. Let $s \in H^0(\mathbb{P}^{p+1}, \mathcal{O}(\delta))$ be the polynomial of degree δ defining Y'. We claim that for any small Stein open set $W \subset \mathbb{P}^{p+1}$ and any L^2 holomorphic p-form u on $Y' \cap W$, there is a L^2 holomorphic $(p+1)$-form $\widetilde{u}$ on W with values in $\mathcal{O}(\delta)$ such that $\widetilde{u}_{\restriction Y' \cap W} = u \wedge ds$. In fact, this is precisely the conclusion of the Ohsawa-Takegoshi extension theorem [OT87; Ohs88] (see also [Man93] for a more general version); one can also invoke more standard local algebra arguments (see Hartshorne [Har77], Theorem III-7.11). As $K_{\mathbb{P}^{p+1}} = \mathcal{O}(-p-2)$, the form $\widetilde{u}$ can be seen as a section of $\mathcal{O}(\delta - p - 2)$ on W, thus the sheaf morphism $u \mapsto u \wedge ds$ extends into a global section of $\mathrm{Hom}\big(\omega_{Y'}, \mathcal{O}_{Y'}(\delta - p - 2)\big)$. The pull-back by π^* yields a section of $\mathrm{Hom}\big(\pi^*\omega_{Y'}, \mathcal{O}_Y((\delta - p - 2)H)\big)$. Since π is finite and generically $1:1$, it is easy to see that $\pi^*\omega_{Y'} = \omega_Y$. The proposition follows. □

By an appropriate induction process based on the above results, we can now improve Siu's effective version of the Big Matsusaka Theorem [Siu93]. Our version depends on a constant λ_n such that $m(K_X + (n+2)L) + G$ is very ample for $m \geqslant \lambda_n$ and every nef line bundle G. Corollary 8.6 shows that $\lambda_n \leqslant \binom{3n+1}{n} - 2n$, and a similar argument involving the recent results of Angehrn-Siu [AS95] implies $\lambda_n \leqslant n^3 - n^2 - n - 1$ for $n \geqslant 2$. Of course, it is expected that $\lambda_n = 1$ in view of the Fujita conjecture.

(9.6) Effective Version of the Big Matsusaka Theorem. *Let L and B be nef line bundles on a projective n-fold X. Assume that L is ample and let H be the very ample line bundle $H = \lambda_n(K_X + (n+2)L)$. Then $mL - B$ is very ample for*

$$m \geqslant (2n)^{(3^{n-1}-1)/2} \frac{(L^{n-1} \cdot (B+H))^{(3^{n-1}+1)/2} (L^{n-1} \cdot H)^{3^{n-2}(n/2-3/4)-1/4}}{(L^n)^{3^{n-2}(n/2-1/4)+1/4}}.$$

In particular mL is very ample for

$$m \geqslant C_n \, (L^n)^{3^{n-2}} \left(n + 2 + \frac{L^{n-1} \cdot K_X}{L^n}\right)^{3^{n-2}(n/2+3/4)+1/4}$$

with $C_n = (2n)^{(3^{n-1}-1)/2} (\lambda_n)^{3^{n-2}(n/2+3/4)+1/4}$.

Proof. We use Proposition 9.4 and Proposition 9.5 to construct inductively a sequence of (non necessarily irreducible) algebraic subvarieties $X = Y_n \supset Y_{n-1} \supset$

$\cdots \supset Y_2 \supset Y_1$ such that $Y_p = \bigcup_j Y_{p,j}$ is p-dimensional, and Y_{p-1} is obtained for each $p \geqslant 2$ as the union of zero sets of sections

$$\sigma_{p,j} \in H^0(Y_{p,j}, \mathcal{O}_{Y_{p,j}}(m_{p,j}L - B))$$

with suitable integers $m_{p,j} \geqslant 1$. We proceed by induction on decreasing values of the dimension p, and find inductively upper bounds m_p for the integers $m_{p,j}$.

By Corollary 9.2, an integer m_n for $m_nL - B$ to have a section σ_n can be found with

$$m_n \leqslant n\frac{L^{n-1}\cdot(B + K_X + (n+1)L)}{L^n} \leqslant n\frac{L^{n-1}\cdot(B+H)}{L^n}.$$

Now suppose that the sections $\sigma_n, \ldots, \sigma_{p+1,j}$ have been constructed. Then we get inductively a p-cycle $\widetilde{Y}_p = \sum \mu_{p,j}Y_{p,j}$ defined by $\widetilde{Y}_p =$ sum of zero divisors of sections $\sigma_{p+1,j}$ in $Y_{p+1,j}$, where the mutiplicity $\mu_{p,j}$ on $Y_{p,j} \subset Y_{p+1,k}$ is obtained by multiplying the corresponding multiplicity $\mu_{p+1,k}$ of $Y_{p+1,j}$ with the vanishing order of $\sigma_{p+1,k}$ along $Y_{p,j}$. As cohomology classes, we find

$$\widetilde{Y}_p \equiv \sum (m_{p+1,k}L - B)\cdot(\mu_{p+1,k}Y_{p+1,k}) \leqslant m_{p+1}L\cdot\widetilde{Y}_{p+1}.$$

Inductively, we thus have the numerical inequality

$$\widetilde{Y}_p \leqslant m_{p+1}\cdots m_nL^{n-p}.$$

Now, for each component $Y_{p,j}$, Proposition 9.4 shows that there exists a section of $\omega_{Y_{p,j}} \otimes \mathcal{O}_{Y_{p,j}}(m_{p,j}L - B)$ for some integer

$$m_{p,j} \leqslant p\frac{L^{p-1}\cdot B\cdot Y_{p,j}}{L^p\cdot Y_{p,j}} + p + 1 \leqslant pm_{p+1}\cdots m_n\,L^{n-1}\cdot B + p + 1.$$

Here, we have used the obvious lower bound $L^{p-1}\cdot Y_{p,j} \geqslant 1$ (this is of course a rather weak point in the argument). The degree of $Y_{p,j}$ with respect to H admits the upper bound

$$\delta_{p,j} := H^p\cdot Y_{p,j} \leqslant m_{p+1}\cdots m_nH^p\cdot L^{n-p}.$$

We use the Hovanski-Teissier concavity inequality ([Hov79; Tei79; Tei82])

$$(L^{n-p}\cdot H^p)^{\frac{1}{p}}(L^n)^{1-\frac{1}{p}} \leq L^{n-1}\cdot H$$

to express our bounds in terms of the intersection numbers L^n and $L^{n-1}\cdot H$ only. We then get

$$\delta_{p,j} \leqslant m_{p+1}\ldots m_n\frac{(L^{n-1}\cdot H)^p}{(L^n)^{p-1}}.$$

By Proposition 9.5, there is a nontrivial section in

$$\mathrm{Hom}\big(\omega_{Y_{p,j}}, \mathcal{O}_{Y_{p,j}}((\delta_{p,j} - p - 2)H)\big).$$

Combining this section with the section in $\omega_{Y_{p,j}} \otimes \mathcal{O}_{Y_{p,j}}(m_{p,j}L - B)$ already constructed, we get a section of $\mathcal{O}_{Y_{p,j}}(m_{p,j}L - B + (\delta_{p,j} - p - 2)H)$ on $Y_{p,j}$. Since we do not want H to appear at this point, we replace B with $B + (\delta_{p,j} - p - 2)H$ and thus get a section $\sigma_{p,j}$ of $\mathcal{O}_{Y_{p,j}}(m_{p,j}L - B)$ with some integer $m_{p,j}$ such that

$$\begin{aligned} m_{p,j} &\leqslant p m_{p+1} \cdots m_n \, L^{n-1} \cdot (B + (\delta_{p,j} - p - 2)H) + p + 1 \\ &\leqslant p\, m_{p+1} \cdots m_n \, \delta_{p,j} \, L^{n-1} \cdot (B + H) \\ &\leqslant p\, (m_{p+1} \cdots m_n)^2 \frac{(L^{n-1} \cdot H)^p}{(L^n)^{p-1}} \, L^{n-1} \cdot (B + H). \end{aligned}$$

Therefore, by putting $M = n\, L^{n-1} \cdot (B + H)$, we get the recursion relation

$$m_p \leqslant M \, \frac{(L^{n-1} \cdot H)^p}{(L^n)^{p-1}} \, (m_{p+1} \cdots m_n)^2 \qquad \text{for } 2 \leqslant p \leqslant n-1,$$

with initial value $m_n \leqslant M/L^n$. If we let $(\overline{m}_p)$ be the sequence obtained by the same recursion formula with equalities instead of inequalities, we get $m_p \leqslant \overline{m}_p$ with $\overline{m}_{n-1} = M^3 (L^{n-1} \cdot H)^{n-1} / (L^n)^n$ and

$$\overline{m}_p = \frac{L^n}{L^{n-1} \cdot H} \, \overline{m}_{p+1}^2 \overline{m}_{p+1}$$

for $2 \leqslant p \leqslant n - 2$. We then find inductively

$$m_p \leqslant \overline{m}_p = M^{3^{n-p}} \frac{(L^{n-1} \cdot H)^{3^{n-p-1}(n-3/2)+1/2}}{(L^n)^{3^{n-p-1}(n-1/2)+1/2}}.$$

We next show that $m_0 L - B$ is nef for

$$m_0 = \max\bigl(m_2\,,\, m_3,\, \ldots\,, m_n\,,\, m_2 \cdots m_n \, L^{n-1} \cdot B\bigr).$$

In fact, let $C \subset X$ be an arbitrary irreducible curve. Either $C = Y_{1,j}$ for some j or there exists an integer $p = 2, \ldots, n$ such that C is contained in Y_p but not in Y_{p-1}. If $C \subset Y_{p,j} \smallsetminus Y_{p-1}$, then $\sigma_{p,j}$ does not vanish identically on C. Hence $(m_{p,j}L - B)_{\restriction C}$ has nonnegative degree and

$$(m_0 L - B) \cdot C \geqslant (m_{p,j} L - B) \cdot C \geq 0.$$

On the other hand, if $C = Y_{1,j}$, then

$$(m_0 L - B) \cdot C \geqslant m_0 - B \cdot \widetilde{Y}_1 \geqslant m_0 - m_2 \cdots m_n \, L^{n-1} \cdot B \geqslant 0.$$

By the definition of λ_n (and by Corollary 8.6 showing that such a constant exists), $H + G$ is very ample for every nef line bundle G, in particular $H + m_0 L - B$ is very ample. We thus replace again B with $B + H$. This has the effect of replacing M with $M = n\,\bigl(L^{n-1} \cdot (B + 2H)\bigr)$ and m_0 with

$$m_0 = \max\bigl(m_n\,,\, m_{n-1},\, \ldots\,, m_2\,,\, m_2 \cdots m_n \, L^{n-1} \cdot (B + H)\bigr).$$

The last term is the largest one, and from the estimate on $\overline{m}_p$, we get

$$m_0 \leqslant M^{(3^{n-1}-1)/2}\frac{(L^{n-1}\cdot H)^{(3^{n-2}-1)(n-3/2)/2+(n-2)/2}(L^{n-1}\cdot(B+H))}{(L^n)^{(3^{n-2}-1)(n-1/2)/2+(n-2)/2+1}}$$

$$\leqslant (2n)^{(3^{n-1}-1)/2}\frac{(L^{n-1}\cdot(B+H))^{(3^{n-1}+1)/2}(L^{n-1}\cdot H)^{3^{n-2}(n/2-3/4)-1/4}}{(L^n)^{3^{n-2}(n/2-1/4)+1/4}}.$$

□

(9.7) Remark. In the surface case $n = 2$, one can take $\lambda_n = 1$ and our bound yields mL very ample for

$$m \geqslant 4\,\frac{(L\cdot(K_X+4L))^2}{L^2}.$$

If one looks more carefully at the proof, the initial constant 4 can be replaced by 2. In fact, it has been shown recently by Fernández del Busto that mL is very ample for

$$m > \frac{1}{2}\left[\frac{(L\cdot(K_X+4L)+1)^2}{L^2}+3\right],$$

and an example of G. Xiao shows that this bound is essentially optimal (see [FdB96]).

Chapter 10

Positivity Concepts for Vector Bundles

In the course of the proof of Skoda's L^2 estimates in the next chapter, we will have to deal with dual bundles and exact sequences of Hermitian vector bundles. The following fundamental differential geometric lemma will be needed.

(10.1) Lemma. *Let E be a Hermitian holomorphic vector bundle of rank r on a complex n-dimensional manifold X. Then the Chern connections of E and E^* are related by $\Theta_{E^*} = -{}^t\Theta_E$ where ^{t} denotes transposition. In other words, the associated Hermitian forms $\widetilde{\Theta}_E$ and $\widetilde{\Theta}_{E^*}$ are related by*

$$\widetilde{\Theta}_E(\tau,\tau) = \sum_{1\leqslant j,k\leqslant n,\,1\leqslant\lambda,\mu\leqslant r} c_{jk\lambda\mu}\tau_{j\lambda}\overline{\tau}_{k\mu}, \qquad \tau = \sum_{j,\lambda}\tau_{j,\lambda}\frac{\partial}{\partial z_j}\otimes e_\lambda,$$

$$\widetilde{\Theta}_{E^*}(\tau,\tau) = -\sum_{1\leqslant j,k\leqslant n,\,1\leqslant\lambda,\mu\leqslant r} c_{jk\mu\lambda}\tau^*_{j\lambda}\overline{\tau}^*_{k\mu}, \qquad \tau^* = \sum_{j,\lambda}\tau^*_{j,\lambda}\frac{\partial}{\partial z_j}\otimes e^*_\lambda.$$

In particular $E >_{\mathrm{Grif}} 0$ if and only if $E^* <_{\mathrm{Grif}} 0$.

Notice that the corresponding duality statement for Nakano positivity is wrong (because of the twist of indices, which is fortunately irrelevant in the case of decomposable tensors).

Proof. The Chern connections of E and E^* are related by the Leibniz rule

$$d(\sigma\wedge s) = (D_{E^*}\sigma)\wedge s + (-1)^{\deg\sigma}\sigma\wedge D_E s$$

whenever s, σ are forms with values in E, E^* respectively, and $\sigma\wedge s$ is computed using the pairing $E^*\otimes E\to\mathbb{C}$. If we differentiate a second time, this yields the identity

$$0 = (D^2_{E^*}\sigma)\wedge s + \sigma\wedge D^2_E s,$$

which is equivalent to the formula $\Theta_{E^*} = -{}^t\Theta_E$. All other assertions follow. □

(10.2) Lemma. *Let*

$$0 \longrightarrow S \xrightarrow{\;j\;} E \xrightarrow{\;g\;} Q \longrightarrow 0$$

be an exact sequence of holomorphic vector bundles. Assume that E is equipped with a smooth Hermitian metric, and that S and Q are endowed with the metrics (restriction-metric and quotient-metric) induced by that of E. Then

$$j^* \oplus g : E \to S \oplus Q, \qquad j \oplus g^* : S \oplus Q \to E \tag{10.3}$$

are C^∞ isomorphisms of bundles, which are inverse of each other. In the $\mathscr{C}^\infty$-splitting $E \simeq S \oplus Q$, the Chern connection of E admits a matrix decomposition

$$D_E = \begin{pmatrix} D_S & -\beta^* \\ \beta & D_Q \end{pmatrix} \tag{10.4}$$

in terms of the Chern connections of S and Q, where

$$\beta \in \mathscr{C}^\infty\big(X, \Lambda^{1,0}T_X^* \otimes \mathrm{Hom}(S,Q)\big), \qquad \beta^* \in \mathscr{C}^\infty\big(X, \Lambda^{0,1}T_X^* \otimes \mathrm{Hom}(Q,S)\big).$$

The form β is called the second fundamental form associated with the exact sequence. It is uniquely defined by each of the two formulas

$$D'_{\mathrm{Hom}(S,E)} j = g^* \circ \beta, \qquad j \circ \beta^* = -D''_{\mathrm{Hom}(Q,E)} g^*. \tag{10.5}$$

We have $D'_{\mathrm{Hom}(S,Q)}\beta = 0$, $D''_{\mathrm{Hom}(Q,S)}\beta^* = 0$, and the curvature form of E splits as

$$\Theta_E = \begin{pmatrix} \Theta_S - \beta^* \wedge \beta & -D'_{\mathrm{Hom}(Q,S)}\beta^* \\ D''_{\mathrm{Hom}(S,Q)}\beta & \Theta_Q - \beta \wedge \beta^* \end{pmatrix}, \tag{10.6}$$

and the curvature forms of S and Q can be expressed as

$$\Theta_S = \Theta_{E\restriction S} + \beta^* \wedge \beta, \qquad \Theta_Q = \Theta_{E\restriction Q} + \beta \wedge \beta^*, \tag{10.7}$$

where $\Theta_{E\restriction S}$, $\Theta_{E\restriction Q}$ stand for $j^* \circ \Theta_E \circ j$ and $g \circ \Theta_E \circ g^*$.

Proof. Because of the uniqueness property of Chern connections, it is easy to see that we have a Leibnitz formula

$$D_F(f \wedge u) = (D_{\mathrm{Hom}(E,F)} f) \wedge u + (-1)^{\deg f} f \wedge D_E u$$

whenever u, f are forms with values in Hermitian vector bundles E and $\mathrm{Hom}(E,F)$ (where $\mathrm{Hom}(E,F) = E^* \otimes F$ is equipped with the tensor product metric and $f \wedge u$ incorporates the evaluation mapping $\mathrm{Hom}(E,F) \otimes E \to F$). In our case, given a form u with values in E, we write $u = ju_S + g^*u_Q$ where $u_S = j^*u$ and $u_Q = gu$ are the projections of u on S and Q. We then get

$$\begin{aligned} D_E u &= D_E(ju_S + g^*u_Q) \\ &= (D_{\mathrm{Hom}(S,E)} j) \wedge u_S + j \cdot D_S u_S + (D_{\mathrm{Hom}(Q,E)} g^*) \wedge u_Q + g^* \cdot D_Q u_Q. \end{aligned}$$

Since j is holomorphic as well as $j^* \circ j = \mathrm{Id}_S$, we find $D''_{\mathrm{Hom}(S,E)} j = 0$ and

$$D''_{\mathrm{Hom}(S,S)} \mathrm{Id}_S = 0 = D''_{\mathrm{Hom}(E,S)} j^* \circ j.$$

By taking the adjoint, we see that $j^* \circ D'_{\mathrm{Hom}(S,E)} j = 0$, hence $D'_{\mathrm{Hom}(S,E)} j$ takes values in g^*Q and we thus have a unique form β as in the lemma such that $D'_{\mathrm{Hom}(S,E)} j = g^* \circ \beta$. Similarly, g and $g \circ g^* = \mathrm{Id}_Q$ are holomorphic, thus

$$D''_{\mathrm{Hom}(Q,Q)} \mathrm{Id}_Q = 0 = g \circ D''_{\mathrm{Hom}(Q,E)} g^*$$

and there is a form $\gamma \in \mathscr{C}^\infty(X, \Lambda^{0,1} T^*_X \otimes \mathrm{Hom}(Q,S))$ such that $D''_{\mathrm{Hom}(Q,E)} g^* = j \circ \gamma$. By adjunction, we get $D'_{\mathrm{Hom}(E,Q)} g = \gamma^* \circ j^*$ and $D''_{\mathrm{Hom}(E,Q)} g = 0$ implies $D'_{\mathrm{Hom}(Q,E)} g^* = 0$. If we differentiate $g \circ j = 0$ we then get

$$0 = D'_{\mathrm{Hom}(E,Q)} g \circ j + g \circ D'_{\mathrm{Hom}(S,E)} j = \gamma^* \circ j^* \circ j + g \circ g^* \circ \beta = \gamma^* + \beta,$$

thus $\gamma = -\beta^*$ and $D''_{\mathrm{Hom}(Q,E)} g^* = -j \circ \beta^*$. Combining all this, we get

$$\begin{aligned} D_E u &= g^* \beta \wedge u_S + j \cdot D_S u_S - j\beta^* \wedge u_Q + g^* \cdot D_Q u_Q \\ &= j(D_S u_S - \beta^* \wedge u_Q) + g^*(\beta \wedge u_S + D_Q u_Q), \end{aligned}$$

and the asserted matrix decomposition formula follows. By squaring the matrix, we get

$$D_E^2 = \begin{pmatrix} D_S^2 - \beta^* \wedge \beta & -D_S \circ \beta^* - \beta^* \circ D_Q \\ D_Q \circ \beta + \beta \circ D_S & D_Q^2 - \beta \wedge \beta^* \end{pmatrix}.$$

As $D_Q \circ \beta + \beta \circ D_S = D_{\mathrm{Hom}(S,Q)} \beta$ and $D_S \circ \beta^* + \beta^* \circ D_Q = D_{\mathrm{Hom}(Q,S)} \beta^*$ by the Leibniz rule, the curvature formulas follow (observe, since the Chern curvature form is of type $(1,1)$, that we must have $D'_{\mathrm{Hom}(S,Q)} \beta = 0$, $D''_{\mathrm{Hom}(Q,S)} \beta^* = 0$). □

(10.8) Corollary. *Let $0 \to S \to E \to Q \to 0$ be an exact sequence of Hermitian vector bundles. Then*

(a) $E \geqslant_{\mathrm{Grif}} 0 \implies Q \geqslant_{\mathrm{Grif}} 0$,

(b) $E \leqslant_{\mathrm{Grif}} 0 \implies S \leqslant_{\mathrm{Grif}} 0$,

(c) $E \leqslant_{\mathrm{Nak}} 0 \implies S \leqslant_{\mathrm{Nak}} 0$,

and analogous implications hold true for strict positivity.

Proof. If β is written $\sum dz_j \otimes \beta_j$, $\beta_j \in \mathrm{Hom}(S,Q)$, then Formulas (10.7) yield

$$\begin{aligned} \mathrm{i}\Theta_S &= \mathrm{i}\Theta_{E \restriction S} - \sum dz_j \wedge d\overline{z}_k \otimes \beta_k^* \beta_j, \\ \mathrm{i}\Theta_Q &= \mathrm{i}\Theta_{E \restriction Q} + \sum dz_j \wedge d\overline{z}_k \otimes \beta_j \beta_k^*. \end{aligned}$$

Since $\beta \cdot (\xi \otimes s) = \sum \xi_j \beta_j \cdot s$ and $\beta^* \cdot (\xi \otimes s) = \sum \overline{\xi}_k \beta_k^* \cdot s$ we get

$$\widetilde{\Theta}_S(\xi \otimes s, \xi' \otimes s') = \widetilde{\Theta}_E(\xi \otimes s, \xi' \otimes s') - \sum_{j,k} \xi_j \overline{\xi}'_k \langle \beta_j \cdot s, \beta_k \cdot s' \rangle,$$

$$\widetilde{\Theta}_S(u,u) = \widetilde{\Theta}_E(u,u) - |\beta \cdot u|^2,$$

$$\widetilde{\Theta}_Q(\xi \otimes s, \xi' \otimes s') = \widetilde{\Theta}_E(\xi \otimes s, \xi' \otimes s') + \sum_{j,k} \xi_j \overline{\xi}'_k \langle \beta_k^* \cdot s, \beta_j^* \cdot s' \rangle,$$

$$\widetilde{\Theta}_Q(\xi \otimes s, \xi \otimes s) = \widetilde{\Theta}_E(\xi \otimes s, \xi \otimes s) = |\beta^* \cdot (\xi \otimes s)|^2. \qquad \square$$

Next, we need positivity properties which somehow interpolate between Griffiths and Nakano positivity. This leads to the concept of m-tensor positivity.

(10.9) Definition. *Let T and E be complex vector spaces of dimensions n, r respectively, and let Θ be a Hermitian form on $T \otimes E$.*

(a) *A tensor $u \in T \otimes E$ is said to be of rank m if m is the smallest $\geqslant 0$ integer such that u can be written*

$$u = \sum_{j=1}^{m} \xi_j \otimes s_j, \qquad \xi_j \in T,\ s_j \in E.$$

(b) *Θ is said to be m-tensor positive (resp. m-tensor semi-positive) if $\Theta(u,u) > 0$ (resp. $\Theta(u,u) \geqslant 0$) for every tensor $u \in T \otimes E$ of rank $\leqslant m$, $u \neq 0$. In this case, we write*

$$\Theta >_m 0 \qquad (\textit{resp. } \Theta \geqslant_m 0).$$

We say that a Hermitian vector bundle E is m-tensor positive if $\widetilde{\Theta}_E >_m 0$. Griffiths positivity corresponds to $m = 1$ and Nakano positivity to $m \geqslant \min(n,r)$. Recall from Theorem (5.2) that we have

$$\langle [\mathrm{i}\Theta_E, \Lambda]u, u \rangle = \sum_{|S|=q-1} \sum_{j,k,\lambda,\mu} c_{jk\lambda\mu}\, u_{jS,\lambda} \overline{u}_{kS,\mu}$$

for every (n,q)-form $u = \sum u_{K,\lambda}\, dz_1 \wedge \cdots \wedge dz_n \wedge d\overline{z}_K \otimes e_\lambda$ with values in E. Since $u_{jS,\lambda} = 0$ for $j \in S$, the rank of the tensor $(u_{jS,\lambda})_{j,\lambda} \in \mathbb{C}^n \otimes \mathbb{C}^r$ is in fact $\leqslant \min\{n-q+1, r\}$. We obtain therefore:

(10.10) Lemma. *Assume that $E \geqslant_m 0$ (resp. $E >_m 0$). Then the Hermitian operator $[\mathrm{i}\Theta_E, \Lambda]$ is semipositive (resp. positive definite) on $\Lambda^{n,q}T^*X \otimes E$ for $q \geqslant 1$ and $m \geqslant \min\{n-q+1, r\}$.*

The Nakano vanishing theorem can then be improved as follows.

(10.11) Theorem. *Let X be a weakly pseudoconvex Kähler manifold of dimension n and let E a Hermitian vector bundle of rank r such that $\widetilde{\Theta}_E >_m 0$ over X. Then*

$$H^{n,q}(X,E) = 0 \qquad \textit{for } q \geqslant 1 \textit{ and } m \geqslant \min\{n-q+1, r\}.$$

We next study some important relations which exist between the various positivity concepts. Our starting point is the following result of [DSk79].

(10.12) Theorem. *For any Hermitian vector bundle E,*

$$E >_{\mathrm{Grif}} 0 \quad \Longrightarrow \quad E \otimes \det E >_{\mathrm{Nak}} 0.$$

To prove this result, we use the fact that

$$\Theta_{\det E} = \mathrm{Tr}_E\, \Theta_E \tag{10.13}$$

where $\mathrm{Tr}_E : \mathrm{Hom}(E, E) \to \mathbb{C}$ is the trace map, together with the identity

$$\Theta_{E \otimes \det E} = \Theta_E + \mathrm{Tr}_E(\Theta_E) \otimes \mathrm{Id}_E,$$

which is itself a consequence of (10.13) and of the standard formula

$$\Theta_{E\otimes F} = \Theta_E \otimes \mathrm{Id}_F + \mathrm{Id}_E \otimes \Theta_F.$$

In order to prove (10.13), for instance, we differentiate twice a wedge product, according to the formula

$$D_{\Lambda^p E}(s_1 \wedge \cdots \wedge s_p) = \sum_{j=1}^{p} (-1)^{\deg s_1 + \cdots + \deg s_{j-1}} s_1 \wedge \cdots \wedge s_{j-1} \wedge D_E s_j \wedge \cdots \wedge s_p.$$

The corresponding Hermitian forms on $T_X \otimes E$ are thus related by

$$\widetilde{\Theta}_{E \otimes \det E} = \widetilde{\Theta}_E + \mathrm{Tr}_E\, \widetilde{\Theta}_E \otimes h,$$

where h denotes the Hermitian metric on E and $\mathrm{Tr}_E\, \widetilde{\Theta}_E$ is the Hermitian form on T_X defined by

$$\mathrm{Tr}_E\, \widetilde{\Theta}_E(\xi, \xi) = \sum_{1 \leqslant \lambda \leqslant r} \widetilde{\Theta}_E(\xi \otimes e_\lambda, \xi \otimes e_\lambda), \qquad \xi \in T_X,$$

for any orthonormal frame $(e_1, \ldots, e_r)$ of E. Theorem 10.12 is now a consequence of the following simple property of Hermitian forms on a tensor product of complex vector spaces.

(10.14) Proposition. *Let T, E be complex vector spaces of respective dimensions n, r, and h a Hermitian metric on E. Then for every Hermitian form Θ on $T \otimes E$*

$$\Theta >_{\mathrm{Grif}} 0 \quad \Longrightarrow \quad \Theta + \mathrm{Tr}_E\, \Theta \otimes h >_{\mathrm{Nak}} 0.$$

We first need a lemma analogous to Fourier inversion formula for discrete Fourier transforms.

(10.15) Lemma. *Let q be an integer $\geqslant 3$, and x_λ, y_μ, $1 \leqslant \lambda, \mu \leqslant r$, be complex numbers. Let σ describe the set U_q^r of r-tuples of q-th roots of unity and put*

$$x'_\sigma = \sum_{1\leqslant\lambda\leqslant r} x_\lambda \overline{\sigma}_\lambda, \qquad y'_\sigma = \sum_{1\leqslant\mu\leqslant r} y_\mu \overline{\sigma}_\mu, \qquad \sigma \in U_q^r.$$

Then for every pair (α, β), $1 \leqslant \alpha, \beta \leqslant r$, the following identity holds:

$$q^{-r} \sum_{\sigma\in U_q^r} x'_\sigma \overline{y}'_\sigma \sigma_\alpha \overline{\sigma}_\beta = \begin{cases} x_\alpha \overline{y}_\beta & \text{if } \alpha \neq \beta, \\ \displaystyle\sum_{1\leqslant\mu\leqslant r} x_\mu \overline{y}_\mu & \text{if } \alpha = \beta. \end{cases}$$

Proof. The coefficient of $x_\lambda \overline{y}_\mu$ in the summation $q^{-r}\sum_{\sigma\in U_q^r} x'_\sigma \overline{y}'_\sigma \sigma_\alpha \overline{\sigma}_\beta$ is given by

$$q^{-r} \sum_{\sigma\in U_q^r} \sigma_\alpha \overline{\sigma}_\beta \overline{\sigma}_\lambda \sigma_\mu.$$

This coefficient equals 1 when the pairs $\{\alpha, \mu\}$ and $\{\beta, \lambda\}$ are equal (in which case $\sigma_\alpha \overline{\sigma}_\beta \overline{\sigma}_\lambda \sigma_\mu = 1$ for any one of the q^r elements of U_q^r). Hence, it is sufficient to prove that

$$\sum_{\sigma\in U_q^r} \sigma_\alpha \overline{\sigma}_\beta \overline{\sigma}_\lambda \sigma_\mu = 0$$

when the pairs $\{\alpha, \mu\}$ and $\{\beta, \lambda\}$ are distinct.

If $\{\alpha, \mu\} \neq \{\beta, \lambda\}$, then one of the elements of one of the pairs does not belong to the other pair. As the four indices $\alpha, \beta, \lambda, \mu$ play the same role, we may suppose for example that $\alpha \notin \{\beta, \lambda\}$. Let us apply to σ the substitution $\sigma \mapsto \tau$, where τ is defined by

$$\tau_\alpha = e^{2\pi \mathrm{i}/q} \sigma_\alpha, \ \tau_\nu = \sigma_\nu \quad \text{for} \quad \nu \neq \alpha.$$

We get

$$\sum_\sigma \sigma_\alpha \overline{\sigma}_\beta \overline{\sigma}_\lambda \sigma_\mu = \sum_\tau = \begin{cases} e^{2\pi \mathrm{i}/q} \displaystyle\sum_\sigma & \text{if } \alpha \neq \mu, \\ e^{4\pi \mathrm{i}/q} \displaystyle\sum_\sigma & \text{if } \alpha = \mu. \end{cases}$$

Since $q \geqslant 3$ by hypothesis, it follows that

$$\sum_\sigma \sigma_\alpha \overline{\sigma}_\beta \overline{\sigma}_\lambda \sigma_\mu = 0.$$

Proof of Proposition. 10.14 Let $(t_j)_{1\leqslant j\leqslant n}$ be a basis of T, $(e_\lambda)_{1\leqslant\lambda\leqslant r}$ an orthonormal basis of E and $\xi = \sum_j \xi_j t_j \in T$, $u = \sum_{j,\lambda} u_{j\lambda} t_j \otimes e_\lambda \in T \otimes E$. The coefficients $c_{jk\lambda\mu}$ of Θ with respect to the basis $t_j \otimes e_\lambda$ satisfy the symmetry relation

$\overline{c}_{jk\lambda\mu} = c_{kj\mu\lambda}$, and we have the formulas

$$\Theta(u,u) = \sum_{j,k,\lambda,\mu} c_{jk\lambda\mu} u_{j\lambda} \overline{u}_{k\mu},$$

$$\mathrm{Tr}_E\,\Theta(\xi,\xi) = \sum_{j,k,\lambda} c_{jk\lambda\lambda} \xi_j \overline{\xi}_k,$$

$$(\Theta + \mathrm{Tr}_E\,\Theta \otimes h)(u,u) = \sum_{j,k,\lambda,\mu} c_{jk\lambda\mu} u_{j\lambda} \overline{u}_{k\mu} + c_{jk\lambda\lambda} u_{j\mu} \overline{u}_{k\mu}.$$

For every $\sigma \in U_q^r$ (cf. Lemma 10.15), put

$$u'_{j\sigma} = \sum_{1 \leqslant \lambda \leqslant r} u_{j\lambda} \overline{\sigma}_\lambda \in \mathbb{C},$$

$$\widehat{u}_\sigma = \sum_j u'_{j\sigma} t_j \in T \quad , \quad \widehat{e}_\sigma = \sum_\lambda \sigma_\lambda e_\lambda \in E.$$

Lemma 10.15 implies

$$q^{-r} \sum_{\sigma \in U_q^r} \Theta(\widehat{u}_\sigma \otimes \widehat{e}_\sigma, \widehat{u}_\sigma \otimes \widehat{e}_\sigma) = q^{-r} \sum_{\sigma \in U_q^r} c_{jk\lambda\mu} u'_{j\sigma} \overline{u}'_{k\sigma} \sigma_\lambda \overline{\sigma}_\mu$$

$$= \sum_{j,k,\lambda \neq \mu} c_{jk\lambda\mu} u_{j\lambda} \overline{u}_{k\mu} + \sum_{j,k,\lambda,\mu} c_{jk\lambda\lambda} u_{j\mu} \overline{u}_{k\mu}.$$

The Griffiths positivity assumption shows that the left hand side is $\geqslant 0$, hence

$$(\Theta + \mathrm{Tr}_E\,\Theta \otimes h)(u,u) \geqslant \sum_{j,k,\lambda} c_{jk\lambda\lambda} u_{j\lambda} \overline{u}_{k\lambda} \geqslant 0$$

with strict positivity if $\Theta >_{\mathrm{Grif}} 0$ and $u \neq 0$. □

We now relate Griffiths positivity to m-tensor positivity. The most useful result is the following

(10.16) Proposition. *Let T be a complex vector space and (E,h) a Hermitian vector space of respective dimensions n, r with $r \geqslant 2$. Then for any Hermitian form Θ on $T \otimes E$ and any integer $m \geqslant 1$*

$$\Theta >_{\mathrm{Grif}} 0 \quad \Longrightarrow \quad m\,\mathrm{Tr}_E\,\Theta \otimes h - \Theta >_m 0.$$

Proof. Let us distinguish two cases.

(a) $m = 1$. Let $u \in T \otimes E$ be a tensor of rank 1. Then u can be written $u = \xi_1 \otimes e_1$ with $\xi_1 \in T$, $\xi_1 \neq 0$, and $e_1 \in E$, $|e_1| = 1$. Complete e_1 into an orthonormal basis $(e_1, \ldots, e_r)$ of E. One gets immediately

$$(\mathrm{Tr}_E\,\Theta \otimes h)(u,u) = \mathrm{Tr}_E\,\Theta(\xi_1,\xi_1) = \sum_{1 \leqslant \lambda \leqslant r} \Theta(\xi_1 \otimes e_\lambda, \xi_1 \otimes e_\lambda)$$

$$> \Theta(\xi_1 \otimes e_1, \xi_1 \otimes e_1) = \Theta(u,u).$$

(b) $m \geqslant 2$. Every tensor $u \in T \otimes E$ of rank $\leqslant m$ can be written

$$u = \sum_{1 \leqslant \lambda \leqslant q} \xi_\lambda \otimes e_\lambda, \quad \xi_\lambda \in T,$$

with $q = \min(m, r)$ and $(e_\lambda)_{1 \leqslant \lambda \leqslant r}$ an orthonormal basis of E. Let F be the vector subspace of E generated by $(e_1, \ldots, e_q)$ and Θ_F the restriction of Θ to $T \otimes F$. The first part shows that

$$\Theta' := \mathrm{Tr}_F \Theta_F \otimes h - \Theta_F >_{\mathrm{Grif}} 0.$$

Proposition 10.14 applied to Θ' on $T \otimes F$ yields

$$\Theta' + \mathrm{Tr}_F \Theta' \otimes h = q \,\mathrm{Tr}_F \Theta_F \otimes h - \Theta_F >_q 0.$$

Since $u \in T \otimes F$ is of rank $\leqslant q \leqslant m$, we get (for $u \neq 0$)

$$\begin{aligned} \Theta(u, u) = \Theta_F(u, u) &< q(\mathrm{Tr}_F \Theta_F \otimes h)(u, u) \\ &= q \sum_{1 \leqslant j, \lambda \leqslant q} \Theta(\xi_j \otimes e_\lambda, \xi_j \otimes e_\lambda) \leqslant m \,\mathrm{Tr}_E \Theta \otimes h(u, u). \end{aligned}$$ □

Proposition 10.16 is of course also true in the semi-positive case. From these facts, we deduce

(10.17) Theorem. *Let E be a Griffiths (semi-)positive bundle of rank $r \geqslant 2$. Then for any integer $m \geqslant 1$*

$$E^* \otimes (\det E)^m >_m 0 \quad (\textit{resp.} \ \geqslant_m 0).$$

Proof. We apply Proposition 10.16 to $\Theta = -\Theta(E^*) = {}^t\Theta_E \geqslant_{\mathrm{Grif}} 0$ on $T_X \otimes E^*$ and observe that

$$\Theta_{\det E} = \mathrm{Tr}_E \Theta_E = \mathrm{Tr}_{E^*} \Theta.$$

(10.18) Theorem. *Let $0 \to S \to E \to Q \to 0$ be an exact sequence of Hermitian vector bundles. Then for any $m \geqslant 1$*

$$E >_m 0 \quad \Longrightarrow \quad S \otimes (\det Q)^m >_m 0.$$

Proof. Formulas 10.7 imply

$$\mathrm{i}\Theta_S >_m \mathrm{i}\beta^* \wedge \beta \quad , \quad \mathrm{i}\Theta_Q >_m \mathrm{i}\beta \wedge \beta^*,$$

$$\mathrm{i}\Theta_{\det Q} = \mathrm{Tr}_Q(\mathrm{i}\Theta_Q) > \mathrm{Tr}_Q(\mathrm{i}\beta \wedge \beta^*).$$

If we write $\beta = \sum dz_j \otimes \beta_j$ as in the proof of Corollary 10.8, then

$$\begin{aligned}\mathrm{Tr}_Q(\mathrm{i}\beta \wedge \beta^*) &= \sum idz_j \wedge d\overline{z}_k \,\mathrm{Tr}_Q(\beta_j \beta_k^*) \\ &= \sum idz_j \wedge d\overline{z}_k \,\mathrm{Tr}_S(\beta_k^* \beta_j) = \mathrm{Tr}_S(-\mathrm{i}\beta^* \wedge \beta).\end{aligned}$$

Furthermore, it has been already proved that $-\mathrm{i}\beta^* \wedge \beta \geqslant_{\mathrm{Nak}} 0$. By Proposition 10.16 applied to the corresponding Hermitian form Θ on $T_X \otimes S$, we get

$$m\,\mathrm{Tr}_S(-\mathrm{i}\beta^* \wedge \beta) \otimes \mathrm{Id}_S + \mathrm{i}\beta^* \wedge \beta \geqslant_m 0,$$

and Theorem 10.18 follows. □

(10.19) Corollary. *Let X be a weakly pseudoconvex Kähler n-dimensional manifold, E a holomorphic vector bundle of rank $r \geqslant 2$ and $m \geqslant 1$ an integer. Then*

(a) *$E >_{\mathrm{Grif}} 0 \Rightarrow H^{n,q}(X, E \otimes \det E) = 0$ for $q \geqslant 1$;*

(b) *$E >_{\mathrm{Grif}} 0 \Rightarrow H^{n,q}\big(X, E^* \otimes (\det E)^m\big) = 0$ for $q \geqslant 1$ and $m \geqslant \min\{n-q+1, r\}$;*

(c) *Let $0 \to S \to E \to Q \to 0$ be an exact sequence of vector bundles and $m = \min\{n-q+1, \mathrm{rk}\, S\}$, $q \geqslant 1$. If $E >_m 0$ and if L is a line bundle such that $L \otimes (\det Q)^{-m} \geqslant 0$, then*

$$H^{n,q}(X, S \otimes L) = 0.$$

Proof. Immediate consequence of Theorem 10.11, in combination with Theorem 10.12 for (a), Theorem 10.17 for (b) and Theorem 10.18 for (c). □

Chapter 11

Skoda's L^2 Estimates for Surjective Bundle Morphisms

11.A. Surjectivity and Division Theorems

Let (X, ω) be a Kähler manifold, $\dim X = n$, and let $g : E \to Q$ a holomorphic morphism of Hermitian vector bundles over X. Assume in the first instance that g is *surjective*. We are interested in conditions insuring that the induced morphisms $g : H^{n,k}(X, E) \longrightarrow H^{n,k}(X, Q)$ are also surjective (dealing with $(n, \bullet)$ bidegrees is always easier, since we have to understand positivity conditions for the curvature term). For that purpose, it is natural to consider the subbundle $S = \operatorname{Ker} g \subset E$ and the exact sequence

$$0 \longrightarrow S \xrightarrow{\ j\ } E \xrightarrow{\ g\ } Q \longrightarrow 0 \tag{11.1}$$

where $j : S \to E$ is the inclusion. In fact, we need a little more flexibility to handle the curvature terms, so we take the tensor product of the exact sequence by a holomorphic line bundle L (whose properties will be specified later):

$$0 \longrightarrow S \otimes L \longrightarrow E \otimes L \xrightarrow{\ g\ } Q \otimes L \longrightarrow 0. \tag{11.2}$$

(11.3) Theorem. *Let k be an integer such that $0 \leqslant k \leqslant n$. Set $r = \operatorname{rk} E$, $q = \mathrm{rkQ}$, $s = \operatorname{rk} S = r - q$ and*

$$m = \min\{n - k, s\} = \min\{n - k, r - q\}.$$

Assume that (X, ω) possesses also a complete Kähler metric $\widehat{\omega}$, that $E \geqslant_m 0$, and that $L \longrightarrow X$ is a Hermitian holomorphic line bundle such that

$$\mathrm{i}\Theta_L - (m + \varepsilon)\mathrm{i}\Theta_{\det Q} \geqslant 0$$

for some $\varepsilon > 0$. Then for every D''-closed form f of type (n, k) with values in $Q \otimes L$ such that $\|f\| < +\infty$, there exists a D''-closed form h of type (n, k) with values in $E \otimes L$ such that $f = g \cdot h$ and

$$\|h\|^2 \leqslant (1 + m/\varepsilon)\, \|f\|^2.$$

The idea of the proof is essentially due to [Sko78], who actually proved the special case $k = 0$. The general case appeared in [Dem82b].

Proof. Let $j : S \to E$ be the inclusion morphism, $g^* : Q \to E$ and $j^* : E \to S$ the adjoints of g, j, and the matrix of D_E with respect to the orthogonal splitting $E \simeq S \oplus Q$ (cf. Lemma 10.2). Then $g^* f$ is a lifting of f in $E \otimes L$. We will try to find h under the form

$$h = g^* f + ju, \qquad u \in L^2(X, \Lambda^{n,k} T_X^* \otimes S \otimes L).$$

As the images of S and Q in E are orthogonal, we have $|h|^2 = |f|^2 + |u|^2$ at every point of X. On the other hand $D''_{Q\otimes L} f = 0$ by hypothesis and $D'' g^* = -j \circ \beta^*$ by (10.5), hence

$$D''_{E\otimes L} h = -j(\beta^* \wedge f) + j\, D''_{S\otimes L} = j(D''_{S\otimes L} - \beta^* \wedge f).$$

We are thus led to solve the equation

$$D''_{S\otimes L} u = \beta^* \wedge f, \tag{11.4}$$

and for that, we apply Theorem 6.1 to the $(n, k+1)$-form $\beta^* \wedge f$. One now observes that the curvature of $S \otimes L$ can be expressed in terms of β. This remark will be used to prove:

(11.5) Lemma. *Let $A_k = [\mathrm{i}\Theta_{S\otimes L}, \Lambda]$ be the curvature operator acting as an Hermitian operator on the bundle of $(n, k+1)$-forms. Then*

$$\langle A_k^{-1}(\beta^* \wedge f), (\beta^* \wedge f)\rangle \leqslant (m/\varepsilon)\, |f|^2.$$

If the lemma is taken for granted, Theorem 9.4 yields a solution u of (11.4) in $L^2(X, \Lambda^{n,q} T_X^* \otimes S \otimes L)$ such that $\|u\|^2 \leqslant (m/\varepsilon)\, \|f\|^2$. As $\|h\|^2 = \|f\|^2 + \|u\|^2$, the proof of Theorem 11.3 is complete.

Proof of Lemma 11.5. Exactly as in the proof of Theorem 10.18, the formulas (10.7) yield

$$\mathrm{i}\Theta_S \geqslant_m \mathrm{i}\beta^* \wedge \beta, \qquad \mathrm{i}\Theta_{\det Q} \geqslant \mathrm{Tr}_Q(\mathrm{i}\beta \wedge \beta^*) = \mathrm{Tr}_S(-\mathrm{i}\beta^* \wedge \beta).$$

Since $\mathscr{C}^\infty(X, \Lambda^{1,1} T_X^* \otimes \mathrm{Herm}\, S) \ni \Theta := -\mathrm{i}\beta^* \wedge \beta \geqslant_{\mathrm{Grif}} 0$, Proposition 10.16 implies

$$m\ \mathrm{Tr}_S(-\mathrm{i}\beta^* \wedge \beta) \otimes \mathrm{Id}_S + \mathrm{i}\beta^* \wedge \beta \geqslant_m 0.$$

From the hypothesis on the curvature of L we get

$$\begin{aligned}
\mathrm{i}\Theta_{S\otimes L} &\geqslant_m \mathrm{i}\Theta_S \otimes \mathrm{Id}_L + (m+\varepsilon)\, \mathrm{i}\Theta_{\det Q} \otimes \mathrm{Id}_{S\otimes L}\\
&\geqslant_m \big(\mathrm{i}\beta^* \wedge \beta + (m+\varepsilon)\ \mathrm{Tr}_S(-\mathrm{i}\beta^* \wedge \beta) \otimes \mathrm{Id}_S\big) \otimes \mathrm{Id}_L\\
&\geqslant_m (\varepsilon/m)\, (-\mathrm{i}\beta^* \wedge \beta) \otimes \mathrm{Id}_S \otimes \mathrm{Id}_L\,.
\end{aligned}$$

For any $v \in \Lambda^{n,k+1}T_X^* \otimes S \otimes L$, Lemma 10.10 implies

$$\langle A_k v, v\rangle \geqslant (\varepsilon/m)\,\langle -\mathrm{i}\beta^* \wedge \beta \wedge \Lambda v, v\rangle, \tag{11.6}$$

because $\mathrm{rk}(S \otimes L) = s$ and $m = \min\{n-k, s\}$. Let $(dz_1, \ldots, dz_n)$ be an orthonormal basis of T_X^* at a given point $x_0 \in X$ and set

$$\beta = \sum_{1\leqslant j\leqslant n} dz_j \otimes \beta_j, \qquad \beta_j \in \mathrm{Hom}(S, Q).$$

The adjoint of the operator $\beta^* \wedge \bullet = \sum d\overline{z}_j \wedge \beta_j^* \bullet$ is the contraction operator $\beta \lrcorner \bullet$ defined by

$$\beta \lrcorner v = \sum \frac{\partial}{\partial \overline{z}_j} \lrcorner (\beta_j v) = \sum -\mathrm{i}dz_j \wedge \Lambda(\beta_j v) = -\mathrm{i}\beta \wedge \Lambda v.$$

Consequently, we get $\langle -\mathrm{i}\beta^* \wedge \beta \wedge \Lambda v, v\rangle = |\beta \lrcorner v|^2$ and (11.6) implies

$$|\langle \beta^* \wedge f, v\rangle|^2 = |\langle f, \beta \lrcorner v\rangle|^2 \leqslant |f|^2\, |\beta \lrcorner v|^2 \leqslant (m/\varepsilon)\langle A_k v, v\rangle\, |f|^2.$$

This is equivalent to the estimate asserted in the lemma. □

If X has a plurisubharmonic exhaustion function ψ, we can select a convex increasing function $\chi \in \mathscr{C}^\infty(\mathbb{R}, \mathbb{R})$ and multiply the metric of L by the weight $\exp(-\chi \circ \psi)$ in order to make the L^2 norm of f converge. Theorem 11.3 implies therefore:

(11.7) Corollary. *Let (X, ω) be a weakly pseudoconvex Kähler manifold, $g : E \to Q$ a surjective bundle morphism with $r = \mathrm{rk}\, E$, $q = \mathrm{rk}\, Q$, and $L \to X$ a Hermitian holomorphic line bundle. We set $m = \min\{n-k, r-q\}$ and assume that $E \geqslant_m 0$ and*

$$\mathrm{i}\Theta_L - (m+\varepsilon)\,\mathrm{i}\Theta_{\det Q} \geqslant 0$$

for some $\varepsilon > 0$. Then g induces a surjective map

$$H^{n,k}(X, E \otimes L) \longrightarrow H^{n,k}(X, Q \otimes L).$$

The most remarkable feature of this result is that it does not require any strict positivity assumption on the curvature (for instance E can be a flat bundle). A careful examination of the proof shows that it amounts to verify that the image of the coboundary morphism

$$-\beta^* \wedge \bullet \ :\ H^{n,k}(X, Q \otimes L) \longrightarrow H^{n,k+1}(X, S \otimes L)$$

vanishes; however the cohomology group $H^{n,k+1}(X, S \otimes L)$ itself does not necessarily vanish, as it would do under a strict positivity assumption.

We want now to get estimates also when Q is endowed with a metric given a priori, that can be distinct from the quotient metric of E by g. Then the map $g^*(gg^*)^{-1}: Q \to E$ is the lifting of Q orthogonal to $S = \operatorname{Ker} g$. The quotient metric $|\bullet|'$ on Q is therefore defined in terms of the original metric $|\bullet|$ by

$$|v|'^2 = |g^*(gg^*)^{-1}v|^2 = \langle (gg^*)^{-1}v, v\rangle = \det(gg^*)^{-1}\,\langle \widetilde{gg^*}v, v\rangle$$

where $\widetilde{gg^*} \in \operatorname{End}(Q)$ denotes the endomorphism of Q whose matrix is the transposed comatrix of gg^*. For every $w \in \det Q$, we find

$$|w|'^2 = \det(gg^*)^{-1}\,|w|^2.$$

If Q' denotes the bundle Q with the quotient metric, we get

$$\mathrm{i}\Theta_{\det Q'} = \mathrm{i}\Theta_{\det Q} + \mathrm{i}d'd'' \log\det(gg^*).$$

In order that the hypotheses of Theorem 11.3 be satisfied, we are led to define a new metric $|\bullet|'$ on L by $|u|'^2 = |u|^2\left(\det(gg^*)\right)^{-m-\varepsilon}$. Then

$$\mathrm{i}\Theta_{L'} = \mathrm{i}\Theta(L) + (m+\varepsilon)\,\mathrm{i}d'd''\log\det(gg^*) \geqslant (m+\varepsilon)\,\mathrm{i}\Theta_{\det Q'}.$$

Theorem 11.3 applied to (E, Q', L') can now be reformulated:

(11.8) Theorem. *Let X be a complete Kähler manifold equipped with a Kähler metric ω on X, let $E \to Q$ be a surjective morphism of Hermitian vector bundles and let $L \to X$ be a Hermitian holomorphic line bundle. Set $r = \operatorname{rk} E$, $q = \operatorname{rk} Q$ and $m = \min\{n-k, r-q\}$, and assume that $E \geqslant_m 0$ and*

$$\mathrm{i}\Theta_L - (m+\varepsilon)\mathrm{i}\Theta_{\det Q} \geqslant 0$$

for some $\varepsilon > 0$. Then for every D''-closed form f of type (n,k) with values in $Q \otimes L$ such that

$$I = \int_X \langle \widetilde{gg^*}f, f\rangle\,(\det gg^*)^{-m-1-\varepsilon}\,dV < +\infty,$$

there exists a D''-closed form h of type (n,k) with values in $E \otimes L$ such that $f = g\cdot h$ and

$$\int_X |h|^2\,(\det gg^*)^{-m-\varepsilon}\,dV \leqslant (1 + m/\varepsilon)\,I. \qquad \square$$

Our next goal is to extend Theorem 11.8 in the case when $g : E \longrightarrow Q$ is only *generically* surjective; this means that the analytic set

$$Y = \{x \in X \;;\; g_x : E_x \longrightarrow Q_x \text{ is not surjective}\}$$

defined by the equation $\Lambda^q g = 0$ is nowhere dense in X. Here $\Lambda^q g$ is a section of the bundle $\operatorname{Hom}(\Lambda^q E, \det Q)$. The idea is to apply the above Theorem 11.8 to $X \smallsetminus Y$. For this, we have to know whether $X \smallsetminus Y$ has a complete Kähler metric.

(11.9) Lemma. *Let (X,ω) be a Kähler manifold, and $Y=\sigma^{-1}(0)$ an analytic subset defined by a section of a Hermitian vector bundle $E\to X$. If X is weakly pseudoconvex and exhausted by $X_c=\{x\in X\ ;\ \psi(x)<c\}$, then $X_c\smallsetminus Y$ has a complete Kähler metric for all $c\in\mathbb{R}$. The same conclusion holds for $X\smallsetminus Y$ if (X,ω) is complete and if for some constant $C\geqslant 0$ we have $\Theta_E\leqslant_{\rm Grif} C\,\omega\otimes\langle\ ,\ \rangle_E$ on X.*

Proof. Set $\tau=\log|\sigma|^2$. Then $d'\tau=\{D'\sigma,\sigma\}/|\sigma|^2$ and $D''D'\sigma=D^2\sigma=\Theta_E\sigma$, thus

$$\mathrm{i}d'd''\tau=\mathrm{i}\frac{\{D'\sigma,D'\sigma\}}{|\sigma|^2}-\mathrm{i}\frac{\{D'\sigma,\sigma\}\wedge\{\sigma,D'\sigma\}}{|\sigma|^4}-\frac{\{\mathrm{i}\Theta_E\sigma,\sigma\}}{|\sigma|^2}.$$

For every $\xi\in T_X$, we find therefore

$$\begin{aligned}H\tau(\xi)&=\frac{|\sigma|^2\,|D'\sigma\cdot\xi|^2-|\langle D'\sigma\cdot\xi,\sigma\rangle|^2}{|\sigma|^4}-\frac{\widetilde\Theta_E(\xi\otimes\sigma,\xi\otimes\sigma)}{|\sigma|^2}\\&\geqslant-\frac{\widetilde\Theta_E(\xi\otimes\sigma,\xi\otimes\sigma)}{|\sigma|^2}\end{aligned}$$

by the Cauchy-Schwarz inequality. If C is a bound for the coefficients of $\widetilde\Theta_E$ on the compact subset $\overline{X}_c$, we get $\mathrm{i}d'd''\tau\geqslant -C\omega$ on X_c. Let $\chi\in\mathscr{C}^\infty(\mathbb{R},\mathbb{R})$ be a convex increasing function. We set

$$\widehat\omega=\omega+\mathrm{i}d'd''(\chi\circ\tau)=\omega+\mathrm{i}\big(\chi'\circ\tau\ d'd''\tau+\chi''\circ\tau\ d'\tau\wedge d''\tau\big).$$

We thus see that $\widehat\omega$ is positive definite if $\chi'\leqslant 1/2C$, and by a computation similar to the one preceding Theorem 6.2, we check that $\widehat\omega$ is complete near $Y=\tau^{-1}(-\infty)$ as soon as

$$\int_{-\infty}^0\sqrt{\chi''(t)}\,dt=+\infty.$$

One can choose for example χ such that $\chi(t)=\frac{1}{5C}(t-\log|t|)$ for $t\leqslant -1$. In order to obtain a complete Kähler metric on $X_c\smallsetminus Y$, we also need the metric to be complete near ∂X_c. If $\widehat\omega$ is not, such a metric can be defined by

$$\begin{aligned}\widetilde\omega&=\widehat\omega+\mathrm{i}d'd''\log(c-\psi)^{-1}=\widehat\omega+\frac{\mathrm{i}d'd''\psi}{c-\psi}+\frac{\mathrm{i}d'\psi\wedge d''\psi}{(c-\psi)^2}\\&\geqslant \mathrm{i}d'\log(c-\psi)^{-1}\wedge d''\log(c-\psi)^{-1}\ ;\end{aligned}$$

$\widetilde\omega$ is complete on $X_c\smallsetminus\Omega$ because $\log(c-\psi)^{-1}$ tends to $+\infty$ on ∂X_c. □

We also need another elementary lemma dealing with the extension of partial differential equalities across analytic sets.

(11.10) Lemma. *Let Ω be an open subset of $\mathbb{C}^n$ and Y an analytic subset of Ω. Assume that v is a $(p,q-1)$-form with $L^2_{\rm loc}$ coefficients and w a (p,q)-form with $L^1_{\rm loc}$ coefficients such that $d''v=w$ on $\Omega\smallsetminus Y$ (in the sense of distribution theory). Then $d''v=w$ on Ω.*

Proof. An induction on the dimension of Y shows that it is sufficient to prove the result in a neighborhood of a regular point $a \in Y$. By using a local analytic isomorphism, the proof is reduced to the case where Y is contained in the hyperplane $z_1 = 0$, with $a = 0$. Let $\lambda \in \mathscr{C}^\infty(\mathbb{R}, \mathbb{R})$ be a function such that $\lambda(t) = 0$ for $t \leqslant \frac{1}{2}$ and $\lambda(t) = 1$ for $t \geqslant 1$. We must show that

$$\int_\Omega w \wedge \alpha = (-1)^{p+q} \int_\Omega v \wedge d''\alpha \tag{11.11}$$

for all $\alpha \in \mathscr{D}(\Omega, \Lambda^{n-p,n-q}T^*_\Omega)$. Set $\lambda_\varepsilon(z) = \lambda(|z_1|/\varepsilon)$ and replace α in the integral by $\lambda_\varepsilon\alpha$. Then $\lambda_\varepsilon\alpha \in \mathscr{D}(\Omega \smallsetminus Y, \Lambda^{n-p,n-q}T^*_\Omega)$ and the hypotheses imply

$$\int_\Omega w \wedge \lambda_\varepsilon\alpha = (-1)^{p+q} \int_\Omega v \wedge d''(\lambda_\varepsilon\alpha) = (-1)^{p+q} \int_\Omega v \wedge (d''\lambda_\varepsilon \wedge \alpha + \lambda_\varepsilon d''\alpha).$$

As w and v have L^1_{loc} coefficients on Ω, the integrals of $w \wedge \lambda_\varepsilon\alpha$ and $v \wedge \lambda_\varepsilon d''\alpha$ converge respectively to the integrals of $w \wedge \alpha$ and $v \wedge d''\alpha$ as ε tends to 0. The remaining term can be estimated by means of the Cauchy-Schwarz inequality:

$$\Big| \int_\Omega v \wedge d''\lambda_\varepsilon \wedge \alpha \Big|^2 \leqslant \int_{|z_1| \leqslant \varepsilon} |v \wedge \alpha|^2 \, dV \int_{\mathrm{Supp}\,\alpha} |d''\lambda_\varepsilon|^2 \, dV\,;$$

as $v \in L^2_{\mathrm{loc}}(\Omega)$, the integral $\int_{|z_1| \leqslant \varepsilon} |v \wedge \alpha|^2 \, dV$ converges to 0 with ε, whereas

$$\int_{\mathrm{Supp}\,\alpha} |d''\lambda_\varepsilon|^2 \, dV \leqslant \frac{C}{\varepsilon^2} \mathrm{Vol}\big(\mathrm{Supp}\,\alpha \cap \{|z_1| \leqslant \varepsilon\}\big) \leqslant C'.$$

Equality (11.11) follows when ε tends to 0. □

(11.12) Theorem. *The existence statement and the estimates of Theorem* 11.8 *remain true for a generically surjective morphism $g : E \to Q$, provided that X is weakly pseudoconvex.*

Proof. Apply Theorem 11.8 to each relatively compact domain $X_c \smallsetminus Y$ (these domains are complete Kähler by Lemma 11.9). From a sequence of solutions on $X_c \smallsetminus Y$ we can extract a subsequence converging weakly on $X \smallsetminus Y$ as c tends to $+\infty$. One gets a form h satisfying the estimates, such that $D''h = 0$ on $X \smallsetminus Y$ and $f = g \cdot h$. In order to see that $D''h = 0$ on X, it suffices to apply Lemma 11.10 and to observe that h has L^2_{loc} coefficients on X by our estimates. □

A very special but interesting case is obtained for the trivial bundles $E = \Omega \times \mathbb{C}^r$, $Q = \Omega \times \mathbb{C}$ over a pseudoconvex open set $\Omega \subset \mathbb{C}^n$. Then the morphism g is given by a r-tuple $(g_1, \ldots, g_r)$ of holomorphic functions on Ω. Let us take $k = 0$ and $L = \Omega \times \mathbb{C}$ with the metric given by a weight $e^{-\varphi}$. If we observe that $\widetilde{gg^*} = \mathrm{Id}$ when $\mathrm{rk}\, Q = 1$, Theorem 11.8 applied on $X = \Omega \smallsetminus g^{-1}(0)$ and Lemmas 11.9, 11.10 give:

(11.13) Theorem (Skoda [Sko72b]). *Let Ω be a complete Kähler open subset of $\mathbb{C}^n$, let φ be a plurisubharmonic function and $g = (g_1, \ldots, g_r)$ be a r-tuple of* *holomorphic functions on* Ω. *Set* $m = \min\{n, r-1\}$. *Then for every holomorphic function* f *on* Ω *such that*

$$I = \int_{\Omega \smallsetminus Z} |f|^2\, |g|^{-2(m+1+\varepsilon)} e^{-\varphi}\, dV < +\infty,$$

where $Z = g^{-1}(0)$, *there exist holomorphic functions* $(h_1, \ldots, h_r)$ *on* Ω *such that* $f = \sum g_j h_j$ *and*

$$\int_{\Omega \smallsetminus Y} |h|^2\, |g|^{-2(m+\varepsilon)} e^{-\varphi}\, dV \leqslant (1 + m/\varepsilon) I. \qquad \square$$

11.B. Applications to Local Algebra: the Briançon-Skoda Theorem

We now show that Theorem 11.13 can be applied to get deep results concerning ideals of the local ring $\mathcal{O}_n = \mathbb{C}\{z_1, \ldots, z_n\}$ of germs of holomorphic functions on $(\mathbb{C}^n, 0)$. Let $\mathcal{I} = (g_1, \ldots, g_r) \neq (0)$ be an ideal of $\mathcal{O}_n$.

(11.14) Definition. *Let* $k \in \mathbb{R}_+$. *We associate to* $\mathcal{I}$ *the following ideals:*

(a) *the ideal* $\overline{\mathcal{I}}^{(k)}$ *of germs* $u \in \mathcal{O}_n$ *such that* $|u| \leqslant C|g|^k$ *for some constant* $C \geqslant 0$, *where* $|g|^2 = |g_1|^2 + \cdots + |g_r|^2$;

(b) *the ideal* $\widehat{\mathcal{I}}^{(k)}$ *of germs* $u \in \mathcal{O}_n$ *such that*

$$\int_\Omega |u|^2\, |g|^{-2(k+\varepsilon)}\, dV < +\infty$$

on a small ball Ω *centered at* 0, *if* $\varepsilon > 0$ *is small enough.*

(11.15) Proposition. *For all* $k, l \in \mathbb{R}_+$ *we have*

(a) $\overline{\mathcal{I}}^{(k)} \subset \widehat{\mathcal{I}}^{(k)}$;

(b) $\mathcal{I}^k \subset \overline{\mathcal{I}}^{(k)}$ *if* $k \in \mathbb{N}$;

(c) $\overline{\mathcal{I}}^{(k)} . \overline{\mathcal{I}}^{(l)} \subset \overline{\mathcal{I}}^{(k+l)}$;

(d) $\overline{\mathcal{I}}^{(k)} . \widehat{\mathcal{I}}^{(l)} \subset \widehat{\mathcal{I}}^{(k+l)}$.

All properties are immediate from the definitions except (a) which is a consequence of the integrability of $|g|^{-\varepsilon}$ for $\varepsilon > 0$ small (exercise to the reader!). Before stating the main result, we need a simple lemma.

(11.16) Lemma. *If $\mathcal{I} = (g_1, \ldots, g_r)$ and $r > n$, we can find elements $\widetilde{g}_1, \ldots, \widetilde{g}_n$ in $\mathcal{I}$ such that $C^{-1}|g| \leqslant |\widetilde{g}| \leqslant C|g|$ on a neighborhood of 0. Each $\widetilde{g}_j$ can be taken to be a linear combination*

$$\widetilde{g}_j = a_j . g = \sum_{1 \leqslant k \leqslant r} a_{jk} g_k, \qquad a_j \in \mathbb{C}^r \smallsetminus \{0\}$$

where the coefficients $([a_1], \ldots, [a_n])$ are chosen in the complement of a proper analytic subset of $(\mathbb{P}^{r-1})^n$.

It follows from the lemma that the ideal $\mathcal{J} = (\widetilde{g}_1, \ldots, \widetilde{g}_n) \subset \mathcal{I}$ satisfies $\overline{\mathcal{J}}^{(k)} = \overline{\mathcal{I}}^{(k)}$ and $\widehat{\mathcal{J}}^{(k)} = \widehat{\mathcal{I}}^{(k)}$ for all k.

Proof. Assume that $g \in \mathcal{O}(\Omega)^r$. Consider the analytic subsets in $\Omega \times (\mathbb{P}^{r-1})^n$ defined by

$$A = \big\{(z, [w_1], \ldots, [w_n])\,;\, w_j \cdot g(z) = 0\big\},$$
$$A^* = \bigcup \text{irreducible components of } A \text{ not contained in } g^{-1}(0) \times (\mathbb{P}^{r-1})^n.$$

For $z \notin g^{-1}(0)$ the fiber $A_z = \{([w_1], \ldots, [w_n])\,;\, w_j . g(z) = 0\} = A^*_z$ is a product of n hyperplanes in $\mathbb{P}^{r-1}$, hence $A \cap (\Omega \smallsetminus g^{-1}(0)) \times (\mathbb{P}^{r-1})^n$ is a fiber bundle with base $\Omega \smallsetminus g^{-1}(0)$ and fiber $(\mathbb{P}^{r-2})^n$. As A^* is the closure of this set in $\Omega \times (\mathbb{P}^{r-1})^n$, we have

$$\dim A^* = n + n(r-2) = n(r-1) = \dim(\mathbb{P}^{r-1})^n.$$

It follows that the zero fiber

$$A^*_0 = A^* \cap \big(\{0\} \times (\mathbb{P}^{r-1})^n\big)$$

is a proper subset of $\{0\} \times (\mathbb{P}^{r-1})^n$. Choose $(a_1, \ldots, a_n) \in (\mathbb{C}^r \smallsetminus \{0\})^n$ such that $(0, [a_1], \ldots, [a_n])$ is not in A^*_0. By compactness the set $A^* \cap \big(\overline{B}(0,\varepsilon) \times (\mathbb{P}^{r-1})^n\big)$ is disjoint from the neighborhood $B(0,\varepsilon) \times \prod [B(a_j,\varepsilon)]$ of $(0, [a_1], \ldots, [a_n])$ for ε small enough. For $z \in B(0,\varepsilon)$ we have $|a_j \cdot g(z)| \geqslant \varepsilon |g(z)|$ for some j, otherwise the inequality $|a_j \cdot g(z)| < \varepsilon |g(z)|$ would imply the existence of $h_j \in \mathbb{C}^r$ with $|h_j| < \varepsilon$ and $a_j\, g(z) = h_j\, g(z)$. Since $g(z) \neq 0$, we would have

$$(z, [a_1 - h_1], \ldots, [a_n - h_n]) \in A^* \cap \big(B(0,\varepsilon) \times (\mathbb{P}^{r-1})^n\big),$$

a contradiction. We obtain therefore

$$\varepsilon |g(z)| \leqslant \max |a_j\, g(z)| \leqslant (\max |a_j|)\, |g(z)| \quad \text{on } B(0,\varepsilon). \qquad \square$$

(11.17) Theorem (Briançon-Skoda [BSk74]). *Set $p = \min\{n-1, r-1\}$. Then*

(a) $\widehat{\mathcal{I}}^{(k+1)} = \mathcal{I}\, \widehat{\mathcal{I}}^{(k)} = \overline{\mathcal{I}}\, \widehat{\mathcal{I}}^{(k)}$ *for $k \geqslant p$.*

(b) $\overline{\mathcal{I}}^{(k+p)} \subset \widehat{\mathcal{I}}^{(k+p)} \subset \mathcal{I}^k$ *for all $k \in \mathbb{N}$.*

Proof. (a) The inclusions $\mathcal{I}.\widehat{\mathcal{I}}^{(k)} \subset \overline{\mathcal{I}}.\widehat{\mathcal{I}}^{(k)} \subset \widehat{\mathcal{I}}^{(k+1)}$ are obvious thanks to Proposition 11.15, so we only have to prove that $\widehat{\mathcal{I}}^{(k+1)} \subset \mathcal{I}.\widehat{\mathcal{I}}^{(k)}$. Assume first that $r \leqslant n$. Let $f \in \widehat{\mathcal{I}}^{(k+1)}$ be such that

$$\int_\Omega |f|^2\, |g|^{-2(k+1+\varepsilon)}\, dV < +\infty.$$

For $k \geqslant p-1$, we can apply Theorem 11.13 with $m = r-1$ and with the weight $\varphi = (k-m)\log|g|^2$. Hence f can be written $f = \sum g_j h_j$ with

$$\int_\Omega |h|^2\, |g|^{-2(k+\varepsilon)}\, dV < +\infty,$$

thus $h_j \in \widehat{\mathcal{I}}^{(k)}$ and $f \in \mathcal{I}.\widehat{\mathcal{I}}^{(k)}$. When $r > n$, Lemma 11.16 shows that there is an ideal $\mathcal{J} \subset \mathcal{I}$ with n generators such that $\widehat{\mathcal{J}}^{(k)} = \widehat{\mathcal{I}}^{(k)}$. We find

$$\widehat{\mathcal{I}}^{(k+1)} = \widehat{\mathcal{J}}^{(k+1)} \subset \mathcal{J}\,\widehat{\mathcal{J}}^{(k)} \subset \mathcal{I}\widehat{\mathcal{I}}^{(k)} \qquad \text{for } \ k \geqslant n-1.$$

(b) Property (a) implies inductively $\widehat{\mathcal{I}}^{(k+p)} = \mathcal{I}^k\,\widehat{\mathcal{I}}^{(p)}$ for all $k \in \mathbb{N}$. This gives in particular $\widehat{\mathcal{I}}^{(k+p)} \subset \mathcal{I}^k$. □

(11.18) Corollary.

(a) *The ideal $\overline{\mathcal{I}}$ is the integral closure of $\mathcal{I}$, i.e. by definition the set of germs $u \in \mathcal{O}_n$ which satisfy an equation*

$$u^d + a_1 u^{d-1} + \cdots + a_d = 0, \qquad a_s \in \mathcal{I}^s, \qquad 1 \leqslant s \leqslant d.$$

(b) *Similarly, $\overline{\mathcal{I}}^{(k)}$ is the set of germs $u \in \mathcal{O}_n$ which satisfy an equation*

$$u^d + a_1 u^{d-1} + \cdots + a_d = 0, \qquad a_s \in \mathcal{I}^{\lceil ks\rceil}, \qquad 1 \leqslant s \leqslant d,$$

where $\lceil t\rceil$ denotes the smallest integer $\geqslant t$.

As the ideal $\overline{\mathcal{I}}^{(k)}$ is finitely generated, property (b) shows that there always exists a rational number $l \geqslant k$ such that $\overline{\mathcal{I}}^{(l)} = \overline{\mathcal{I}}^{(k)}$.

Proof. (a) If $u \in \mathcal{O}_n$ satisfies a polynomial equation with coefficients $a_s \in \mathcal{I}^s$, then clearly $|a_s| \leqslant C_s\, |g|^s$ and the usual elementary bound

$$|\text{roots}| \leqslant 2 \max_{1\leqslant s\leqslant d} |a_s|^{1/s}$$

for the roots of a monic polynomial implies $|u| \leqslant C\,|g|$.

Conversely, assume that $u \in \overline{\mathcal{J}}$. The ring $\mathcal{O}_n$ is Noetherian, so the ideal $\widehat{\mathcal{J}}^{(p)}$ has a finite number of generators $v_1, \ldots, v_N$. For every j we have $uv_j \in \overline{\mathcal{J}}\,\widehat{\mathcal{J}}^{(p)} = \mathcal{J}\,\widehat{\mathcal{J}}^{(p)}$, hence there exist elements $b_{jk} \in \mathcal{J}$ such that

$$uv_j = \sum_{1\leqslant k\leqslant N} b_{jk} v_k.$$

The matrix $(u\delta_{jk} - b_{jk})$ has the non zero vector (v_j) in its kernel, thus u satisfies the equation $\det(u\delta_{jk} - b_{jk}) = 0$, which is of the required type.

(b) Observe that $v_1, \ldots, v_N$ satisfy simultaneously some integrability condition $\int_\Omega |v_j|^{-2(p+\varepsilon)} < +\infty$, thus $\widehat{\mathcal{J}}^{(p)} = \widehat{\mathcal{J}}^{(p+\eta)}$ for $\eta \in [0, \varepsilon[$. Let $u \in \overline{\mathcal{J}}^{(k)}$. For every integer $m \in \mathbb{N}$ we have

$$u^m v_j \in \overline{\mathcal{J}}^{(km)}\,\widehat{\mathcal{J}}^{(p+\eta)} \subset \widehat{\mathcal{J}}^{(km+\eta+p)}.$$

If $k \notin \mathbb{Q}$, we can find m such that $d(km + \varepsilon/2, \mathbb{Z}) < \varepsilon/2$, thus $km + \eta \in \mathbb{N}$ for some $\eta \in]0, \varepsilon[$. If $k \in \mathbb{Q}$, we take m such that $km \in \mathbb{N}$ and $\eta = 0$. Then

$$u^m v_j \in \widehat{\mathcal{J}}^{(N+p)} = \mathcal{J}^N\,\widehat{\mathcal{J}}^{(p)} \qquad \text{with} \quad N = km + \eta \in \mathbb{N},$$

and the reasoning made in (a) gives $\det(u^m\delta_{jk} - b_{jk}) = 0$ for some $b_{jk} \in \mathcal{J}^N$. This is an equation of the type described in (b), where the coefficients a_s vanish when s is not a multiple of m and $a_{ms} \in \mathcal{J}^{Ns} \subset \mathcal{J}^{\lceil kms\rceil}$. □

Let us mention that Briançon and Skoda's result 11.17 (b) is optimal for $k = 1$. Take for example $\mathcal{J} = (g_1, \ldots, g_r)$ with $g_j(z) = z_j^r$, $1 \leqslant j \leqslant r$, and $f(z) = z_1 \cdots z_r$. Then $|f| \leqslant C|g|$ and 11.17 (b) yields $f^r \in \mathcal{J}$; however, it is easy to verify that $f^{r-1} \notin \mathcal{J}$. The theorem also gives an answer to the following conjecture made by J. Mather.

(11.19) Corollary. *Let $f \in \mathcal{O}_n$ and $\mathcal{J}_f = (z_1\partial f/\partial z_1, \ldots, z_n\partial f/\partial z_n)$. Then $f \in \overline{\mathcal{J}}_f$, and for every integer $k \geqslant 0$, $f^{k+n-1} \in \mathcal{J}_f^k$.*

The Corollary is also optimal for $k = 1$: for example, one can verify that the function $f(z) = (z_1 \cdots z_n)^3 + z_1^{3n-1} + \cdots + z_n^{3n-1}$ is such that $f^{n-1} \notin \mathcal{J}_f$.

Proof. Set $g_j(z) = z_j\,\partial f/\partial z_j$, $1 \leqslant j \leqslant n$. By 11.17 (b), it suffices to show that $|f| \leqslant C|g|$. For every germ of analytic curve $\mathbb{C} \ni t \longmapsto \gamma(t)$, $\gamma \not\equiv 0$, the vanishing order of $f \circ \gamma(t)$ at $t = 0$ is the same as that of

$$t\,\frac{d(f\circ\gamma)}{dt} = \sum_{1\leqslant j\leqslant n} t\,\gamma_j'(t)\,\frac{\partial f}{\partial z_j}(\gamma(t)).$$

We thus obtain

$$|f\circ\gamma(t)| \leqslant C_1\,|t|\,\left|\frac{d(f\circ\gamma)}{dt}\right| \leqslant C_2 \sum_{1\leqslant j\leqslant n} |t\,\gamma_j'(t)|\,\left|\frac{\partial f}{\partial z_j}(\gamma(t))\right| \leqslant C_3\,|g\circ\gamma(t)|$$

and conclude by the following elementary lemma. □

(11.20) Curve selection lemma. *Let $f, g_1, \ldots, g_r \in \mathcal{O}_n$ be germs of holomorphic functions vanishing at 0. Then we have $|f| \leqslant C|g|$ for some constant C if and only if for every germ of analytic curve γ through 0 there exists a constant C_γ such that $|f \circ \gamma| \leqslant C_\gamma |g \circ \gamma|$.*

Proof. If the inequality $|f| \leqslant C|g|$ does not hold on any neighborhood of 0, the germ of analytic set $(A, 0) \subset (\mathbb{C}^{n+r}, 0)$ defined by

$$g_j(z) = f(z) z_{n+j}, \qquad 1 \leqslant j \leqslant r,$$

contains a sequence of points $\big(z_\nu, g_j(z_\nu)/f(z_\nu)\big)$ converging to 0 as ν tends to $+\infty$, with $f(z_\nu) \neq 0$. Hence $(A, 0)$ contains an irreducible component on which $f \not\equiv 0$ and there is a germ of curve $\widetilde{\gamma} = (\gamma, \gamma_{n+j}) : (\mathbb{C}, 0) \to (\mathbb{C}^{n+r}, 0)$ contained in $(A, 0)$ such that $f \circ \gamma \not\equiv 0$. We get $g_j \circ \gamma = (f \circ \gamma)\gamma_{n+j}$, hence $|g \circ \gamma(t)| \leqslant C|t|\,|f \circ \gamma(t)|$ and the inequality $|f \circ \gamma| \leqslant C_\gamma |g \circ \gamma|$ does not hold. □

Chapter 12

The Ohsawa-Takegoshi L^2 Extension Theorem

The Ohsawa-Takegoshi theorem addresses the following extension problem: let Y be a complex analytic submanifold of a complex manifold X ; given a holomorphic function f on Y satisfying suitable L^2 conditions on Y, find a holomorphic extension F of f to X, together with a good L^2 estimate for F on X. The first satisfactory solution has been obtained in the fundamental papers [OT87; Ohs88]. We follow here a more geometric approach due to Manivel [Man93], which provides a generalized extension theorem in the general framework of vector bundles. As in Ohsawa-Takegoshi's fundamental paper, the main idea is to use a modified Bochner-Kodaira-Nakano inequality. Such inequalities were originally introduced in the work of Donnelly-Fefferman [DF83] and Donnelly-Xavier [DX84].

12.A. The Basic a Priori Inequality

The main a priori inequality we are going to use is a simplified (and slightly extended) version of the original Ohsawa-Takegoshi a priori inequality, along the lines proposed by Ohsawa [Ohs95].

(12.1) Lemma (Ohsawa [Ohs95]). *Let E be a Hermitian vector bundle on a complex manifold X equipped with a Kähler metric ω. Let η, $\lambda > 0$ be smooth functions on X. Then for every form $u \in \mathscr{D}(X, \Lambda^{p,q}T^*_X \otimes E)$ with compact support we have*

$$\begin{aligned}\|(\eta^{\frac12} + \lambda^{\frac12})D''^* u\|^2 + \|\eta^{\frac12} D'' u\|^2 + \|\lambda^{\frac12} D' u\|^2 + 2\|\lambda^{-\frac12} d'\eta \wedge u\|^2 \\ \geqslant \langle\langle [\eta\, \mathrm{i}\Theta_E - \mathrm{i}\, d'd''\eta - \mathrm{i}\lambda^{-1} d'\eta \wedge d''\eta, \Lambda]u, u\rangle\rangle.\end{aligned}$$

Proof. Let us consider the "twisted" Laplace-Beltrami operators

$$\begin{aligned}D'\eta D'^* + D'^*\eta D' &= \eta[D', D'^*] + [D', \eta]D'^* + [D'^*, \eta]D' \\ &= \eta\Delta' + (d'\eta)D'^* - (d'\eta)^* D', \\ D''\eta D''^* + D''^*\eta D'' &= \eta[D'', D''^*] + [D'', \eta]D''^* + [D''^*, \eta]D'' \\ &= \eta\Delta'' + (d''\eta)D''^* - (d''\eta)^* D'',\end{aligned}$$

where η, $(d'\eta)$, $(d''\eta)$ are abbreviated notations for the multiplication operators $\eta\bullet$, $(d'\eta) \wedge \bullet$, $(d''\eta) \wedge \bullet$. By subtracting the above equalities and taking into account the Bochner-Kodaira-Nakano identity $\Delta'' - \Delta' = [\mathrm{i}\Theta_E, \Lambda]$, we get

$$\begin{aligned}&D''\eta D''^* + D''^*\eta D'' - D'\eta D'^* - D'^*\eta D'\\ &\quad = \eta[\mathrm{i}\Theta_E, \Lambda] + (d''\eta)D''^* - (d''\eta)^*D'' + (d'\eta)^*D' - (d'\eta)D'^*.\end{aligned}\tag{12.2}$$

Moreover, the Jacobi identity yields

$$[D'', [d'\eta, \Lambda]] - [d'\eta, [\Lambda, D'']] + [\Lambda, [D'', d'\eta]] = 0,$$

whilst $[\Lambda, D''] = -\mathrm{i}D'^*$ by the basic commutation relations 4.5. A straightforward computation shows that $[D'', d'\eta] = -(d'd''\eta)$ and $[d'\eta, \Lambda] = \mathrm{i}(d''\eta)^*$. Therefore we get

$$\mathrm{i}[D'', (d''\eta)^*] + \mathrm{i}[d'\eta, D'^*] - [\Lambda, (d'd''\eta)] = 0,$$

that is,

$$[\mathrm{i}\, d'd''\eta, \Lambda] = [D'', (d''\eta)^*] + [D'^*, d'\eta] = D''(d''\eta)^* + (d''\eta)^*D'' + D'^*(d'\eta) + (d'\eta)D'^*.$$

After adding this to (12.2), we find

$$\begin{aligned}&D''\eta D''^* + D''^*\eta D'' - D'\eta D'^* - D'^*\eta D' + [\mathrm{i}\, d'd''\eta, \Lambda]\\ &\quad = \eta[\mathrm{i}\Theta_E, \Lambda] + (d''\eta)D''^* + D''(d''\eta)^* + (d'\eta)^*D' + D'^*(d'\eta).\end{aligned}$$

We apply this identity to a form $u \in \mathscr{D}(X, \Lambda^{p,q}T_X^* \otimes E)$ and take the inner bracket with u. Then

$$\langle\!\langle (D''\eta D''^*)u, u\rangle\!\rangle = \langle\!\langle \eta D''^*u, D''^*u\rangle\!\rangle = \|\eta^{\frac{1}{2}}D''^*u\|^2,$$

and likewise for the other similar terms. The above equalities imply

$$\begin{aligned}&\|\eta^{\frac{1}{2}}D''^*u\|^2 + \|\eta^{\frac{1}{2}}D''u\|^2 - \|\eta^{\frac{1}{2}}D'u\|^2 - \|\eta^{\frac{1}{2}}D'^*u\|^2\\ &\quad = \langle\!\langle [\eta\,\mathrm{i}\Theta_E - \mathrm{i}\, d'd''\eta, \Lambda]u, u\rangle\!\rangle + 2\,\mathrm{Re}\,\langle\!\langle D''^*u, (d''\eta)^*u\rangle\!\rangle + 2\,\mathrm{Re}\,\langle\!\langle D'u, d'\eta \wedge u\rangle\!\rangle.\end{aligned}$$

By neglecting the negative terms $-\|\eta^{\frac{1}{2}}D'u\|^2 - \|\eta^{\frac{1}{2}}D'^*u\|^2$ and adding the squares

$$\|\lambda^{\frac{1}{2}}D''^*u\|^2 + 2\,\mathrm{Re}\,\langle\!\langle D''^*u, (d''\eta)^*u\rangle\!\rangle + \|\lambda^{-\frac{1}{2}}(d''\eta)^*u\|^2 \geqslant 0,$$
$$\|\lambda^{\frac{1}{2}}D'u\|^2 + 2\,\mathrm{Re}\,\langle\!\langle D'u, d'\eta \wedge u\rangle\!\rangle + \|\lambda^{-\frac{1}{2}}d'\eta \wedge u\|^2 \geqslant 0$$

we get

$$\begin{aligned}&\|(\eta^{\frac{1}{2}} + \lambda^{\frac{1}{2}})D''^*u\|^2 + \|\eta^{\frac{1}{2}}D''u\|^2 + \|\lambda^{\frac{1}{2}}D'u\|^2 + \|\lambda^{-\frac{1}{2}}d'\eta \wedge u\|^2 + \|\lambda^{-\frac{1}{2}}(d''\eta)^*u\|^2\\ &\quad \geqslant \langle\!\langle [\eta\,\mathrm{i}\Theta_E - \mathrm{i}\, d'd''\eta, \Lambda]u, u\rangle\!\rangle.\end{aligned}$$

Finally, we use the identities

$$(d'\eta)^*(d'\eta) - (d''\eta)(d''\eta)^* = \mathrm{i}[d''\eta, \Lambda](d'\eta) + \mathrm{i}(d''\eta)[d'\eta, \Lambda] = [\mathrm{i}d''\eta \wedge d'\eta, \Lambda],$$
$$\|\lambda^{-\frac{1}{2}}d'\eta \wedge u\|^2 - \|\lambda^{-\frac{1}{2}}(d''\eta)^*u\|^2 = -\langle\!\langle [\mathrm{i}\lambda^{-1}d'\eta \wedge d''\eta, \Lambda]u, u\rangle\!\rangle,$$

The inequality asserted in Lemma 12.1 follows by adding the second identity to our last inequality. □

In the special case of (n,q)-forms, the forms $D'u$ and $d'\eta \wedge u$ are of bidegree $(n+1,q)$, hence the estimate takes the simpler form

$$\|(\eta^{\frac{1}{2}} + \lambda^{\frac{1}{2}})D''^*u\|^2 + \|\eta^{\frac{1}{2}}D''u\|^2 \geqslant \langle\!\langle [\eta\,\mathrm{i}\Theta_E - \mathrm{i}\, d'd''\eta - \mathrm{i}\lambda^{-1}\, d'\eta \wedge d''\eta, \Lambda]u, u\rangle\!\rangle.\tag{12.3}$$

12.B. Abstract L^2 Existence Theorem for Solutions of $\overline{\partial}$-Equations

Using standard arguments from functional analysis – actually just basic properties of Hilbert spaces along the lines already explained in section 5 – the a priori inequality (12.3) implies a very strong L^2 existence theorem for solutions of $\overline{\partial}$-equations.

(12.4) Proposition. *Let X be a complete Kähler manifold equipped with a (non necessarily complete) Kähler metric ω, and let E be a Hermitian vector bundle over X. Assume that there are smooth and bounded functions η, $\lambda > 0$ on X such that the (Hermitian) curvature operator*

$$B = B^{n,q}_{E,\omega,\eta} = [\eta\,\mathrm{i}\Theta_E - \mathrm{i}\,d'd''\eta - \mathrm{i}\lambda^{-1}d'\eta\wedge d''\eta, \Lambda_\omega]$$

is positive definite everywhere on $\Lambda^{n,q}T^*_X\otimes E$, for some $q\geqslant 1$. Then for every form $g\in L^2(X,\Lambda^{n,q}T^*_X\otimes E)$ such that $D''g=0$ and $\int_X\langle B^{-1}g,g\rangle\,dV_\omega < +\infty$, there exists $f\in L^2(X,\Lambda^{n,q-1}T^*_X\otimes E)$ such that $D''f=g$ and

$$\int_X(\eta+\lambda)^{-1}|f|^2\,dV_\omega \leqslant 2\int_X\langle B^{-1}g,g\rangle\,dV_\omega.$$

Proof. The proof is almost identical to the proof of standard L^2 estimates for $\overline{\partial}$ (see Theorem 5.1), except that we use (12.3) instead of (4.7). Assume first that ω is complete. With the same notation as in 7.4, we get for every $v=v_1+v_2\in(\operatorname{Ker}D'')\oplus(\operatorname{Ker}D'')^\perp$ the inequalities

$$|\langle g,v\rangle|^2 = |\langle g,v_1\rangle|^2 \leqslant \int_X\langle B^{-1}g,g\rangle\,dV_\omega\int_X\langle Bv_1,v_1\rangle\,dV_\omega,$$

and

$$\int_X\langle Bv_1,v_1\rangle\,dV_\omega \leqslant \|(\eta^{\frac12}+\lambda^{\frac12})D''^*v_1\|^2+\|\eta^{\frac12}D''v_1\|^2 = \|(\eta^{\frac12}+\lambda^{\frac12})D''^*v\|^2$$

provided that $v\in\operatorname{Dom}D''^*$. Combining both, we find

$$|\langle g,v\rangle|^2 \leqslant \Big(\int_X\langle B^{-1}g,g\rangle\,dV_\omega\Big)\|(\eta^{\frac12}+\lambda^{\frac12})D''^*v\|^2.$$

This shows the existence of an element $w\in L^2(X,\Lambda^{n,q}T^*_X\otimes E)$ such that

$$\|w\|^2 \leqslant \int_X\langle B^{-1}g,g\rangle\,dV_\omega \qquad\text{and}$$
$$\langle\langle v,g\rangle\rangle = \langle\langle(\eta^{\frac12}+\lambda^{\frac12})D''^*v,w\rangle\rangle \qquad \forall g\in\operatorname{Dom}D''\cap\operatorname{Dom}D''^*.$$

As $(\eta^{1/2}+\lambda^{\frac12})^2\leqslant 2(\eta+\lambda)$, it follows that $f=(\eta^{1/2}+\lambda^{\frac12})w$ satisfies $D''f=g$ as well as the desired L^2 estimate. If ω is not complete, we set $\omega_\varepsilon=\omega+\varepsilon\widehat{\omega}$ with some complete Kähler metric $\widehat{\omega}$. The final conclusion is then obtained by passing to the limit and using a monotonicity argument (the integrals are monotonic with respect to ε). □

(12.5) Remark. We will also need a variant of the L^2-estimate, so as to obtain approximate solutions with weaker requirements on the data: given $\delta > 0$ and $g \in L^2(X, \Lambda^{n,q}T_X^* \otimes E)$ such that $D''g = 0$ and $\int_X \langle (B+\delta I)^{-1}g, g\rangle\, dV_\omega < +\infty$, there exists an approximate solution $f \in L^2(X, \Lambda^{n,q-1}T_X^* \otimes E)$ and a correcting term $h \in L^2(X, \Lambda^{n,q}T_X^* \otimes E)$ such that $D''f + \delta^{1/2}h = g$ and

$$\int_X (\eta+\lambda)^{-1}|f|^2\, dV_\omega + \int_X |h|^2\, dV_\omega \leqslant 2\int_X \langle (B+\delta I)^{-1}g, g\rangle\, dV_\omega.$$

The proof is almost unchanged, we rely instead on the estimates

$$|\langle g, v_1\rangle|^2 \leqslant \int_X \langle (B+\delta I)^{-1}g, g\rangle\, dV_\omega \int_X \langle (B+\delta I)v_1, v_1\rangle\, dV_\omega,$$

and

$$\int_X \langle (B+\delta I)v_1, v_1\rangle\, dV_\omega \leqslant \|(\eta^{\frac{1}{2}} + \lambda^{\frac{1}{2}})D''^* v\|^2 + \delta\|v\|^2.$$ □

12.C. The L^2 Extension Theorem

According to a concept already widely used in Chapter 5, a (non necessarily compact) complex manifold will be said to be weakly pseudoconvex if it possesses a smooth weakly plurisubharmonic exhaustion function.

(12.6) Theorem. *Let X be a weakly pseudoconvex complex n-dimensional manifold possessing a Kähler metric ω, and let L (resp. E) be a Hermitian holomorphic line bundle (resp. a Hermitian holomorphic vector bundle of rank r over X), and s a global holomorphic section of E. Assume that s is generically transverse to the zero section, and let*

$$Y = \{x \in X\,;\ s(x) = 0, \Lambda^r ds(x) \neq 0\}, \qquad p = \dim Y = n - r.$$

Moreover, assume that the $(1,1)$-form $\mathrm{i}\Theta_L + r\,\mathrm{i}\,d'd''\log|s|^2$ is semi-positive and that there is a continuous function $\alpha \geqslant 1$ such that the following two inequalities hold everywhere on X:

(a) $\mathrm{i}\Theta_L + r\,\mathrm{i}\,d'd''\log|s|^2 \geqslant \alpha^{-1}\dfrac{\{\mathrm{i}\Theta_E s, s\}}{|s|^2}$,

(b) $|s| \leqslant e^{-\alpha}$.

Then for every smooth D''-closed $(0,q)$-form f over Y with values in the line bundle $\Lambda^n T_X^* \otimes L$ (restricted to Y), such that $\int_Y |f|^2|\Lambda^r(ds)|^{-2}dV_\omega < +\infty$, there *exists a D''-closed $(0,q)$-form F over X with values in $\Lambda^n T_X^* \otimes L$, such that F is smooth over $X \smallsetminus \{s = \Lambda^r(ds) = 0\}$, satisfies $F_{\restriction Y} = f$ and*

$$\int_X \frac{|F|^2}{|s|^{2r}(-\log|s|)^2}\, dV_{X,\omega} \leqslant C_r \int_Y \frac{|f|^2}{|\Lambda^r(ds)|^2} dV_{Y,\omega}\,,$$

where C_r is a numerical constant depending only on r.

Observe that the differential ds (which is intrinsically defined only at points where s vanishes) induces a vector bundle isomorphism $ds : T_X/T_Y \to E$ along Y, hence a non vanishing section $\Lambda^r(ds)$, taking values in

$$\Lambda^r(T_X/T_Y)^* \otimes \det E \subset \Lambda^r T_X^* \otimes \det E.$$

The norm $|\Lambda^r(ds)|$ is computed here with respect to the metrics on $\Lambda^r T_X^*$ and $\det E$ induced by the Kähler metric ω and by the given metric on E. Also notice that if hypothesis (a) is satisfied for some α, one can always achieve (b) by multiplying the metric of E with a sufficiently small weight $e^{-\chi\circ\psi}$ (with ψ a psh exhaustion on X and χ a convex increasing function; property (a) remains valid after we multiply the metric of L by $e^{-(r+\alpha_0^{-1})\chi\circ\psi}$, where $\alpha_0 = \inf_{x\in X}\alpha(x)$.

Proof. Let us first assume that the singularity set $\Sigma = \{s = 0\} \cap \{\Lambda^r(ds) = 0\}$ is empty, so that Y is closed and nonsingular. We claim that there exists a smooth section

$$F_\infty \in \mathscr{C}^\infty(X, \Lambda^{n,q} T_X^* \otimes L) = \mathscr{C}^\infty(X, \Lambda^{0,q} T_X^* \otimes \Lambda^n T_X^* \otimes L)$$

such that

(a) F_∞ coincides with f in restriction to Y,

(b) $|F_\infty| = |f|$ at every point of Y,

(c) $D'' F_\infty = 0$ at every point of Y.

For this, consider coordinates patches $U_j \subset X$ biholomorphic to polydiscs such that

$$U_j \cap Y = \{z \in U_j\,;\ z_1 = \cdots = z_r = 0\}$$

in the corresponding coordinates. We can certainly find a section

$$\widetilde{f} \in \mathscr{C}^\infty(X, \Lambda^{n,q} T_X^* \otimes L)$$

which achieves (a) and (b), since the restriction map $(\Lambda^{0,q} T_X^*)_{\restriction Y} \to \Lambda^{0,q} T_Y^*$ can be viewed as an orthogonal projection onto a $\mathscr{C}^\infty$-subbundle of $(\Lambda^{0,q} T_X^*)_{\restriction Y}$. It is enough to extend this subbundle from $U_j \cap Y$ to U_j (e.g. by extending each component of a frame), and then to extend f globally via local smooth extensions and a partition of unity. For any such extension $\widetilde{f}$ we have

$$(D''\widetilde{f})_{\restriction Y} = (D''\widetilde{f}_{\restriction Y}) = D'' f = 0.$$

It follows that we can divide $D''\widetilde{f} = \sum_{1\leqslant\lambda\leqslant r} g_{j,\lambda}(z)\wedge d\overline{z}_\lambda$ on $U_j\cap Y$, with suitable smooth $(0,q)$-forms $g_{j,\lambda}$ which we also extend arbitrarily from $U_j\cap Y$ to U_j. Then

$$F_\infty := \widetilde{f} - \sum_j \theta_j(z) \sum_{1\leqslant\lambda\leqslant r} \overline{z}_\lambda g_{j,\lambda}(z)$$

coincides with $\widetilde{f}$ on Y and satisfies (c). Since we do not know about F_∞ except in an infinitesimal neighborhood of Y, we will consider a truncation F_ε of F_∞ with support in a small tubular neighborhood $|s|<\varepsilon$ of Y, and solve the equation $D''u_\varepsilon = D''F_\varepsilon$ with the constraint that u_ε should be 0 on Y. As $\operatorname{codim} Y = r$, this will be the case if we can guarantee that $|u_\varepsilon|^2|s|^{-2r}$ is locally integrable near Y. For this, we will apply Proposition 12.4 with a suitable choice of the functions η and λ, and an additional weight $|s|^{-2r}$ in the metric of L.

Let us consider the smooth strictly convex function $\chi_0 :]-\infty,0] \to]-\infty,0]$ defined by $\chi_0(t) = t - \log(1-t)$ for $t\leqslant 0$, which is such that $\chi_0(t)\leqslant t$, $1\leqslant \chi_0'\leqslant 2$ and $\chi_0''(t) = 1/(1-t)^2$. We set

$$\sigma_\varepsilon = \log(|s|^2+\varepsilon^2), \qquad \eta_\varepsilon = \varepsilon - \chi_0(\sigma_\varepsilon).$$

As $|s|\leqslant e^{-\alpha}\leqslant e^{-1}$, we have $\sigma_\varepsilon\leqslant 0$ for ε small, and

$$\eta_\varepsilon \geqslant \varepsilon - \sigma_\varepsilon \geqslant \varepsilon - \log(e^{-2\alpha}+\varepsilon^2).$$

Given a relatively compact subset $X_c = \{\psi < c\} \subset\subset X$, we thus have $\eta_\varepsilon \geqslant 2\alpha$ for $\varepsilon < \varepsilon(c)$ small enough. Simple calculations yield

$$\begin{aligned}
\mathrm{i}\, d'\sigma_\varepsilon &= \frac{\mathrm{i}\{D's,s\}}{|s|^2+\varepsilon^2},\\
\mathrm{i}\, d'd''\sigma_\varepsilon &= \frac{\mathrm{i}\{D's,D's\}}{|s|^2+\varepsilon^2} - \frac{\mathrm{i}\{D's,s\}\wedge\{s,D's\}}{(|s|^2+\varepsilon^2)^2} - \frac{\{\mathrm{i}\Theta_E s,s\}}{|s|^2+\varepsilon^2}\\
&\geqslant \frac{\varepsilon^2}{|s|^2}\,\frac{\mathrm{i}\{D's,s\}\wedge\{s,D's\}}{(|s|^2+\varepsilon^2)^2} - \frac{\{\mathrm{i}\Theta_E s,s\}}{|s|^2+\varepsilon^2}\\
&\geqslant \frac{\varepsilon^2}{|s|^2}\mathrm{i}d'\sigma_\varepsilon\wedge d''\sigma_\varepsilon - \frac{\{\mathrm{i}\Theta_E s,s\}}{|s|^2+\varepsilon^2},
\end{aligned}$$

thanks to Lagrange's inequality $\mathrm{i}\{D's,s\}\wedge\{s,D's\}\leqslant |s|^2\mathrm{i}\{D's,D's\}$. On the other hand, we have $d'\eta_\varepsilon = -\chi_0'(\sigma_\varepsilon)d\sigma_\varepsilon$ with $1\leqslant\chi_0'(\sigma_\varepsilon)\leqslant 2$, hence

$$\begin{aligned}
-\mathrm{i}d'd''\eta_\varepsilon &= \chi_0'(\sigma_\varepsilon)\mathrm{i}d'd''\sigma_\varepsilon + \chi_0''(\sigma_\varepsilon)\mathrm{i}d'\sigma_\varepsilon\wedge d''\sigma_\varepsilon\\
&\geqslant \Big(\frac{1}{\chi_0'(\sigma_\varepsilon)}\,\frac{\varepsilon^2}{|s|^2} + \frac{\chi_0''(\sigma_\varepsilon)}{\chi_0'(\sigma_\varepsilon)^2}\Big)\mathrm{i}d'\eta_\varepsilon\wedge d''\eta_\varepsilon - \chi_0'(\sigma_\varepsilon)\frac{\{\mathrm{i}\Theta_E s,s\}}{|s|^2+\varepsilon^2}.
\end{aligned}$$

We consider the original metric of L multiplied by the weight $|s|^{-2r}$. In this way, we get a curvature form

$$\mathrm{i}\Theta_L + r\,\mathrm{i}d'd''\log|s|^2 \geqslant \frac{1}{2}\chi_0'(\sigma_\varepsilon)\alpha^{-1}\frac{\{\mathrm{i}\Theta_E s,s\}}{|s|^2+\varepsilon^2}$$

by hypothesis (a), thanks to the semipositivity of the left hand side and the fact that $\frac{1}{2}\chi_0'(\sigma_\varepsilon)\frac{1}{|s|^2+\varepsilon^2} \leqslant \frac{1}{|s|^2}$. As $\eta_\varepsilon \geqslant 2\alpha$ on X_c for ε small, we infer

$$\eta_\varepsilon(\mathrm{i}\Theta_L + \mathrm{i}d'd''\log|s|^2) - \mathrm{i}d'd''\eta_\varepsilon - \frac{\chi_0''(\sigma_\varepsilon)}{\chi_0'(\sigma_\varepsilon)^2}\mathrm{i}d'\eta_\varepsilon \wedge d''\eta_\varepsilon \geqslant \frac{\varepsilon^2}{\chi_0'(\sigma_\varepsilon)|s|^2}\mathrm{i}d'\eta_\varepsilon \wedge d''\eta_\varepsilon$$

on X_c. Hence, if $\lambda_\varepsilon = \chi_0'(\sigma_\varepsilon)^2/\chi_0''(\sigma_\varepsilon)$, we obtain

$$\begin{aligned} B_\varepsilon &:= \big[\eta_\varepsilon(\mathrm{i}\Theta_L + \mathrm{i}d'd''\log|s|^2) - \mathrm{i}d'd''\eta_\varepsilon - \lambda_\varepsilon^{-1}\mathrm{i}d'\eta_\varepsilon \wedge d''\eta_\varepsilon\,,\ \Lambda\big] \\ &\geqslant \Big[\frac{\varepsilon^2}{\chi_0'(\sigma_\varepsilon)|s|^2}\mathrm{i}d'\eta_\varepsilon \wedge d''\eta_\varepsilon\,,\ \Lambda\Big] = \frac{\varepsilon^2}{\chi_0'(\sigma_\varepsilon)|s|^2}(d''\eta_\varepsilon)(d''\eta_\varepsilon)^* \end{aligned}$$

as an operator on (n,q)-forms (see the proof of Lemma 12.1).

Let $\theta : \mathbb{R} \to [0,1]$ be a smooth cut-off function such that $\theta(t) = 1$ on $]-\infty, 1/2]$, $\operatorname{Supp}\theta \subset\ ]-\infty, 1[$ and $|\theta'| \leqslant 3$. For $\varepsilon > 0$ small, we consider the (n,q)-form $F_\varepsilon = \theta(\varepsilon^{-2}|s|^2)F_\infty$ and its D''-derivative

$$g_\varepsilon = D''F_\varepsilon = (1+\varepsilon^{-2}|s|^2)\theta'(\varepsilon^{-2}|s|^2)d''\sigma_\varepsilon \wedge F_\infty + \theta(\varepsilon^{-2}|s|^2)D''F_\infty$$

(as is easily seen from the equality $1+\varepsilon^{-2}|s|^2 = \varepsilon^{-2}e^{\sigma_\varepsilon}$). We observe that g_ε has its support contained in the tubular neighborhood $|s| < \varepsilon$; moreover, as $\varepsilon \to 0$, the second term in the right hand side converges uniformly to 0 on every compact set; it will therefore produce no contribution in the limit. On the other hand, the first term has the same order of magnitude as $d''\sigma_\varepsilon$ and $d''\eta_\varepsilon$, and can be controlled in terms of B_ε. In fact, for any (n,q)-form u and any $(n,q+1)$-form v we have

$$\begin{aligned} |\langle d''\eta_\varepsilon \wedge u, v\rangle|^2 &= |\langle u, (d''\eta_\varepsilon)^*v\rangle|^2 \leqslant |u|^2|(d''\eta_\varepsilon)^*v|^2 \\ &= |u|^2\langle (d''\eta_\varepsilon)(d''\eta_\varepsilon)^*v, v\rangle \leqslant \frac{\chi_0'(\sigma_\varepsilon)|s|^2}{\varepsilon^2}|u|^2\langle B_\varepsilon v, v\rangle. \end{aligned}$$

This implies

$$\langle B_\varepsilon^{-1}(d''\eta_\varepsilon \wedge u), (d''\eta_\varepsilon \wedge u)\rangle \leqslant \frac{\chi_0'(\sigma_\varepsilon)|s|^2}{\varepsilon^2}|u|^2.$$

The main term in g_ε can be written

$$g_\varepsilon^{(1)} := (1+\varepsilon^{-2}|s|^2)\theta'(\varepsilon^{-2}|s|^2)\chi_0'(\sigma_\varepsilon)^{-1}d''\eta_\varepsilon \wedge F_\infty.$$

On $\operatorname{Supp} g_\varepsilon^{(1)} \subset \{|s| < \varepsilon\}$, since $\chi_0'(\sigma_\varepsilon) \geqslant 1$, we thus find

$$\langle B_\varepsilon^{-1}g_\varepsilon^{(1)}, g_\varepsilon^{(1)}\rangle \leqslant (1+\varepsilon^{-2}|s|^2)^2\,\theta'(\varepsilon^{-2}|s|^2)^2|F_\infty|^2.$$

Instead of working on X itself, we will work rather on the relatively compact subset $X_c \smallsetminus Y_c$, where $Y_c = Y \cap X_c = Y \cap \{\psi < c\}$. We know that $X_c \smallsetminus Y_c$ is again complete Kähler by a standard lemma (see [Dem82b], Theorem 1.5). In this way, we avoid the singularity of the weight $|s|^{-2r}$ along Y. We find

$$\int_{X_c\smallsetminus Y_c}\langle B_\varepsilon^{-1}g_\varepsilon^{(1)}, g_\varepsilon^{(1)}\rangle\,|s|^{-2r}dV_\omega \leqslant \int_{X_c\smallsetminus Y_c}|F_\infty|^2(1+\varepsilon^{-2}|s|^2)^2\theta'(\varepsilon^{-2}|s|^2)^2|s|^{-2r}dV_\omega.$$

Now, we let $\varepsilon \to 0$ and view s as "transverse local coordinates" around Y. As F_∞ coincides with f on Y, it is not hard to see that the right hand side converges to $c_r \int_{Y_c} |f|^2 |\Lambda^r(ds)|^{-2} dV_{Y,\omega}$ where c_r is the "universal" constant

$$c_r = \int_{z \in \mathbb{C}^r,\, |z| \leqslant 1} (1+|z|^2)^2 \theta'(|z|^2)^2 \frac{\mathrm{i}^{r^2} \Lambda^r(dz) \wedge \Lambda^r(d\bar{z})}{|z|^{2r}} < +\infty$$

depending only on r. The second term

$$g_\varepsilon^{(2)} = \theta(\varepsilon^{-2}|s|^2) d'' F_\infty$$

in g_ε satisfies $\mathrm{Supp}(g_\varepsilon^{(2)}) \subset \{|s| < \varepsilon\}$ and $|g_\varepsilon^{(2)}| = O(|s|)$ (just look at the Taylor expansion of $d''F_\infty$ near Y). From this we easily conclude that

$$\int_{X_c \smallsetminus Y_c} \langle B_\varepsilon^{-1} g_\varepsilon^{(2)}, g_\varepsilon^{(2)} \rangle \, |s|^{-2r} dV_{X,\omega} = O(\varepsilon^2),$$

provided that B_ε remains locally uniformly bounded below near Y (this is the case for instance if we have strict inequalities in the curvature assumption (a)). If this holds true, we apply Proposition 12.4 on $X_c \smallsetminus Y_c$ with the additional weight factor $|s|^{-2r}$. Otherwise, we use the modified estimate stated in Remark 12.5 in order to solve the approximate equation $D''u + \delta^{1/2} h = g_\varepsilon$ with $\delta > 0$ small. This yields sections $u = u_{c,\varepsilon,\delta}$, $h = h_{c,\varepsilon,\delta}$ such that

$$\int_{X_c \smallsetminus Y_c} (\eta_\varepsilon + \lambda_\varepsilon)^{-1} |u_{c,\varepsilon,\delta}|^2 |s|^{-2r} \, dV_\omega + \int_{X_c \smallsetminus Y_c} |h_{c,\varepsilon,\delta}|^2 |s|^{-2r} \, dV_\omega$$
$$\leqslant 2 \int_{X_c \smallsetminus Y_c} \langle (B_\varepsilon + \delta I)^{-1} g_\varepsilon, g_\varepsilon \rangle |s|^{-2r} \, dV_\omega,$$

and the right hand side is under control in all cases. The extra error term $\delta^{1/2} h$ can be removed at the end by letting δ tend to 0. Since there is essentially no additional difficulty involved in this process, we will assume for simplicity of exposition that we do have the required lower bound for B_ε and the estimates of $g_\varepsilon^{(1)}$ and $g_\varepsilon^{(2)}$ as above. For $\delta = 0$, the above estimate provides a solution $u_{c,\varepsilon}$ of the equation $D'' u_{c,\varepsilon} = g_\varepsilon = D'' F_\varepsilon$ on $X_c \smallsetminus Y_c$, such that

$$\int_{X_c \smallsetminus Y_c} (\eta_\varepsilon + \lambda_\varepsilon)^{-1} |u_{c,\varepsilon}|^2 |s|^{-2r} dV_{X,\omega} \leqslant 2 \int_{X_c \smallsetminus Y_c} \langle B_\varepsilon^{-1} g_\varepsilon, g_\varepsilon \rangle \, |s|^{-2r} dV_{X,\omega}$$
$$\leqslant 2\, c_r \int_{Y_c} \frac{|f|^2}{|\Lambda^r(ds)|^2} dV_{Y,\omega} + O(\varepsilon).$$

Here we have

$$\sigma_\varepsilon = \log(|s|^2 + \varepsilon^2) \leqslant \log(e^{-2\alpha} + \varepsilon^2) \leqslant -2\alpha + O(\varepsilon^2) \leqslant -2 + O(\varepsilon^2),$$
$$\eta_\varepsilon = \varepsilon - \chi_0(\sigma_\varepsilon) \leqslant (1 + O(\varepsilon)) \sigma_\varepsilon^2,$$
$$\lambda_\varepsilon = \frac{\chi_0'(\sigma_\varepsilon)^2}{\chi_0''(\sigma_\varepsilon)} = (1 - \sigma_\varepsilon)^2 + (1 - \sigma_\varepsilon) \leqslant (3 + O(\varepsilon)) \sigma_\varepsilon^2,$$
$$\eta_\varepsilon + \lambda_\varepsilon \leqslant (4 + O(\varepsilon)) \sigma_\varepsilon^2 \leqslant (4 + O(\varepsilon)) \big(-\log(|s|^2 + \varepsilon^2) \big)^2.$$

As F_ε is uniformly bounded with support in $\{|s| < \varepsilon\}$, we conclude from an obvious volume estimate that

$$\int_{X_c} \frac{|F_\varepsilon|^2 dV_{X,\omega}}{(|s|^2+\varepsilon^2)^r(-\log(|s|^2+\varepsilon^2))^2} \leqslant \frac{\text{Const.}}{(\log\varepsilon)^2}.$$

Therefore, thanks to the usual inequality $|t+u|^2 \leqslant (1+k)|t|^2 + (1+k^{-1})|u|^2$ applied to the sum $F_{c,\varepsilon} = \widetilde{f}_\varepsilon - u_{c,\varepsilon}$ with $k = |\log\varepsilon|$, we obtain from our previous estimates

$$\int_{X_c \smallsetminus Y_c} \frac{|F_{c,\varepsilon}|^2 dV_{X,\omega}}{(|s|^2+\varepsilon^2)^r(-\log(|s|^2+\varepsilon^2))^2} \leqslant 8\, c_r \int_{Y_c} \frac{|f|^2 dV_{Y,\omega}}{|\Lambda^r(ds)|^2} + O(|\log\varepsilon|^{-1}).$$

In addition to this, we have $d''F_{c,\varepsilon} = 0$ by construction, and this equation can be seen to extend from $X_c \smallsetminus Y_c$ to X_c by the L^2 estimate ([Dem82b], Lemma 6.9).

If $q = 0$, then $u_{c,\varepsilon}$ must also be smooth, and the non integrability of the weight $|s|^{-2r}$ along Y shows that $u_{c,\varepsilon}$ vanishes on Y, therefore

$$F_{c,\varepsilon\restriction Y} = F_{\varepsilon\restriction Y} = F_{\infty\restriction Y} = f.$$

The theorem and its final estimate are thus obtained by extracting weak limits, first as $\varepsilon \to 0$, and then as $c \to +\infty$. The initial assumption that $\Sigma = \{s = \Lambda^r(ds) = 0\}$ is empty can be easily removed in two steps: i) the result is true if X is Stein, since we can always find a complex hypersurface Z in X such that $\Sigma \subset \overline{Y} \cap Z \subsetneq \overline{Y}$, and then apply the extension theorem on the Stein manifold $X \smallsetminus Z$, in combination with L^2 extension; ii) the whole procedure still works when Σ is nowhere dense in $\overline{Y}$ (and possibly nonempty). Indeed local L^2 extensions $\widetilde{f}_j$ still exist by step i) applied on small coordinate balls U_j; we then set $F_\infty = \sum \theta_j \widetilde{f}_j$ and observe that $|D''F_\infty|^2|s|^{-2r}$ is locally integrable, thanks to the estimate $\int_{U_j} |\widetilde{f}_j|^2|s|^{-2r}(\log|s|)^{-2}dV < +\infty$ and the fact that $|\sum d''\theta_j \wedge \widetilde{f}_j| = O(|s|^\delta)$ for suitable $\delta > 0$ [as follows from Hilbert's Nullstensatz applied to $\widetilde{f}_j - \widetilde{f}_k$ at singular points of $\overline{Y}$].

When $q \geqslant 1$, the arguments needed to get a smooth solution involve more delicate considerations, and we will skip the details, which are extremely technical and not very enlightening.

(12.7) Remarks.

(a) When $q = 0$, the estimates provided by Theorem 12.6 are independent of the Kähler metric ω. In fact, if f and F are holomorphic sections of $\Lambda^n T^*_X \otimes L$ over Y (resp. X), viewed as $(n,0)$-forms with values in L, we can "divide" f by $\Lambda^r(ds) \in \Lambda^r(TX/TY)^* \otimes \det E$ to get a section $f/\Lambda^r(ds)$ of $\Lambda^p T^*_Y \otimes L \otimes (\det E)^{-1}$ over Y. We then find

$$|F|^2 dV_{X,\omega} = \mathrm{i}^{n^2}\{F, F\},$$

$$\frac{|f|^2}{|\Lambda^r(ds)|^2} dV_{Y,\omega} = \mathrm{i}^{p^2}\{f/\Lambda^r(ds), f/\Lambda^r(ds)\},$$

where $\{\bullet, \bullet\}$ is the canonical bilinear pairing described in (3.3).

(b) The Hermitian structure on E is not really used in depth. In fact, one only needs E to be equipped with a *Finsler metric*, that is, a smooth complex homogeneous function of degree 2 on E (or equivalently, a smooth Hermitian metric on the tautological bundle $\mathcal{O}_{P(E)}(-1)$ of lines of E over the projectivized bundle $P(E)$, see (4.12)). The section s of E induces a section $[s]$ of $P(E)$ over $X \smallsetminus s^{-1}(0)$ and a corresponding section $\widetilde{s}$ of the pull-back line bundle $[s]^*\mathcal{O}_{P(E)}(-1)$. A trivial check shows that Theorem 12.6 as well as its proof extend to the case of a Finsler metric on E, if we replace everywhere $\{\mathrm{i}\Theta_E s, s\}$ by $\{\mathrm{i}\Theta_{[s]^*\mathcal{O}_{P(E)}(-1)}\widetilde{s}, \widetilde{s}\}$ (especially in hypothesis 12.6 (b)). A minor issue is that $|\Lambda^r(ds)|$ is (a priori) no longer defined, since no obvious Hermitian norm exists on $\det E$. A posteriori, we have the following ad hoc definition of a metric on $(\det E)^*$ which makes the L^2 estimates work as before: for $x \in X$ and $\xi \in \Lambda^r E^*_x$, we set

$$|\xi|^2_x = \frac{1}{c_r}\int_{z\in E_x}(1+|z|^2)^2\theta'(|z|^2)^2\,\frac{\mathrm{i}^{r^2}\xi\wedge\overline{\xi}}{|z|^{2r}}$$

where $|z|$ is the Finsler norm on E_x [the constant c_r is there to make the result agree with the Hermitian case; it is not hard to see that this metric does not depend on the choice of θ].

(c) Even when $q = 0$, the regularity of $u_{c,\varepsilon,\delta}$ requires some explanations, in case $\delta > 0$. In fact, the equation

$$D''u_{c,\varepsilon,\delta} + \delta^{1/2}h_{c,\varepsilon,\delta} = g_\varepsilon = D''F_\varepsilon$$

does not immediately imply smoothness of $u_{c,\varepsilon,\delta}$ (since $h_{c,\varepsilon,\delta}$ need not be smooth in general). However, if we take the pair $(u_{c,\varepsilon,\delta}, h_{c,\varepsilon,\delta})$ to be the minimal L^2 solution orthogonal to the kernel of $D'' \oplus \delta^{1/2}\,\mathrm{Id}$, then it must be in the closure of the image of the adjoint operator $D''^* \oplus \delta^{1/2}\,\mathrm{Id}$, i.e. it must satisfy the additional condition $D''^*h_{c,\varepsilon,\delta} = \delta^{1/2}u_{c,\varepsilon,\delta}$, whence $(\Delta''+\delta\,\mathrm{Id})h_{c,\varepsilon,\delta} = (D''D''^*+\delta\,\mathrm{Id})h_{c,\varepsilon,\delta} = \delta^{1/2}D''F_\varepsilon$, and therefore $h_{c,\varepsilon,\delta}$ is smooth by the ellipticity of Δ''. □

We now present a few interesting corollaries. The first one is a surjectivity theorem for restriction morphisms in Dolbeault cohomology.

(12.8) Corollary. *Let X be a projective algebraic manifold and E a holomorphic vector bundle of rank r over X, s a holomorphic section of E which is everywhere transverse to the zero section, $Y = s^{-1}(0)$, and let L be a holomorphic line bundle such that $F = L^{1/r}\otimes E^*$ is Griffiths positive (by this, we just mean formally that $\frac{1}{r}\mathrm{i}\Theta_L\otimes \mathrm{Id}_E - \mathrm{i}\Theta_E >_{\mathrm{Grif}} 0$). Then the restriction morphism*

$$H^{0,q}(X, \Lambda^n T^*_X \otimes L) \to H^{0,q}(Y, \Lambda^n T^*_X \otimes L)$$

is surjective for every $q \geqslant 0$.

Proof. A short computation gives

$$\begin{aligned}\mathrm{i}\,d'd''\log|s|^2 &= \mathrm{i}\,d'\Big(\frac{\{s,D's\}}{|s|^2}\Big)\\ &= \mathrm{i}\Big(\frac{\{D's,D's\}}{|s|^2}-\frac{\{D's,s\}\wedge\{s,D's\}}{|s|^4}+\frac{\{s,\Theta_E s\}}{|s|^2}\Big)\geqslant -\frac{\{\mathrm{i}\Theta_E s,s\}}{|s|^2}\end{aligned}$$

thanks to Lagrange's inequality and the fact that Θ_E is antisymmetric. Hence, if δ is a small positive constant such that

$$-\mathrm{i}\Theta_E+\frac{1}{r}\mathrm{i}\Theta_L\otimes\mathrm{Id}_E\geqslant_{\mathrm{Grif}}\delta\,\omega\otimes\mathrm{Id}_E>0,$$

we find

$$\mathrm{i}\Theta_L+r\,\mathrm{i}\,d'd''\log|s|^2\geqslant r\delta\,\omega.$$

The compactness of X implies $\mathrm{i}\Theta_E\leqslant C\omega\otimes\mathrm{Id}_E$ for some $C>0$. Theorem 12.6 can thus be applied with $\alpha=r\delta/C$ and Corollary 12.8 follows. By Remark 12.7 (b), the above surjectivity property even holds if $L^{1/r}\otimes E^*$ is just assumed to be ample (in the sense that the associated line bundle $\pi^*L^{1/r}\otimes\mathcal{O}_{P(E)}(1)$ is positive on the projectivized bundle $\pi: P(E)\to X$ of lines of E). □

Another interesting corollary is the following special case, dealing with bounded pseudoconvex domains $\Omega\subset\subset\mathbb{C}^n$. Even this simple version retains highly interesting information on the behavior of holomorphic and plurisubharmonic functions.

(12.9) Corollary. *Let $\Omega\subset\mathbb{C}^n$ be a bounded pseudoconvex domain, and let $Y\subset X$ be a nonsingular complex submanifold defined by a section s of some Hermitian vector bundle E with* bounded *curvature tensor on Ω. Assume that s is everywhere transverse to the zero section and that $|s|\leqslant e^{-1}$ on Ω. Then there is a constant $C>0$ (depending only on E), with the following property: for every psh function φ on Ω, every holomorphic function f on Y with $\int_Y|f|^2|\Lambda^r(ds)|^{-2}e^{-\varphi}dV_Y<+\infty$, there exists an extension F of f to Ω such that*

$$\int_\Omega\frac{|F|^2}{|s|^{2r}(-\log|s|)^2}e^{-\varphi}dV_\Omega\leqslant C\int_Y\frac{|f|^2}{|\Lambda^r(ds)|^2}e^{-\varphi}dV_Y.$$

Proof. We apply essentially the same idea as for the previous corollary, in the special case when $L=\Omega\times\mathbb{C}$ is the trivial bundle equipped with a weight function $e^{-\varphi-A|z|^2}$. The choice of a sufficiently large constant $A>0$ guarantees that the curvature assumption 12.6 (a) is satisfied (A just depends on the presupposed bound for the curvature tensor of E). □

(12.10) Remark. The special case when $Y=\{z_0\}$ is a point is especially interesting. In that case, we just take $s(z)=(e\,\mathrm{diam}\,\Omega)^{-1}(z-z_0)$, viewed as a section of the rank $r = n$ trivial vector bundle $\Omega \times \mathbb{C}^n$ with $|s| \leqslant e^{-1}$. We take $\alpha = 1$ and replace $|s|^{2n}(-\log|s|)^2$ in the denominator by $|s|^{2(n-\varepsilon)}$, using the inequality

$$-\log|s| = \frac{1}{\varepsilon}\log|s|^{-\varepsilon} \leqslant \frac{1}{\varepsilon}|s|^{-\varepsilon}, \qquad \forall\varepsilon > 0.$$

For any given value f_0, we then find a holomorphic function f such that $f(z_0) = f_0$ and

$$\int_\Omega \frac{|f(z)|^2}{|z - z_0|^{2(n-\varepsilon)}} e^{-\varphi(z)} dV_\Omega \leqslant \frac{C_n}{\varepsilon^2(\operatorname{diam}\Omega)^{2(n-\varepsilon)}}|f_0|^2 e^{-\varphi(z_0)}.$$

12.D. Skoda's Division Theorem for Ideals of Holomorphic Functions

Following a strategy inpired by T. Ohsawa [Ohs02; Ohs04], we reprove here a version of Skoda's division theorem for ideals of holomorphic functions, by reducing it to an extension problem. Our approach uses Manivel's version of the extension theorem presented above, and leads to results very close to those of Skoda [Sko80], albeit somewhat weaker.

Let (X, ω) be a Kähler manifold, $\dim X = n$, and let $g : E \to Q$ a holomorphic morphism of Hermitian vector bundles over X. Assume for a moment that g is *everywhere surjective*. Given a holomorphic line bundle $L \to X$, we are interested in conditions insuring that the induced morphism $g : H^0(X, K_X \otimes E \otimes L) \to H^0(X, K_X \otimes Q \otimes L)$ is also surjective (as is observed frequently in algebraic geometry, it will be easier to twist by an adjoint line bundle $K_X \otimes L$ than by L alone). For that purpose, it is natural to consider the subbundle $S = \operatorname{Ker} g \subset E$ and the exact sequence

(12.11) $$0 \longrightarrow S \xrightarrow{\ j\ } E \xrightarrow{\ g\ } Q \longrightarrow 0$$

where $j : S \to E$ is the inclusion, as well as the dual exact sequence

(12.11′) $$0 \longrightarrow Q^* \xrightarrow{\ g^*\ } E^* \xrightarrow{\ j^*\ } S^* \longrightarrow 0,$$

which we will twist by suitable line bundles. The main idea of [Ohs02; Ohs04] is that finding a lifting of a section by g is essentially equivalent to extending the related section on $\mathcal{Y} = P(Q^*) = \mathbb{P}(Q)$ to $\mathcal{X} = P(E^*) = \mathbb{P}(E)$, using the obvious embedding $\mathcal{Y} \subset \mathcal{X}$ of the projectivized bundles. In fact, if $r_S = r_E - r_Q$ are the respective ranks of our vector bundles, we have the classical formula

(12.12) $$K_{\mathcal{X}} = K_{\mathbb{P}(E)} = \pi^*(K_X \otimes \det E) \otimes \mathcal{O}_{\mathbb{P}(E)}(-r_S)$$

where $\pi : \mathbb{P}(E) \to X$ is the canonical projection. Therefore, since E coincides with the direct image sheaf $\pi_*\mathcal{O}_{\mathbb{P}(E)}(1)$, a section of $H^0(X, K_X \otimes E \otimes L)$ can also be seen as a section of

(12.13) $$H^0(\mathcal{X}, K_{\mathcal{X}} \otimes \mathcal{O}_{\mathcal{X}}(r_S + 1) \otimes \pi^*(L \otimes \det E^{-1})).$$

Now, since $\mathcal{O}_{\mathcal{X}}(1)_{\restriction cY} = \mathcal{O}_{\mathcal{Y}}(1) = \mathcal{O}_{\mathbb{P}(Q)}(1)$, the lifting problem is equivalent to extending to $\mathcal{X}$ a section of the line bundle $(K_{\mathcal{X}} \otimes \mathcal{L})_{\restriction \mathcal{Y}}$ where $\mathcal{L} = \mathcal{O}_{\mathcal{X}}(r_S + 1) \otimes \pi^*(L \otimes \det E^{-1})$. As a submanifold, $\mathcal{Y}$ is the zero locus of the bundle morphism

$$\mathcal{O}_{\mathbb{P}(E)}(-1) \hookrightarrow \pi^* E^* \to \pi^*(E^*/Q^*) = \pi^* S^*,$$

hence it is the (transverse) zero locus of a naturally defined section

(12.14) $$s \in H^0(\mathcal{X}, \mathcal{E}) \qquad \text{where} \quad \mathcal{E} := \pi^* S^* \otimes \mathcal{O}_{\mathbb{P}(E)}(1).$$

Let us assume that E is endowed with a smooth Hermitian metric h such that $\Theta_{E,h}$ is Griffiths semi-positive. We equip Q with the quotient metric and S, $\mathcal{O}_{\mathbb{P}(E)}(1)$, $\det E$, $\mathcal{E}$ (...) with the induced metrics. A sufficient curvature condition needed to apply the Ohsawa-Takegoshi-Manivel extension theorem is

$$\mathrm{i}\Theta_{\mathcal{L}} + r_S \mathrm{i}d'd'' \log|s|^2 \geqslant \varepsilon \frac{\{\mathrm{i}\Theta_{\mathcal{E}} s, s\}}{|s|^2}$$

for $\varepsilon > 0$ small enough (i.e. in some range $\varepsilon \in [0, \varepsilon_0]$, $\varepsilon_0 \leqslant 1$). Since $\mathrm{i}d'd'' \log|s|^2 \geqslant -\mathrm{i}\Theta_{\mathcal{O}(\mathbb{P}(E)}(1) - \frac{\{\mathrm{i}\Theta_{\pi^* S^*} s, s\}}{|s|^2}$, we obtain the sufficient condition

(12.15)
$$\pi^* \mathrm{i}\Theta_{L \otimes \det E^{-1}} + (1 - \varepsilon)\mathrm{i}\Theta_{\mathcal{O}_{\mathbb{P}(E)}(1)} - (r_S + \varepsilon)\frac{\{\mathrm{i}\Theta_{\pi^* S^*} s, s\}}{|s|^2} \geqslant 0, \qquad \varepsilon \in [0, \varepsilon_0].$$

The assumption that E is Griffiths semi-positive implies $\mathrm{i}\Theta_{\det E} \geqslant 0$, $\mathrm{i}\Theta_{\mathcal{O}_{\mathbb{P}(E)}(1)} \geqslant 0$ and also

(12.16) $$\frac{\{\mathrm{i}\Theta_{\pi^* S^*} s, s\}}{|s|^2} \leqslant \mathrm{i}\Theta_{\det Q}.$$

In fact this is equivalent to proving that $S \otimes \det Q$ is Griffiths semi-positive, but we have in fact $S \otimes \det Q = S \otimes \det S^{-1} \otimes \det E = \Lambda^{r_S - 1} S^* \otimes \det E$, which is a quotient of $\Lambda^{r_S - 1} E^* \otimes \det E = \Lambda^{r_E - r_S + 1} E \geqslant 0$. This shows that (12.15) is implied by the simpler condition

(12.17) $$\mathrm{i}\Theta_L \geqslant \mathrm{i}\Theta_{\det E} + (r_S + \varepsilon_0)\mathrm{i}\Theta_{\det Q},$$

in particular $L = \det E \otimes (\det Q)^k$, $k > r_S$, satisfies the curvature condition. We derive from there:

(12.18) Theorem. *Assume that (X, ω) is a Kähler manifold possessing a complete Kähler metric $\widehat{\omega}$, and let $g : E \to Q$ be a surjective morphism of holomorphic vector bundles, where (E, h_E) is a Griffiths semi-positive Hermitian bundle. Consider a Hermitian holomorphic line bundle (L, h_L) such that*

$$\mathrm{i}\Theta_L - (r_S + \varepsilon)\mathrm{i}\Theta_{\det Q} - \mathrm{i}\Theta_{\det E} \geqslant 0, \qquad r_S = r_E - r_Q, \quad \varepsilon > 0.$$

Then for every L^2 holomorphic section $f \in H^0(X, K_X \otimes Q \otimes L)$ there exists a L^2 holomorphic section $h \in H^0(X, K_X \otimes E \otimes L)$ such that $f = g \cdot h$ and $\|h\|^2 \leqslant C_{n,r_E,\varepsilon}\|f\|^2$.

Proof. We apply Theorem 12.6 with respect to the data $(\mathcal{X}, \mathcal{Y}, \mathcal{E}, \mathcal{L})$ and $\alpha = \varepsilon^{-1}$, $r = r_S$. Since $|s| \leqslant 1$, we have to multiply s by $\delta = \exp(-1/\varepsilon)$ to enforce hypothesis 12.6 (b). This affects the final estimate only as far as the term $\log|s|$ is concerned, since both $|s|^{2r}$ and $|\Lambda^r(ds)|^2 = 1$ are multiplied by δ^{2r}. Finally, we apply Fubini's theorem to reduce integrals over $\mathcal{X}$ or $\mathcal{Y}$ to integrals over X, observing that all fibers of $\mathcal{X} = \mathbb{P}(E) \to X$ are isometric and therefore produce the same fiber integral. Theorem 12.18 follows. By exercising a little more care in the estimates, one sees that the constant $C_{n,r_E,\varepsilon}$ is actually bounded by $C_{n,r_E}\varepsilon^{-2}$, where the ε^{-2} comes from the term $(-\log|s|)^2$, after s has been multiplied by $\exp(-1/\varepsilon)$. □

Skoda's original method is slightly more accurate. It shows that one can take $C_{n,r_E,\varepsilon} = \varepsilon^{-1}$, and, more importantly, replaces the curvature hypothesis by the weaker one

$$(12.19)\qquad \begin{aligned} &\mathrm{i}\Theta_L - (k+\varepsilon)\mathrm{i}\Theta_{\det Q} - \mathrm{i}\Theta_{\det E} \geqslant 0, \\ &\text{where } k = \min(r_S, n), \quad r_S = r_E - r_Q, \quad n = \dim X, \quad \varepsilon > 0, \end{aligned}$$

which does not seem so easy to obtain with the present method. It is however possible to get estimates also when Q is endowed with a metric given a priori, that can be distinct from the quotient metric of E by g. Then the map $g^*(gg^*)^{-1} : Q \longrightarrow E$ is the lifting of Q orthogonal to $S = \operatorname{Ker} g$. The quotient metric $|\bullet|'$ on Q is therefore defined in terms of the original metric $|\bullet|$ by

$$|v|'^2 = |g^*(gg^*)^{-1}v|^2 = \langle (gg^*)^{-1}v, v\rangle = \det(gg^*)^{-1}\,\langle \widetilde{gg^*}v, v\rangle$$

where $\widetilde{gg^*} \in \operatorname{End}(Q)$ denotes the endomorphism of Q whose matrix is the transposed comatrix of gg^*. For every $w \in \det Q$, we find

$$|w|'^2 = \det(gg^*)^{-1}\,|w|^2.$$

If Q' denotes the bundle Q with the quotient metric, we get

$$\mathrm{i}\Theta_{\det Q'} = \mathrm{i}\Theta_{\det Q} + \mathrm{i}d'd''\log\det(gg^*).$$

In order that the hypotheses of Theorem 12.18 be satisfied, we are led to define a new metric $|\bullet|'$ on L by $|u|'^2 = |u|^2\left(\det(gg^*)\right)^{-m-\varepsilon}$. Then

$$\mathrm{i}\Theta_{L'} = \mathrm{i}\Theta_L + (m+\varepsilon)\,\mathrm{i}d'd''\log\det(gg^*) \geqslant (m+\varepsilon)\,\mathrm{i}\Theta_{\det Q'}.$$

Theorem 12.18 applied to (E, Q', L') can now be reformulated:

(12.20) Theorem. *Let X be a weakly pseudoconvex manifold equipped with a Kähler metric ω, let $E \to Q$ be a generically surjective morphism of Hermitian* *vector bundles with E Griffiths semi-positive, and let $L \to X$ be a Hermitian holomorphic line bundle. Assume that*

$$\mathrm{i}\Theta_L - (r_S + \varepsilon)\mathrm{i}\Theta_{\det Q} - \mathrm{i}\Theta_{\det E} \geqslant 0, \qquad r_S = r_E - r_Q, \;\; \varepsilon > 0.$$

Then for every holomorphic section f of $K_X \otimes Q \otimes L$ such that

$$I = \int_X \langle \widetilde{gg^*} f, f\rangle\,(\det gg^*)^{-r_S-1-\varepsilon}\,dV < +\infty,$$

there exists a holomorphic section of $K_X \otimes E \otimes L$ such that $f = g \cdot h$ and

$$\int_X |h|^2\,(\det gg^*)^{-r_S-\varepsilon}\,dV \leqslant C_{n,r_E,\varepsilon}\,I.$$

In case Q is of rank 1, the estimate reduces to

$$\int_X |h|^2\,|g|^{-2r_S-2\varepsilon}\,dV \leqslant C_{n,r_E,\varepsilon} \int_X |f|^2\,|g|^{-2(r_S+1)-2\varepsilon}\,dV.$$

Proof. if $Z \subset X$ is the analytic locus where $g : E \to Q$ is not surjective and $X_c = \{\psi < c\}$ is an exhaustion of X by weakly pseudoconvex relatively compact open subsets, we exploit here the fact that $X_c \smallsetminus Z$ carries a complete metric (see [Dem82b]). It is easy to see that the L^2 conditions forces a section defined a priori only on $X \smallsetminus Z$ to extend to X. □

The special case where $E = \mathcal{O}_\Omega^{\oplus p}$ and $Q = \mathcal{O}_\Omega$ are trivial bundles over a weakly pseudocovex open set $\Omega \subset \mathbb{C}^n$ is already a quite substantial theorem, which goes back to [Sko72b]. In this case, we take L to be the Hermitian line bundle $(\mathcal{O}_\Omega, e^{-\varphi})$ associated with an arbitrary plurisubharmonic function φ on Ω.

(12.21) Corollary (Skoda's division theorem)**.** *Let $f, g_1, \ldots, g_p$ be holomorphic functions on a weakly pseudoconvex open set $\Omega \subset \mathbb{C}^n$ such that*

$$\int_\Omega |f|^2 |g|^{-2(p+1)-2\varepsilon} e^{-\varphi} dV < +\infty$$

for some plurisubharmonic function φ. Then there exist holomorphic functions h_j, $1 \leqslant j \leqslant p$, such that $f = \sum g_j h_j$ on Ω, and

$$\int_X |h|^2\,|g|^{-2(p-1)-2\varepsilon} e^{-\varphi}\,dV \leqslant C_{n,p,\varepsilon} \int_X |f|^2\,|g|^{-2p-2\varepsilon} e^{-\varphi}\,dV.$$

Chapter 13

Approximation of Closed Positive Currents by Analytic Cycles

13.A. Approximation of Plurisubharmonic Functions Via Bergman kernels

We prove here, as an application of the Ohsawa-Takegoshi extension theorem, that every psh function on a pseudoconvex open set $\Omega \subset \mathbb{C}^n$ can be approximated very accurately by functions of the form $c \log |f|$, where $c > 0$ and f is a holomorphic function. The main idea is taken from [Dem92]. For other applications to algebraic geometry, see [Dem93b] and Demailly-Kollár [DK01]. Recall that the Lelong number of a function $\varphi \in \mathrm{Psh}(\Omega)$ at a point x_0 is defined to be

$$\nu(\varphi, x_0) = \liminf_{z \to x_0} \frac{\varphi(z)}{\log |z - x_0|} = \lim_{r \to 0_+} \frac{\sup_{B(x_0,r)} \varphi}{\log r}. \tag{13.1}$$

In particular, if $\varphi = \log |f|$ with $f \in \mathcal{O}(\Omega)$, then $\nu(\varphi, x_0)$ is equal to the vanishing order

$$\mathrm{ord}_{x_0}(f) = \sup\{k \in \mathbb{N} \,;\, D^\alpha f(x_0) = 0,\ \forall |\alpha| < k\}.$$

(13.2) Theorem. *Let φ be a plurisubharmonic function on a bounded pseudoconvex open set $\Omega \subset \mathbb{C}^n$. For every $m > 0$, let $\mathcal{H}_\Omega(m\varphi)$ be the Hilbert space of holomorphic functions f on Ω such that $\int_\Omega |f|^2 e^{-2m\varphi} d\lambda < +\infty$ and let $\varphi_m = \frac{1}{2m} \log \sum |\sigma_\ell|^2$ where (σ_ℓ) is an orthonormal basis of $\mathcal{H}_\Omega(m\varphi)$. Then there are constants $C_1, C_2 > 0$ independent of m such that*

(a) $\displaystyle \varphi(z) - \frac{C_1}{m} \leqslant \varphi_m(z) \leqslant \sup_{|\zeta - z| < r} \varphi(\zeta) + \frac{1}{m} \log \frac{C_2}{r^n}$ *for every $z \in \Omega$ and $r < d(z, \partial\Omega)$. In particular, φ_m converges to φ pointwise and in L^1_{loc} topology on Ω when $m \to +\infty$ and*

(b) $\displaystyle \nu(\varphi, z) - \frac{n}{m} \leqslant \nu(\varphi_m, z) \leqslant \nu(\varphi, z)$ *for every $z \in \Omega$.*

Proof. (a) Note that $\sum |\sigma_\ell(z)|^2$ is the square of the norm of the evaluation linear form $\mathrm{ev}_z : f \mapsto f(z)$ on $\mathcal{H}_\Omega(m\varphi)$, since $\sigma_\ell(z) = \mathrm{ev}_z(\sigma_\ell)$ is the ℓ-th coordinate of ev_z in the orthonormal basis (σ_ℓ). In other words, we have

$$\sum |\sigma_\ell(z)|^2 = \sup_{f \in B(1)} |f(z)|^2$$

where $B(1)$ is the unit ball of $\mathcal{H}_\Omega(m\varphi)$ (The sum is called the *Bergman kernel* associated with $\mathcal{H}_\Omega(m\varphi)$). As φ is locally bounded from above, the L^2 topology is actually stronger than the topology of uniform convergence on compact subsets of Ω. It follows that the series $\sum |\sigma_\ell|^2$ converges uniformly on Ω and that its sum is real analytic. Moreover, by what we just explained, we have

$$\varphi_m(z) = \sup_{f\in B(1)} \frac{1}{m}\log|f(z)|.$$

For $z_0 \in \Omega$ and $r < d(z_0, \partial\Omega)$, the mean value inequality applied to the psh function $|f|^2$ implies

$$\begin{aligned} |f(z_0)|^2 &\leqslant \frac{1}{\pi^n r^{2n}/n!}\int_{|z-z-0|<r} |f(z)|^2 d\lambda(z) \\ &\leqslant \frac{1}{\pi^n r^{2n}/n!}\exp\Big(2m \sup_{|z-z_0|<r}\varphi(z)\Big)\int_\Omega |f|^2 e^{-2m\varphi}d\lambda. \end{aligned}$$

If we take the supremum over all $f \in B(1)$ we get

$$\varphi_m(z_0) \leqslant \sup_{|z-z_0|<r}\varphi(z) + \frac{1}{2m}\log\frac{1}{\pi^n r^{2n}/n!}$$

and the second inequality in (a) is proved — as we see, this is an easy consequence of the mean value inequality. Conversely, the Ohsawa-Takegoshi extension theorem (Corollary 12.9) applied to the 0-dimensional subvariety $\{z_0\} \subset \Omega$ shows that for any $a \in \mathbb{C}$ there is a holomorphic function f on Ω such that $f(z_0) = a$ and

$$\int_\Omega |f|^2 e^{-2m\varphi}d\lambda \leqslant C_3|a|^2 e^{-2m\varphi(z_0)},$$

where C_3 only depends on n and $\operatorname{diam}\Omega$. We fix a such that the right hand side is 1. Then $\|f\| \leqslant 1$ and so we get

$$\varphi_m(z_0) \geqslant \frac{1}{m}\log|f(z_0)| = \frac{1}{m}\log|a| = \varphi(z) - \frac{\log C_3}{2m}.$$

The inequalities given in (a) are thus proved. Taking $r = 1/m$, we find that $\lim_{m\to+\infty}\sup_{|\zeta-z|<1/m}\varphi(\zeta) = \varphi(z)$ by the upper semicontinuity of φ, and therefore $\lim\varphi_m(z) = \varphi(z)$, since $\lim\frac{1}{m}\log(C_2 m^n) = 0$.

(b) The above estimates imply

$$\sup_{|z-z_0|<r}\varphi(z) - \frac{C_1}{m} \leqslant \sup_{|z-z_0|<r}\varphi_m(z) \leqslant \sup_{|z-z_0|<2r}\varphi(z) + \frac{1}{m}\log\frac{C_2}{r^n}.$$

After dividing by $\log r < 0$ when $r \to 0$, we infer

$$\frac{\sup_{|z-z_0|<2r}\varphi(z) + \frac{1}{m}\log\frac{C_2}{r^n}}{\log r} \leqslant \frac{\sup_{|z-z_0|<r}\varphi_m(z)}{\log r} \leqslant \frac{\sup_{|z-z_0|<r}\varphi(z) - \frac{C_1}{m}}{\log r},$$

and from this and definition (13.1), it follows immediately that

$$\nu(\varphi, x) - \frac{n}{m} \leqslant \nu(\varphi_m, z) \leqslant \nu(\varphi, z). \qquad \square$$

Theorem 13.2 implies in a straightforward manner the deep result of [Siu74] on the analyticity of the Lelong number upperlevel sets.

(13.3) Corollary ([Siu74]). *Let φ be a plurisubharmonic function on a complex manifold X. Then, for every $c > 0$, the Lelong number upperlevel set*

$$E_c(\varphi) = \{z \in X\,;\ \nu(\varphi, z) \geqslant c\}$$

is an analytic subset of X.

Proof. Since analyticity is a local property, it is enough to consider the case of a psh function φ on a pseudoconvex open set $\Omega \subset \mathbb{C}^n$. The inequalities obtained in Theorem 13.2 (b) imply that

$$E_c(\varphi) = \bigcap_{m \geqslant m_0} E_{c-n/m}(\varphi_m).$$

Now, it is clear that $E_c(\varphi_m)$ is the analytic set defined by the equations $\sigma_\ell^{(\alpha)}(z) = 0$ for all multi-indices α such that $|\alpha| < mc$. Thus $E_c(\varphi)$ is analytic as a (countable) intersection of analytic sets. $\square$

(13.4) Remark. It can be easily shown that the Lelong numbers of any closed positive (p,p)-current coincide (at least locally) with the Lelong numbers of a suitable plurisubharmonic potential φ (see [Sko72a]). Hence Siu's theorem also holds true for the Lelong number upperlevel sets $E_c(T)$ of any closed positive (p,p)-current T.

13.B. Global Approximation of Closed (1,1)-Currents on a Compact Complex Manifold

We take here X to be an arbitrary compact complex manifold (no Kähler assumption is needed). Now, let T be a closed $(1,1)$-current on X. We assume that T is *almost positive*, i.e. that there exists a $(1,1)$-form γ with continuous coefficients such that $T \geqslant \gamma$; the case of positive currents ($\gamma = 0$) is of course the most important.

(13.5) Lemma. *There exists a smooth closed* $(1,1)$*-form α representing the same $\partial\overline{\partial}$-cohomology class as T and an* almost psh *function φ on X such that $T = \alpha + \frac{\mathrm{i}}{\pi}\partial\overline{\partial}\varphi$. (We say that a function φ is almost psh if its complex Hessian is bounded below by a* $(1,1)$*-form with locally bounded coefficients, that is, if $\mathrm{i}\partial\overline{\partial}\varphi$ is almost positive).*

Proof. Select an open covering (U_j) of X by coordinate balls such that $T = \frac{i}{\pi}\partial\overline{\partial}\varphi_j$ over U_j, and construct a global function $\varphi = \sum \theta_j\varphi_j$ by means of a partition of unity $\{\theta_j\}$ subordinate to U_j. Now, we observe that $\varphi - \varphi_k$ is smooth on U_k because all differences $\varphi_j - \varphi_k$ are smooth in the intersections $U_j \cap U_k$ and $\varphi - \varphi_k = \sum \theta_j(\varphi_j - \varphi_k))$. Therefore $\alpha := T - \frac{i}{\pi}\partial\overline{\partial}\varphi$ is smoth. □

By replacing T with $T - \alpha$ and γ with $\gamma - \alpha$, we can assume without loss of generality that $\{T\} = 0$, i.e. that $T = \frac{i}{\pi}\partial\overline{\partial}\varphi$ with an almost psh function φ on X such that $\frac{i}{\pi}\partial\overline{\partial}\varphi \geqslant \gamma$.

Our goal is to approximate T in the weak topology by currents $T_m = \frac{rmi}{\pi}\partial\overline{\partial}\varphi_m$ such their potentials φ_m have analytic singularities in the sense of Definition 1.10, more precisely, defined on a neighborhood V_{x_0} of any point $x_0 \in X$ in the form $\varphi_m(z) = c_m \log \sum_j |\sigma_{j,m}|^2 + O(1)$, where $c_m > 0$ and the $\sigma_{j,m}$ are holomorphic functions on V_{x_0}.

We select a finite covering (W_ν) of X with open coordinate charts. Given $\delta > 0$, we take in each W_ν a maximal family of points with (coordinate) distance to the boundary $\geqslant 3\delta$ and mutual distance $\geqslant \delta/2$. In this way, we get for $\delta > 0$ small a finite covering of X by open balls U'_j of radius δ (actually every point is even at distance $\leqslant \delta/2$ of one of the centers, otherwise the family of points would not be maximal), such that the concentric ball U_j of radius 2δ is relatively compact in the corresponding chart W_ν. Let $\tau_j : U_j \longrightarrow B(a_j, 2\delta)$ be the isomorphism given by the coordinates of W_ν. Let $\varepsilon(\delta)$ be a modulus of continuity for γ on the sets U_j, such that $\lim_{\delta\to 0}\varepsilon(\delta) = 0$ and $\gamma_x - \gamma_{x'} \leqslant \frac{1}{2}\varepsilon(\delta)\,\omega_x$ for all $x, x' \in U_j$. We denote by γ_j the $(1,1)$-form with constant coefficients on $B(a_j, 2\delta)$ such that $\tau_j^*\gamma_j$ coincides with $\gamma - \varepsilon(\delta)\,\omega$ at $\tau_j^{-1}(a_j)$. Then we have

$$0 \leqslant \gamma - \tau_j^*\gamma_j \leqslant 2\varepsilon(\delta)\,\omega \quad \text{on } U'_j \tag{13.6}$$

for $\delta > 0$ small. We set $\varphi_j = \varphi \circ \tau_j^{-1}$ on $B(a_j, 2\delta)$ and let q_j be the homogeneous quadratic function in $z - a_j$ such that $\frac{i}{\pi}\partial\overline{\partial}q_j = \gamma_j$ on $B(a_j, 2\delta)$. Finally, we set

$$\psi_j(z) = \varphi_j(z) - q_j(z) \quad \text{on } B(a_j, 2\delta). \tag{13.7}$$

Then ψ_j is plurisubharmonic, since

$$\frac{i}{\pi}\partial\overline{\partial}(\psi_j \circ \tau_j) = T - \tau_j^*\gamma_j \geqslant \gamma - \tau_j^*\gamma_j \geqslant 0.$$

We let $U'_j \subset\subset U''_j \subset\subset U_j$ be concentric balls of radii δ, $1.5\,\delta$, 2δ respectively. On each open set U_j the function $\psi_j := \varphi - q_j \circ \tau_j$ defined in (13.7) is plurisubharmonic, so Theorem 13.2 applied with $\Omega = U_j$ produces functions

$$\psi_{j,m} = \frac{1}{2m}\log\sum_\ell |\sigma_{j,\ell}|^2, \quad (\sigma_{j,\ell}) = \text{basis of } \mathcal{H}_{U_j}(m\psi_j). \tag{13.8}$$

These functions approximate ψ_j as m tends to $+\infty$ and satisfy the inequalities

$$\psi_j(x) - \frac{C_1}{m} \leqslant \psi_{j,m}(x) \leqslant \sup_{|\zeta - x| < r} \psi_j(\zeta) + \frac{1}{m}\log\frac{C_2}{r^n}. \tag{13.9}$$

The functions $\psi_{j,m} + q_j \circ \tau_j$ on U_j then have to be glued together by a partition of unity technique. For this, we rely on the following "discrepancy" lemma, estimating the variation of the approximating functions on overlapping balls.

(13.10) Lemma. *There are constants $C_{j,k}$ independent of m and δ such that the almost psh functions $w_{j,m} = 2m(\psi_{j,m} + q_j \circ \tau_j)$, i.e.*

$$w_{j,m}(x) = 2m\, q_j \circ \tau_j(x) + \log \sum_\ell \big|\sigma_{j,\ell}(x)\big|^2, \qquad x \in U_j'',$$

satisfy

$$|w_{j,m} - w_{k,m}| \leqslant C_{j,k}\big(\log \delta^{-1} + m\varepsilon(\delta)\delta^2\big) \quad \text{on } U_j'' \cap U_k''.$$

Proof. The details will be left as an exercise to the reader. The main idea is the following: for any holomorphic function $f_j \in \mathcal{H}_{U_j}(m\psi_j)$, a $\overline{\partial}$ equation $\overline{\partial} u = \overline{\partial}(\theta f_j)$ can be solved on U_k, where θ is a cut-off function with support in $U_j'' \cap U_k''$, on a ball of radius $< \delta/4$, equal to 1 on the ball of radius $\delta/8$ centered at a given point $x_0 \in U_j'' \cap U_k''$. We apply the L^2 estimate with respect to the weight $(n+1)\log|x - x_0|^2 + 2m\psi_k$, where the first term is picked up so as to force the solution u to vanish at x_0, in such a way that $F_k = u - \theta f_j$ is holomorphic and $F_k(x_0) = f_j(x_0)$. The discrepancy between the weights on U_j'' and U_k'' is

$$\psi_j(x) - \psi_k(x) = -\big(q_j \circ \tau_j(x) - q_k \circ \tau_k(x)\big)$$

and the $\partial\overline{\partial}$ of this difference is $O(\varepsilon(\delta))$, so it is easy to correct the discrepancy up to a $O(\varepsilon(\delta)\delta^2)$ term by multiplying our functions by an invertible holomorphic function G_{jk}. In this way, we get a uniform L^2 control on the L^2 norm of the solution $f_k = G_{jk}F_k = G_{jk}(u - \theta f_j)$ of the form

$$\int_{U_k} |f_k|^2 e^{-2m\psi_k} \leqslant C_{j,k}\delta^{-2n-4} e^{mO(\varepsilon(\delta)\delta^2)} \int_{U_j} |f_j|^2 e^{-2m\psi_j}.$$

The required estimate follows, using the fact that

$$e^{2m\psi_{j,m}(x)} = \sum_\ell |\sigma_{j,\ell}(x)|^2 = \sup_{f \in \mathcal{H}_{U_j}(m\psi_j),\, \|f\| \leqslant 1} |f(x)|^2 \quad \text{on } U_j,$$

and the analogous equality on U_k. □

Now, the actual glueing of our almost psh functions is performed using the following elementary partition of unity calculation.

(13.11) Lemma. *Let $U_j' \subset\subset U_j''$ be locally finite open coverings of a complex manifold X by relatively compact open sets, and let θ_j be smooth nonnegative functions with support in U_j'', such that $\theta_j \leqslant 1$ on U_j'' and $\theta_j = 1$ on U_j'. Let $A_j \geqslant 0$ be such that*

$$\mathrm{i}(\theta_j \partial\overline{\partial}\theta_j - \partial\theta_j \wedge \overline{\partial}\theta_j) \geqslant -A_j\omega \quad \text{on } U_j'' \smallsetminus U_j'$$

for some positive $(1,1)$*-form* ω. *Finally, let* w_j *be almost psh functions on* U_j *with the property that* $i\partial\overline{\partial} w_j \geqslant \gamma$ *for some real* $(1,1)$*-form* γ *on* M, *and let* C_j *be constants such that*

$$w_j(x) \leqslant C_j + \sup_{k\neq j,\, U'_k \ni x} w_k(x) \quad \text{on } U''_j \smallsetminus U'_j.$$

Then the function $w = \log\big(\sum \theta_j^2 e^{w_j}\big)$ *is almost psh and satisfies*

$$i\partial\overline{\partial} w \geqslant \gamma - 2\Big(\sum_j 1\!\!1_{U''_j \smallsetminus U'_j} A_j e^{C_j}\Big)\omega.$$

Proof. If we set $\alpha_j = \theta_j \partial w_j + 2\partial\theta_j$, a straightforward computation shows that

$$\partial w = \frac{\sum(\theta_j^2 \partial w_j + 2\theta_j\partial\theta_j)e^{w_j}}{\sum \theta_j^2 e^{w_j}} = \frac{\sum \theta_j e^{w_j}\alpha_j}{\sum\theta_j^2 e^{w_j}},$$

$$\partial\overline{\partial} w = \frac{\sum\big(\alpha_j \wedge\overline{\alpha}_j + \theta_j^2\partial\overline{\partial}w_j + 2\theta_j\partial\overline{\partial}\theta_j - 2\partial\theta_j\wedge\overline{\partial}\theta_j\big)e^{w_j}}{\sum\theta_j^2 e^{w_j}} - \frac{\sum_{j,k}\theta_j e^{w_j}\theta_k e^{w_k}\alpha_j\wedge\overline{\alpha}_k}{\big(\sum\theta_j^2 e^{w_j}\big)^2}$$

$$= \frac{\sum_{j<k}\big|\theta_j\alpha_k - \theta_k\alpha_j\big|^2 e^{w_j}e^{w_k}}{\big(\sum\theta_j^2 e^{w_j}\big)^2} + \frac{\sum\theta_j^2 e^{w_j}\partial\overline{\partial}w_j}{\sum\theta_j^2 e^{w_j}} + \frac{\sum\big(2\theta_j\partial\overline{\partial}\theta_j - 2\partial\theta_j\wedge\overline{\partial}\theta_j\big)e^{w_j}}{\sum\theta_j^2 e^{w_j}}$$

by using the Legendre identity. The first term in the last line is nonnegative and the second one is $\geqslant \gamma$. In the third term, if x is in the support of $\theta_j\partial\overline{\partial}\theta_j - \partial\theta_j\wedge\overline{\partial}\theta_j$, then $x \in U''_j \smallsetminus U'_j$ and so $w_j(x) \leqslant C_j + w_k(x)$ for some $k \neq j$ with $U'_k \ni x$ and $\theta_k(x) = 1$. This gives

$$i\frac{\sum\big(2\theta_j\partial\overline{\partial}\theta_j - 2\partial\theta_j\wedge\overline{\partial}\theta_j\big)e^{w_j}}{\sum\theta_j^2 e^{w_j}} \geqslant -2\sum_j 1\!\!1_{U''_j \smallsetminus U'_j} e^{C_j} A_j\omega.$$

The expected lower bound follows. □

We apply Lemma 13.11 to functions $\widetilde{w}_{j,m}$ which are just slight modifications of the functions $w_{j,m} = 2m(\psi_{j,m} + q_j \circ \tau_j)$ occurring in (13.10) :

$$\begin{aligned}\widetilde{w}_{j,m}(x) &= w_{j,m}(x) + 2m\Big(\frac{C_1}{m} + C_3\varepsilon(\delta)(\delta^2/2 - |\tau_j(x)|^2)\Big)\\ &= 2m\Big(\psi_{j,m}(x) + q_j\circ\tau_j(x) + \frac{C_1}{m} + C_3\varepsilon(\delta)(\delta^2/2 - |\tau_j(x)|^2)\Big)\end{aligned}$$

where $x \mapsto z = \tau_j(x)$ is a local coordinate identifying U_j to $B(0, 2\delta)$, C_1 is the constant occurring in (13.9) and C_3 is a sufficiently large constant. It is easy to see that we can take $A_j = C_4\delta^{-2}$ in Lemma 13.11. We have

$$\widetilde{w}_{j,m} \geqslant w_{j,m} + 2C_1 + m\frac{C_3}{2}\varepsilon(\delta)\delta^2 \quad \text{on } B(x_j, \delta/2) \subset U'_j,$$

since $|\tau_j(x)| \leqslant \delta/2$ on $B(x_j, \delta/2)$, while

$$\widetilde{w}_{j,m} \leqslant w_{j,m} + 2C_1 - mC_3\varepsilon(\delta)\delta^2 \quad \text{on } U''_j \smallsetminus U'_j.$$

For $m \geqslant m_0(\delta) = (\log \delta^{-1}/(\varepsilon(\delta)\delta^2)$, Lemma 13.10 implies $|w_{j,m} - w_{k,m}| \leqslant C_5 m\varepsilon(\delta)\delta^2$ on $U''_j \cap U''_k$. Hence, for C_3 large enough, we get

$$\widetilde{w}_{j,m}(x) \leqslant \sup_{k\neq j,\, B(x_k,\delta/2)\ni x} w_{k,m}(x) \leqslant \sup_{k\neq j,\, U'_k \ni x} w_{k,m}(x) \quad \text{on } U''_j \smallsetminus U'_j,$$

and we can take $C_j = 0$ in the hypotheses of Lemma 13.11. The associated function $w = \log\big(\sum \theta_j^2 e^{\widetilde{w}_{j,m}}\big)$ is given by

$$w = \log \sum_j \theta_j^2 \exp\Big(2m\big(\psi_{j,m} + q_j \circ \tau_j + \frac{C_1}{m} + C_3\varepsilon(\delta)(\delta^2/2 - |\tau_j|^2)\big)\Big).$$

If we define $\varphi_m = \dfrac{1}{2m} w$, we get

$$\varphi_m(x) := \frac{1}{2m} w(x) \geqslant \psi_{j,m}(x) + q_j \circ \tau_j(x) + \frac{C_1}{m} + \frac{C_3}{4}\varepsilon(\delta)\delta^2 > \varphi(x)$$

in view of (13.9), by picking an index j such that $x \in B(x_j, \delta/2)$. In the opposite direction, the maximum number N of overlapping balls U_j does not depend on δ, and we thus get

$$w \leqslant \log N + 2m\Big(\max_j \{\psi_{j,m}(x) + q_j \circ \tau_j(x)\} + \frac{C_1}{m} + \frac{C_3}{2}\varepsilon(\delta)\delta^2\Big).$$

By definition of ψ_j we have $\sup_{|\zeta - x|<r} \psi_j(\zeta) \leqslant \sup_{|\zeta - x|<r} \varphi(\zeta) - q_j \circ \tau_j(x) + C_5 r$ thanks to the uniform Lipschitz continuity of $q_j \circ \tau_j$, thus by (13.9) again we find

$$\varphi_m(x) \leqslant \frac{\log N}{2m} + \sup_{|\zeta - x|<r} \varphi(\zeta) + \frac{C_1}{m} + \frac{1}{m}\log\frac{C_2}{r^n} + \frac{C_3}{2}\varepsilon(\delta)\delta^2 + C_5 r.$$

By taking for instance $r = 1/m$ and $\delta = \delta_m \to 0$, we see that φ_m converges to φ. On the other hand (13.6) implies $\frac{\mathrm{i}}{\pi}\partial\overline{\partial} q_j \circ \tau_j(x) = \tau_j^* \gamma_j \geqslant \gamma - 2\varepsilon(\delta)\omega$, thus

$$\frac{\mathrm{i}}{\pi}\partial\overline{\partial}\widetilde{w}_{j,m} \geqslant 2m\big(\gamma - C_6\varepsilon(\delta)\omega\big).$$

Lemma 13.11 then produces the lower bound

$$\frac{\mathrm{i}}{\pi}\partial\overline{\partial} w \geqslant 2m\big(\gamma - C_6\varepsilon(\delta)\omega\big) - C_7\delta^{-2}\omega,$$

whence

$$\frac{\mathrm{i}}{\pi}\partial\overline{\partial}\varphi_m \geqslant \gamma - C_8\varepsilon(\delta)\omega$$

for $m \geqslant m_0(\delta) = (\log \delta^{-1})/(\varepsilon(\delta)\delta^2)$. We can fix $\delta = \delta_m$ to be the smallest value of $\delta > 0$ such that $m_0(\delta) \leqslant m$, then $\delta_m \to 0$ and we have obtained a sequence of quasi psh functions φ_m satisfying the following properties.

(13.12) Theorem. *Let φ be an almost psh function on a compact complex manifold X such that $\frac{i}{\pi}\partial\overline{\partial}\varphi \geqslant \gamma$ for some continuous $(1,1)$-form γ. Then there is a sequence of almost psh functions φ_m such that φ_m has the same singularities as a logarithm of a sum of squares of holomorphic functions and a decreasing sequence $\varepsilon_m > 0$ converging to 0 such that*

(a) $\varphi(x) < \varphi_m(x) \leqslant \sup_{|\zeta - x| < r} \varphi(\zeta) + C\Big(\frac{|\log r|}{m} + r + \varepsilon_m\Big)$
with respect to coordinate open sets covering X. In particular, φ_m converges to φ pointwise and in $L^1(X)$ and

(b) $\nu(\varphi, x) - \frac{n}{m} \leqslant \nu(\varphi_m, x) \leqslant \nu(\varphi, x)$ *for every $x \in X$;*

(c) $\frac{i}{\pi}\partial\overline{\partial}\varphi_m \geqslant \gamma - \varepsilon_m\omega$.

In particular, we can apply this to an arbitrary positive or almost positive closed $(1,1)$-current $T = \alpha + \frac{i}{\pi}\partial\overline{\partial}\varphi$.

(13.13) Corollary. *Let T be an almost positive closed $(1,1)$-current on a compact complex manifold X such that $T \geqslant \gamma$ for some continuous $(1,1)$-form γ. Then there is a sequence of currents T_m whose local potentials have the same singularities as $1/m$ times a logarithm of a sum of squares of holomorphic functions and a decreasing sequence $\varepsilon_m > 0$ converging to 0 such that*

(a) *T_m converges weakly to T,*

(b) $\nu(T, x) - \frac{n}{m} \leqslant \nu(T_m, x) \leqslant \nu(T, x)$ *for every $x \in X$;*

(c) $T_m \geqslant \gamma - \varepsilon_m\omega$.

We say that our currents T_m are approximations of T possessing logarithmic poles.

By using blow-ups of X, the structure of the currents T_m can be better understood. In fact, consider the coherent ideals $\mathcal{J}_m$ generated locally by the holomorphic functions $(\sigma_{j,m}^{(k)})$ on U_k in the local approximations

$$\varphi_{k,m} = \frac{1}{2m} \log \sum_j |\sigma_{j,m}^{(k)}|^2 + O(1)$$

of the potential φ of T on U_k. These ideals are in fact globally defined, because the local ideals $\mathcal{J}_m^{(k)} = (\sigma_{j,m}^{(k)})$ are integrally closed, and they coincide on the intersections $U_k \cap U_\ell$ as they have the same order of vanishing by the proof of Lemma 13.10. By Hironaka [Hir64], we can find a composition of blow-ups with smooth centers $\mu_m : \widetilde{X}_m \to X$ such that $\mu_m^* \mathcal{J}_m$ is an invertible ideal sheaf associated with a normal crossing divisor D_m. Now, we can write

$$\mu_m^* \varphi_{k,m} = \varphi_{k,m} \circ \mu_m = \frac{1}{m} \log |s_{D_m}| + \widetilde{\varphi}_{k,m}$$

where s_{D_m} is the canonical section of $\mathcal{O}(-D_m)$ and $\widetilde{\varphi}_{k,m}$ is a smooth potential. This implies

(13.14) $$\mu_m^* T_m = \frac{1}{m}[D_m] + \beta_m$$

where $[D_m]$ is the current of integration over D_m and β_m is a smooth closed $(1,1)$-form which satisfies the lower bound $\beta_m \geqslant \mu_m^*(\gamma - \varepsilon_m\omega)$. (Recall that the pull-back of a closed $(1,1)$-current by a holomorphic map f is always well-defined, by taking a local plurisubharmonic potential φ such that $T = \mathrm{i}\partial\overline{\partial}\varphi$ and writing $f^*T = \mathrm{i}\partial\overline{\partial}(\varphi \circ f)$). In the remainder of this section, we derive from this a rather important geometric consequence, first appeared in [DP04]). We need two related definitions.

(13.15) Definition. *A Kähler current on a compact complex space X is a closed positive current T of bidegree $(1,1)$ which satisfies $T \geqslant \varepsilon\omega$ for some $\varepsilon > 0$ and some smooth positive Hermitian form ω on X.*

(13.16) Definition. *A compact complex manifold is said to be in the* Fujiki class $\mathcal{C}$) *if it is bimeromorphic to a Kähler manifold (or equivalently, using Hironaka's desingularization theorem, if it admits a proper Kähler modification).*

(13.17) Theorem. *A compact complex manifold X is bimeromorphic to a Kähler manifold (i.e. $X \in \mathcal{C}$) if and only if it admits a Kähler current.*

Proof. If X is bimeromorphic to a Kähler manifold Y, Hironaka's desingularization theorem implies that there exists a blow-up $\widetilde{Y}$ of Y (obtained by a sequence of blow-ups with smooth centers) such that the bimeromorphic map from Y to X can be resolved into a modification $\mu : \widetilde{Y} \to X$. Then $\widetilde{Y}$ is Kähler and the push-forward $T = \mu_*\widetilde{\omega}$ of a Kähler form $\widetilde{\omega}$ on $\widetilde{Y}$ provides a Kähler current on X. In fact, if ω is a smooth Hermitian form on X, there is a constant C such that $\mu^*\omega \leqslant C\widetilde{\omega}$ (by compactness of $\widetilde{Y}$), hence

$$T = \mu_*\widetilde{\omega} \geqslant \mu_*(C^{-1}\mu^*\omega) = C^{-1}\omega.$$

Conversely, assume that X admits a Kähler current $T \geqslant \varepsilon\omega$. By Theorem 13.13 (c), there exists a Kähler current $\widetilde{T} = T_m \geqslant \frac{\varepsilon}{2}\omega$ (with $m \gg 1$ so large that $\varepsilon_m \leqslant \varepsilon/2$) in the same $\partial\overline{\partial}$-cohomology class as T, possessing logarithmic poles. Observation (13.14) implies the existence of a composition of blow-ups $\mu : \widetilde{X} \to X$ such that

$$\mu^*\widetilde{T} = [\widetilde{D}] + \widetilde{\beta} \qquad \text{on } \widetilde{X},$$

where $\widetilde{D}$ is a $\mathbb{Q}$-divisor with normal crossings and $\widetilde{\beta}$ a smooth closed $(1,1)$-form such that $\widetilde{\beta} \geqslant \frac{\varepsilon}{2}\mu^*\omega$. In particular $\widetilde{\beta}$ is positive outside the exceptional locus of μ. This is not enough yet to produce a Kähler form on $\widetilde{X}$, but we are not very far. Suppose that $\widetilde{X}$ is obtained as a tower of blow-ups

$$\widetilde{X} = X_N \to X_{N-1} \to \cdots \to X_1 \to X_0 = X,$$

where X_{j+1} is the blow-up of X_j along a smooth center $Y_j \subset X_j$. Denote by $E_{j+1} \subset X_{j+1}$ the exceptional divisor, and let $\mu_j : X_{j+1} \to X_j$ be the blow-up map. Now, we use the following simple

(13.18) Lemma. *For every Kähler current T_j on X_j, there exists $\varepsilon_{j+1} > 0$ and a smooth form u_{j+1} in the $\partial\overline{\partial}$-cohomology class of $[E_{j+1}]$ such that*

$$T_{j+1} = \mu_j^* T_j - \varepsilon_{j+1} u_{j+1}$$

is a Kähler current on X_{j+1}.

Proof. The line bundle $\mathcal{O}(-E_{j+1})|E_{j+1}$ is equal to $\mathcal{O}_{P(N_j)}(1)$ where N_j is the normal bundle to Y_j in X_j. Pick an arbitrary smooth Hermitian metric on N_j, use this metric to get an induced Fubini-Study metric on $\mathcal{O}_{P(N_j)}(1)$, and finally extend this metric as a smooth Hermitian metric on the line bundle $\mathcal{O}(-E_{j+1})$. Such a metric has positive curvature along tangent vectors of X_{j+1} which are tangent to the fibers of $E_{j+1} = P(N_j) \to Y_j$. Assume furthermore that $T_j \geqslant \delta_j \omega_j$ for some Hermitian form ω_j on X_j and a suitable $0 < \delta_j \ll 1$. Then

$$\mu_j^* T_j - \varepsilon_{j+1} u_{j+1} \geqslant \delta_j \mu_j^* \omega_j - \varepsilon_{j+1} u_{j+1}$$

where $\mu_j^* \omega_j$ is semi-positive on X_{j+1}, positive definite on $X_{j+1} \smallsetminus E_{j+1}$, and also positive definite on tangent vectors of $T_{X_{j+1}|E_{j+1}}$ which are not tangent to the fibers of $E_{j+1} \to Y_j$. The statement is then easily proved by taking $\varepsilon_{j+1} \ll \delta_j$ and by using an elementary compactness argument on the unit sphere bundle of $T_{X_{j+1}}$ associated with any given Hermitian metric. □

End of proof of Theorem 13.17. If $\widetilde{u}_j$ is the pull-back of u_j to the final blow-up $\widetilde{X}$, we conclude inductively that $\mu^* \widetilde{T} - \sum \varepsilon_j \widetilde{u}_j$ is a Kähler current. Therefore the smooth form

$$\widetilde{\omega} := \widetilde{\beta} - \sum \varepsilon_j \widetilde{u}_j = \mu^* \widetilde{T} - \sum \varepsilon_j \widetilde{u}_j - [\widetilde{D}]$$

is Kähler and we see that $\widetilde{X}$ is a Kähler manifold. □

(13.19) Remark. A special case of Theorem 13.16 is the following characterization of Moishezon varieties (i.e. manifolds which are bimeromorphic to projective algebraic varieties or, equivalently, whose algebraic dimension is equal to their complex dimension):

A compact complex manifold X is Moishezon if and only if X possesses a Kähler current T such that the De Rham cohomology class $\{T\}$ is rational, i.e. $\{T\} \in H^2(X, \mathbb{Q})$.

In fact, in the above proof, we get an integral current T if we take the push forward $T = \mu_* \widetilde{\omega}$ of an integral ample class $\{\widetilde{\omega}\}$ on Y, where $\mu : Y \to X$ is a projective model of Y. Conversely, if $\{T\}$ is rational, we can take the $\varepsilon_j' s$ to be rational in Lemma 3.5. This produces at the end a Kähler metric $\widetilde{\omega}$ with rational De Rham cohomology class on $\widetilde{X}$. Therefore $\widetilde{X}$ is projective by the Kodaira embedding theorem. This result was already observed in [JS93] (see also [Bon93; Bon98] for a more general perspective based on a singular version of holomorphic Morse inequalities).

13.C. Global Approximation by Divisors

We now translate our previous approximation theorems into a more algebro-geometric setting. Namely, we assume that T is a closed positive $(1,1)$-current which belongs to the first Chern class $c_1(L)$ of a holomorphic line bundle L, and we assume here X to be algebraic (i.e. projective or at the very least Moishezon).

Our goal is to show that T can be approximated by divisors which have roughly the same Lelong numbers as T. The existence of weak approximations by divisors has already been proved in [Lel72] for currents defined on a pseudoconvex open set $\Omega \subset \mathbb{C}^n$ with $H^2(\Omega, \mathbb{R}) = 0$, and in [Dem92, 93b] in the situation considered here (cf. also [Dem82b], although the argument given there is somewhat incorrect). We take the opportunity to present here a slightly simpler derivation.

Let X be a projective manifold and L a line bundle over X. A singular Hermitian metric h on L is a metric such that the weight function φ of h is L^1_{loc} in any local trivialization (such that $L_{|U} \simeq U \times \mathbb{C}$ and $\|\xi\|_h = |\xi| e^{-\varphi(x)}$, $\xi \in L_x \simeq \mathbb{C}$). The curvature of L can then be computed in the sense of distributions

$$T = \frac{\mathrm{i}}{2\pi}\Theta_{L,h} = \frac{\mathrm{i}}{\pi}\partial\overline{\partial}\varphi,$$

and L is said to be pseudo-effective if L admits a singular Hermitian metric h such that the curvature current $T = \frac{\mathrm{i}}{2\pi}\Theta_{L,h}$ is semi-positive [The weight functions φ of L are thus plurisubharmonic]. In what follows, we sometimes use an additive notation for $\mathrm{Pic}(X)$, i.e. kL is meant for the line bundle $L^{\otimes k}$.

We will also make use of the concept of complex singularity exponent, following e.g. [Var82, 83], [ArGV85] and [DK01]. A quasi-plurisubharmonic (quasi-psh) function is by definition a function φ which is locally equal to the sum of a psh function and of a smooth function, or equivalently, a locally integrable function φ such that $\mathrm{i}\partial\overline{\partial}\varphi$ is locally bounded below by $-C\omega$ where ω is a Hermitian metric and C a constant.

(13.20) Definition. *If K is a compact subset of X and φ is a quasi-psh function defined near K, we define*

(a) *The complex singularity exponent $c_K(\varphi)$ to be the supremum of all positive numbers c such that $e^{-2c\varphi}$ is integrable in a neighborhood of every point $z_0 \in K$, with respect to the Lebesgue measure in holomorphic coordinates centered at z_0. In particular $c_K(\varphi) = \inf_{z_0 \in K}(\varphi)$.*

(b) *The concept is easily extended to Hermitian metrics $h = e^{-2\varphi}$ by putting $c_K(h) = c_K(\varphi)$, to holomorphic functions f by $c_K(f) = c_K(\log|f|)$, to coherent ideals $\mathcal{J} = (g_1, \ldots, g_N)$ by $c_K(\mathcal{J}) = c_K(\varphi)$ where $\varphi = \frac{1}{2}\log\sum|g_j|^2$. Also* *for an effective $\mathbb{R}$-divisor D, we put $c_K(D) = c_K(\log|\sigma_D|)$ where σ_D is the canonical section.*

The main technical result of this chapter can be stated as follows, in the case of big line bundles (cf. Proposition 6.14 (f)).

(13.21) Theorem. *Let L be a line bundle on a compact complex manifold X possessing a singular Hermitian metric h with $\Theta_{L,h} \geqslant \varepsilon\omega$ for some $\varepsilon > 0$ and some smooth positive definite Hermitian $(1,1)$-form ω on X. For every real number $m > 0$, consider the space $\mathcal{H}_m = H^0(X, L^{\otimes m} \otimes \mathcal{I}(h^m))$ of holomorphic sections σ of $L^{\otimes m}$ on X such that*

$$\int_X |\sigma|^2_{h^m} dV_\omega = \int_X |\sigma|^2 e^{-2m\varphi} dV_\omega < +\infty,$$

where $dV_\omega = \frac{1}{m!}\omega^m$ is the Hermitian volume form. Then for $m \gg 1$, $\mathcal{H}_m$ is a non zero finite dimensional Hilbert space and we consider the closed positive $(1,1)$-current

$$T_m = \frac{\mathrm{i}}{\pi}\partial\overline{\partial}\Big(\frac{1}{2m}\log\sum_k |g_{m,k}|^2\Big) = \frac{\mathrm{i}}{\pi}\partial\overline{\partial}\Big(\frac{1}{2m}\log\sum_k |g_{m,k}|^2_h\Big) + \Theta_{L,h}$$

where $(g_{m,k})_{1\leqslant k\leqslant N(m)}$ is an orthonormal basis of $\mathcal{H}_m$. Then :

(a) *For every trivialization $L_{|U} \simeq U \times \mathbb{C}$ on a cordinate open set U of X and every compact set $K \subset U$, there are constants $C_1, C_2 > 0$ independent of m and φ such that*

$$\varphi(z) - \frac{C_1}{m} \leqslant \psi_m(z) := \frac{1}{2m}\log\sum_k |g_{m,k}(z)|^2 \leqslant \sup_{|x-z|<r}\varphi(x) + \frac{1}{m}\log\frac{C_2}{r^n}$$

for every $z \in K$ and $r \leqslant \frac{1}{2}d(K, \partial U)$. In particular, ψ_m converges to φ pointwise and in L^1_{loc} topology on Ω when $m \to +\infty$, hence T_m converges weakly to $T = \Theta_{L,h}$.

(b) *The Lelong numbers $\nu(T,z) = \nu(\varphi,z)$ and $\nu(T_m,z) = \nu(\psi_m,z)$ are related by*

$$\nu(T,z) - \frac{n}{m} \leqslant \nu(T_m,z) \leqslant \nu(T,z) \quad \text{for every } z \in X.$$

(c) *For every compact set $K \subset X$, the complex singularity exponents of the metrics given locally by $h = e^{-2\varphi}$ and $h_m = e^{-2\psi_m}$ satisfy*

$$c_K(h)^{-1} - \frac{1}{m} \leqslant c_K(h_m)^{-1} \leqslant c_K(h)^{-1}.$$

Proof. The major part of the proof is a variation of the arguments already explained in Section 13.A.

(a) We note that $\sum |g_{m,k}(z)|^2$ is the square of the norm of the evaluation linear form $\sigma \mapsto \sigma(z)$ on $\mathcal{H}_m$, hence

$$\psi_m(z) = \sup_{\sigma \in B(1)} \frac{1}{m} \log |\sigma(z)|$$

where $B(1)$ is the unit ball of $\mathcal{H}_m$. For $r \leqslant \frac{1}{2} d(K, \partial\Omega)$, the mean value inequality applied to the plurisubharmonic function $|\sigma|^2$ implies

$$\begin{aligned} |\sigma(z)|^2 &\leqslant \frac{1}{\pi^n r^{2n}/n!} \int_{|x-z|<r} |\sigma(x)|^2 d\lambda(x) \\ &\leqslant \frac{1}{\pi^n r^{2n}/n!} \exp\Big(2m \sup_{|x-z|<r} \varphi(x)\Big) \int_\Omega |\sigma|^2 e^{-2m\varphi} d\lambda. \end{aligned}$$

If we take the supremum over all $\sigma \in B(1)$ we get

$$\psi_m(z) \leqslant \sup_{|x-z|<r} \varphi(x) + \frac{1}{2m} \log \frac{1}{\pi^n r^{2n}/n!}$$

and the right hand inequality in (a) is proved. Conversely, the Ohsawa-Takegoshi extension theorem [OhT87], [Ohs88] applied to the 0-dimensional subvariety $\{z\} \subset U$ shows that for any $a \in \mathbb{C}$ there is a holomorphic function f on U such that $f(z) = a$ and

$$\int_U |f|^2 e^{-2m\varphi} d\lambda \leqslant C_3 |a|^2 e^{-2m\varphi(z)},$$

where C_3 only depends on n and $\operatorname{diam} U$. Now, provided a remains in a compact set $K \subset U$, we can use a cut-off function θ with support in U and equal to 1 in a neighborhood of a, and solve the $\overline{\partial}$-equation $\overline{\partial} g = \overline{\partial}(\theta f)$ in the L^2 space associated with the weight $2m\varphi + 2(n+1)|\log|z-a|$, that is, the singular Hermitian metric $h(z)^m |z-a|^{-2(n+1)}$ on $L^{\otimes m}$. For this, we apply the standard Andreotti-Vesentini-Hörmander L^2 estimates (see e.g. [Dem82b] for the required version). This is possible for $m \geqslant m_0$ thanks to the hypothesis that $\Theta_{L,h} \geqslant \varepsilon\omega > 0$, even if X is non Kähler (X is in any event a Moishezon variety from our assumptions). The bound m_0 depends only on ε and the geometry of a finite covering of X by compact sets $K_j \subset U_j$, where U_j are coordinate balls (say); it is independent of the point a and even of the metric h. It follows that $g(a) = 0$ and therefore $\sigma = \theta f - g$ is a holomorphic section of $L^{\otimes m}$ such that

$$\int_X |\sigma|^2_{h^m} dV_\omega = \int_X |\sigma|^2 e^{-2m\varphi} dV_\omega \leqslant C_4 \int_U |f|^2 e^{-2m\varphi} dV_\omega \leqslant C_5 |a|^2 e^{-2m\varphi(z)},$$

in particular $\sigma \in \mathcal{H}_m = H^0(X, L^{\otimes m} \otimes \mathcal{I}(h^m))$. We fix a such that the right hand side is 1. This gives the inequality

$$\psi_m(z) \geqslant \frac{1}{m} \log |a| = \varphi(z) - \frac{\log C_5}{2m}$$

which is the left hand part of statement (a).

(b) The first inequality in (a) implies $\nu(\psi_m, z) \leqslant \nu(\varphi, z)$. In the opposite direction, we find

$$\sup_{|x-z|<r} \psi_m(x) \leqslant \sup_{|x-z|<2r} \varphi(x) + \frac{1}{m} \log \frac{C_2}{r^n}.$$

Divide by $\log r < 0$ and take the limit as r tends to 0. The quotient by $\log r$ of the supremum of a psh function over $B(x, r)$ tends to the Lelong number at x. Thus we obtain

$$\nu(\psi_m, x) \geqslant \nu(\varphi, x) - \frac{n}{m}.$$

(c) Again, the first inequality (in (a) immediately yields $h_m \leqslant C_6 h$, hence $c_K(h_m) \geqslant c_K(h)$. For the converse inequality, since we have

$$c_{\cup K_j}(h) = \min_j c_{K_j}(h),$$

we can assume without loss of generality that K is contained in a trivializing open patch U of L. Let us take $c < c_K(\psi_m)$. Then, by definition, if $V \subset X$ is a sufficiently small open neighborhood of K, the Hölder inequality for the conjugate exponents $p = 1 + mc^{-1}$ and $q = 1 + m^{-1}c$ implies, thanks to equality $\frac{1}{p} = \frac{c}{mq}$,

$$\begin{aligned}
\int_V e^{-2(m/p)\varphi} dV_\omega &= \int_V \Big(\sum_{1 \leqslant k \leqslant N(m)} |g_{m,k}|^2 e^{-2m\varphi} \Big)^{1/p} \Big(\sum_{1 \leqslant k \leqslant N(m)} |g_{m,k}|^2 \Big)^{-c/mq} dV_\omega \\
&\leqslant \bigg(\int_X \sum_{1 \leqslant k \leqslant N(m)} |g_{m,k}|^2 e^{-2m\varphi} dV_\omega \bigg)^{1/p} \bigg(\int_V \Big(\sum_{1 \leqslant k \leqslant N(m)} |g_{m,k}|^2 \Big)^{-c/m} dV_\omega \bigg)^{1/q} \\
&= N(m)^{1/p} \bigg(\int_V \Big(\sum_{1 \leqslant k \leqslant N(m)} |g_{m,k}|^2 \Big)^{-c/m} dV_\omega \bigg)^{1/q} < +\infty.
\end{aligned}$$

From this we infer $c_K(h) \geqslant m/p$, i.e., $c_K(h)^{-1} \leqslant p/m = 1/m + c^{-1}$. As $c < c_K(\psi_m)$ was arbitrary, we get $c_K(h)^{-1} \leqslant 1/m + c_K(h_m)^{-1}$ and the inequalities of (c) are proved. □

(13.22) Remark. The proof would also work, with a few modifications, when X is a Stein manifold and L is an arbitrary holomorphic line bundle.

(13.23) Corollary. *Let $L \to X$ be a holomorphic line bundle and $T = \frac{i}{2\pi}\Theta_{L,h}$ the curvature current of some singular Hermitian metric h on L.*

(a) *If L is big and $\Theta_{L,h} \geqslant \varepsilon\omega > 0$, there exists a sequence of holomorphic sections $h_s \in H^0(X, q_s L)$ with $\lim q_s = +\infty$ such that the $\mathbb{Q}$-divisors $D_s = \frac{1}{q_s}\mathrm{div}(h_s)$ satisfy $T = \lim D_s$ in the weak topology and $\sup_{x \in X} |\nu(D_s, x) - \nu(T, x)| \to 0$ as $s \to +\infty$.*

(b) *If L is just pseudo-effective and $\Theta_{L,h} \geqslant 0$, for any ample line bundle A, there exists a sequence of non zero sections $h_s \in H^0(X, p_sA + q_sL)$ with $p_s, q_s > 0$, $\lim q_s = +\infty$ and $\lim p_s/q_s = 0$, such that the divisors $D_s = \frac{1}{q_s}\mathrm{div}(h_s)$ satisfy $T = \lim D_s$ in the weak topology and $\sup_{x\in X} |\nu(D_s, x) - \nu(T, x)| \to 0$ as $s \to +\infty$.*

Proof. Part (b) is a rather straightforward consequence of part (a) applied to $mL + A$ and $T_m = \frac{1}{m}\Theta_{mL+A,h^m h_A} = T + \frac{1}{m}\Theta_{A,h_A} \to T$ when m tends to infinity. Therefore, it suffices to prove (a).

(a) By Theorem 13.20, we can find sections $g_1, \ldots, g_N \in H^0(X, mL)$ (omitting the index m for simplicity of notation), such that

$$T_m = \frac{\mathrm{i}}{\pi}\partial\overline{\partial}\Big(\frac{1}{2m}\log \sum_{1\leqslant j\leqslant N} |g_j|_h^2\Big) + \Theta_{L,h} = \frac{\mathrm{i}}{\pi}\partial\overline{\partial}\Big(\frac{1}{2m}\log \sum_{1\leqslant j\leqslant N} |g_j|^2\Big)$$

converges weakly to T and satisfies $\nu(T, x) - n/m \leqslant \nu(T_m, x) \leqslant \nu(T, x)$. In fact, since the number N of sections grows at most as $O(m^n)$, we can replace $\sum_{1\leqslant j\leqslant N} |g_j|^2$ by $\max_{1\leqslant j\leqslant N} |g_j|^2$, as the difference of the potentials tends uniformly to 0 with the help of the renormalizing constant $\frac{1}{2m}$. Hence, we can use instead the approximating currents

$$\widetilde{T}_m = \frac{\mathrm{i}}{\pi}\partial\overline{\partial}u_m, \qquad u_m = \frac{1}{m}\log \max_{1\leqslant j\leqslant N} |g_j|.$$

Now, as L is big, by the proof of (6.17b) we can write $k_0L = A + D$ where A is an ample divisor and D is an effective divisor, for some $k_0 > 0$. By enlarging k_0, we can even assume that A is very ample. Let σ_D be the canonical section of D and let $h_1, \ldots, h_N$ be sections of $H^0(X, A)$. We get a section of $H^0(X, (m\ell + k_0)L)$ by considering

$$u_{\ell,m} = (g_1^\ell h_1 + \cdots + g_N^\ell h_N)\sigma_D$$

By enlarging N if necessary and putting e.g. $g_j = g_N$ for $j > N$, we can assume that the sections h_j generate all 1-jets of sections of A at every point (actually, one can always achieve this with $n + 1$ sections only, so this is not really a big demand). Then, for almost every N-tuple $(h_1, \ldots, h_N)$, Lemma 13.24 and the weak continuity of $\partial\overline{\partial}$ imply that

$$\Delta_{\ell,m} = \frac{1}{\ell m}\frac{\mathrm{i}}{\pi}\partial\overline{\partial}\log|u_{\ell,m}| = \frac{1}{\ell m}\mathrm{div}(u_{\ell,m})$$

converges weakly to $\widetilde{T}_m = \frac{\mathrm{i}}{\pi}\partial\overline{\partial}u_m$ as ℓ tends to $+\infty$, and that

$$\nu(T_m, x) \leqslant \nu\Big(\frac{1}{\ell m}\Delta_{\ell,m}, x\Big) \leqslant \nu(T, x) + \frac{\mu + 1}{\ell m},$$

where $\mu = \max_{x\in X} \mathrm{ord}_x(\sigma_D)$. This, together with the first step, implies the proposition for some subsequence $D_s = \Delta_{\ell(s),s}$, $\ell(s) \gg s \gg 1$. We even obtain the more explicit inequality

$$\nu(T, x) - \frac{n}{m} \leqslant \nu\Big(\frac{1}{\ell m}\Delta_{\ell,m}, x\Big) \leqslant \nu(T, x) + \frac{\mu + 1}{\ell m}. \qquad \square$$

(13.24) Lemma. *Let Ω be an open subset in $\mathbb{C}^n$ and let $g_1, \ldots, g_N \in H^0(\Omega, \mathcal{O}_\Omega)$ be non zero functions. Let $S \subset H^0(\Omega, \mathcal{O}_\Omega)$ be a finite dimensional subspace whose elements generate all 1-jets at any point of Ω. Finally, set $u = \log \max_j |g_j|$ and*

$$u_\ell = g_1^\ell h_1 + \cdots + g_N^\ell h_N, \qquad h_j \in S \smallsetminus \{0\}.$$

Then for all $(h_1, \ldots, h_N)$ in $(S \smallsetminus \{0\})^N$ except a set of measure 0, the sequence $\frac{1}{\ell}\log|u_\ell|$ converges to u in $L^1_{\mathrm{loc}}(\Omega)$ and

$$\nu(u,x) \leqslant \nu\Big(\frac{1}{\ell}\log|u_\ell|\Big) \leqslant \nu(u,x) + \frac{1}{\ell}, \qquad \forall x \in X, \ \ \forall \ell \geqslant 1.$$

Proof. The sequence $\frac{1}{\ell}\log|u_\ell|$ is locally uniformly bounded above and we have

$$\lim_{\ell\to+\infty} \frac{1}{\ell}\log\big|u_\ell(z)\big| = u(z)$$

at every point z where all absolute values $|g_j(z)|$ are distinct and all $h_j(z)$ are nonzero. This is a set of full measure in Ω because the sets $\{|g_j|^2 = |g_l|^2,\ j \neq l\}$ and $\{h_j = 0\}$ are real analytic and thus of zero measure (without loss of generality, we may assume that Ω is connected and that the g_j's are not pairwise proportional). The well-known uniform integrability properties of plurisubharmonic functions then show that $\frac{1}{\ell}\log|u_\ell|$ converges to u in $L^1_{\mathrm{loc}}(\Omega)$. It is easy to see that $\nu(u,x)$ is the minimum of the vanishing orders $\mathrm{ord}_x(g_j)$, hence

$$\nu(\log|u_\ell|, x) = \mathrm{ord}_x(u_\ell) \geqslant \ell\,\nu(u,x).$$

In the opposite direction, consider the set $\mathcal{E}_\ell$ of all $(N+1)$-tuples

$$(x, h_1, \ldots, h_N) \in \Omega \times S^N$$

for which $\nu(\log|u_\ell|, x) \geqslant \ell\,\nu(u,x) + 2$. Then $\mathcal{E}_\ell$ is a constructible set in $\Omega \times S^N$: it has a locally finite stratification by analytic sets, since

$$\mathcal{E}_\ell = \bigcup_{s\geqslant 0}\Big(\bigcup_{j,\,|\alpha|=s} \{x\,;\ D^\alpha g_j(x) \neq 0\} \times S^N\Big) \bigcup \bigcap_{|\beta|\leqslant \ell s+1} \{(x,(h_j))\,;\ D^\beta u_\ell(x) = 0\}.$$

The fiber $\mathcal{E}_\ell \cap (\{x\}\times S^N)$ over a point $x \in \Omega$ where $\nu(u,x) = \min \mathrm{ord}_x(g_j) = s$ is the vector space of N-tuples $(h_j) \in S^N$ satisfying the equations $D^\beta\big(\sum g_j^\ell h_j(x)\big) = 0$, $|\beta| \leqslant \ell s + 1$. However, if $\mathrm{ord}_x(g_j) = s$, the linear map

$$(0, \ldots, 0, h_j, 0, \ldots, 0) \longmapsto \big(D^\beta(g_j^\ell h_j(x))\big)_{|\beta|\leqslant \ell s+1}$$

has rank $n+1$, because it factorizes into an injective map $J^1_x h_j \mapsto J^{\ell s+1}_x(g_j^\ell h_j)$. It follows that the fiber $\mathcal{E}_\ell \cap (\{x\} \times S^N)$ has codimension at least $n+1$. Therefore

$$\dim \mathcal{E}_\ell \leqslant \dim(\Omega \times S^N) - (n+1) = \dim S^N - 1$$

and the projection of $\mathcal{E}_\ell$ on S^N has measure zero by Sard's theorem. By definition of $\mathcal{E}_\ell$, any choice of $(h_1, \ldots, h_N) \in S^N \setminus \bigcup_{\ell \geqslant 1} \mathrm{pr}(\mathcal{E}_\ell)$ produces functions u_ℓ such that $\nu(\log|u_\ell|, x) \leqslant \ell\,\nu(u,x) + 1$ on Ω. □

(13.25) Exercise. When L is ample and h is a smooth metric with $T = \frac{i}{2\pi}\Theta_{L,h} > 0$, show that the approximating divisors can be taken smooth (and thus irreducible if X is connected).

Hint. In the above proof of Corollary 13.23, the sections g_j have no common zeroes and one can take $\sigma_D = 1$. Moreover, a smooth divisor Δ in an ample linear system is always connected, otherwise two disjoint parts Δ', Δ'' would be big and nef and $\Delta' \cdot \Delta'' = 0$ would contradict the Hovanskii-Teissier inequality when X is connected.

(13.26) Corollary. *On a projective manifold X, effective $\mathbb{Q}$-divisors are dense in the weak topology in the cone $P^{1,1}_{\mathrm{NS}}(X)$ of closed positive $(1,1)$-currents T whose cohomology class $\{T\}$ belongs to the Neron-Severi space $\mathrm{NS}_{\mathbb{R}}(X)$.*

Proof. We may add ε times a Kähler metric ω so as to get $T + \varepsilon\omega > 0$, and then perturb by a small combination $\sum \delta_j \alpha_j$ of classes α_j in a $\mathbb{Z}$-basis of $\mathrm{NS}(X)$ so that $\Theta = T + \varepsilon\omega + \sum \delta_j\alpha_j \geqslant \frac{\varepsilon}{2}\omega$ and $\{\Theta\} \in H^2(X,\mathbb{Q})$. Then Θ can be approximated by $\mathbb{Q}$-divisors by Corollary 13.23, and the conclusion follows. □

(13.27) Comments. We can rephrase the above results by saying that the cone of closed positive currents $P^{1,1}_{\mathrm{NS}}(X)$ is a completion of the cone of effective $\mathbb{Q}$-divisors. A considerable advantage of using currents is that the cone of currents is locally compact in the weak topology, namely the section of the cone consisting of currents T of mass $\int_X T \wedge \omega^{n-1} = 1$ is compact. This provides a very strong tool for the study of the asymptotic behavior of linear systems, as required for instance in the Minimal Model Program of Kawamata-Mori-Shokurov. One should be aware, however, that the cone of currents is really huge and contains objects which are very far from being algebraic in any reasonable sense. This occurs very frequently in the realm of complex dynamics. For instance, if $P_m(z,c)$ denotes the m-th iterate of the quadratic polynomial $z \mapsto z^2 + c$, then $P_m(z,c)$ defines a polynomial of degree 2^m on $\mathbb{C}^2$, and the sequence of $\mathbb{Q}$-divisors $D_m = \frac{1}{m}\frac{i}{\pi}\partial\overline{\partial}\log|P_m(z,c)|$ which have mass 1 on $\mathbb{C}^2 \subset \mathbb{P}^2_{\mathbb{C}}$ can be shown to converge to a closed positive current T of mass 1 on $\mathbb{P}^2_{\mathbb{C}}$. The support of this current T is extremely complicated : its slices $c = c_0$ are the Julia sets J_c of the quadratic polynomial $z \mapsto z^2 + c$, and the slice $z = 0$ is the famous Mandelbrot set M. Therefore, in general, limits of divisors in asymptotic linear systems may exhibit a fractal behavior. □

13.D. Singularity Exponents and log Canonical Thresholds

The goal of this section to relate "log canonical thresholds" with the α invariant introduced by G. Tian [Tia87] for the study of the existence of Kähler-Einstein metrics. The approximation technique of closed positive $(1,1)$-currents introduced above can be used to show that the α invariant actually coincides with the log canonical threshold (see also [DK01]; [JK01]; [BGK05]; [Dem08]).

Usually, in these applications, only the case of the anticanonical line bundle $L = -K_X$ is considered. Here we will consider more generally the case of an arbitrary line bundle L (or $\mathbb{Q}$-line bundle L) on a complex manifold X, with some additional restrictions which will be stated later. We introduce a generalized version of Tian's invariant α, as defined in [Tia87] (see also [Siu88]).

(13.28) Definition. *Assume that X is a compact manifold and that L is a pseudo-effective line bundle, i.e. L admits a singular Hermitian metric h_0 with $\Theta_{L,h_0} \geqslant 0$. If K is a compact subset of X, we put*

$$\alpha_K(L) = \inf_{\{h,\,\Theta_{L,h}\geqslant 0\}} c_K(h)$$

where h runs over all singular Hermitian metrics on L such that $\Theta_{L,h} \geqslant 0$.

In algebraic geometry, it is more usual to look instead at linear systems defined by a family of linearly independent sections $\sigma_0, \sigma_1, \ldots, \sigma_N \in H^0(X, L^{\otimes m})$. We denote by Σ the vector subspace generated by these sections and by

$$|\Sigma| := P(\Sigma) \subset |mL| := P(H^0(X, L^{\otimes m}))$$

the corresponding linear system. Such an $(N+1)$-tuple of sections $\sigma = (\sigma_j)_{0\leqslant j\leqslant N}$ defines a singular Hermitian metric h on L by putting in any trivialization

$$|\xi|_h^2 = \frac{|\xi|^2}{\left(\sum_j |\sigma_j(z)|^2\right)^{1/m}} = \frac{|\xi|^2}{|\sigma(z)|^{2/m}}, \qquad \xi \in L_z,$$

hence $h(z) = |\sigma(z)|^{-2/m}$ with $\varphi(z) = \frac{1}{m}\log|\sigma(z)| = \frac{1}{2m}\log\sum_j |\sigma_j(z)|^2$ as the associated weight function. Therefore, we are interested in the number $c_K(|\sigma|^{-2/m})$. In the case of a single section σ_0 (corresponding to a one-point linear system), this is the same as the log canonical threshold $\mathrm{lct}_K(X, \frac{1}{m}D) = c_K(\frac{1}{m}D)$ of the associated divisor D, in the notation of Section 1 of [CS08]. We will also use the formal notation $c_K(\frac{1}{m}|\Sigma|)$ in the case of a higher dimensional linear system $|\Sigma| \subset |mL|$. The main result of this section is

(13.29) Theorem. *Let L be a big line bundle on a compact complex manifold X. Then for every compact set K in X we have*

$$\alpha_K(L) = \inf_{\{h,\,\Theta_{L,h}\geqslant 0\}} c_K(h) = \inf_{m\in\mathbb{Z}_{>0}} \inf_{D\in|mL|} c_K\left(\frac{1}{m}D\right).$$

Proof. Observe that the inequality

$$\inf_{m\in\mathbb{Z}_{>0}} \inf_{D\in|mL|} c_K\left(\frac{1}{m}D\right) \geqslant \inf_{\{h,\,\Theta_{L,h}\geqslant 0\}} c_K(h)$$

is trivial, since any divisor $D \in |mL|$ gives rise to a singular Hermitian metric h.

The converse inequality will follow from the approximation techniques discussed above. Given a big line bundle L on X, there exists a modification $\mu : \widetilde{X} \to X$ of X such that $\widetilde{X}$ is projective and $\mu^* L = \mathcal{O}(A+E)$ where A is an ample divisor and E an effective divisor with rational coefficients. By pushing forward by μ a smooth metric h_A with positive curvature on A, we get a singular Hermitian metric h_1 on L such that $\Theta_{L,h_1} \geqslant \mu_* \Theta_{A,h_A} \geqslant \varepsilon\omega$ on X. Then for any $\delta > 0$ and any singular Hermitian metric h on L with $\Theta_{L,h} \geqslant 0$, the interpolated metric $h_\delta = h_1^\delta h^{1-\delta}$ satisfies $\Theta_{L,h_\delta} \geqslant \delta\varepsilon\omega$. Since h_1 is bounded away from 0, it follows that $c_K(h) \geqslant (1-\delta) c_K(h_\delta)$ by monotonicity. By Theorem 13.21(c) applied to h_δ, we infer

$$c_K(h_\delta) = \lim_{m \to +\infty} c_K(h_{\delta,m}),$$

and we also have

$$c_K(h_{\delta,m}) \geqslant c_K\Big(\frac{1}{m} D_{\delta,m}\Big)$$

for any divisor $D_{\delta,m}$ associated with a section $\sigma \in H^0(X, L^{\otimes m} \otimes \mathcal{I}(h_\delta^m))$, since the metric $h_{\delta,m}$ is given by $h_{\delta,m} = (\sum_k |g_{m,k}|^2)^{-1/m}$ for an orthornormal basis of such sections. This clearly implies

$$c_K(h) \geqslant \liminf_{\delta \to 0} \liminf_{m \to +\infty} c_K\Big(\frac{1}{m} D_{\delta,m}\Big) \geqslant \inf_{m \in \mathbb{Z}_{>0}} \inf_{D \in |mL|} c_K\Big(\frac{1}{m} D\Big). \qquad \square$$

In the applications, it is frequent to have a finite or compact group G of automorphisms of X and to look at G-invariant objects, namely G-equivariant metrics on G-equivariant line bundles L ; in the case of a reductive algebraic group G we simply consider a compact real form $G^{\mathbb{R}}$ instead of G itself.

One then gets an α invariant $\alpha_{K,G}(L)$ by looking only at G-equivariant metrics in Definition 13.28. All contructions made are then G-equivariant, especially $\mathcal{H}_m \subset |mL|$ is a G-invariant linear system. For every G-invariant compact set K in X, we thus infer

(13.30)

$$\alpha_{K,G}(L) := \inf_{\{h\ G\text{-equiv.},\, \Theta_{L,h} \geqslant 0\}} c_K(h) = \inf_{m \in \mathbb{Z}_{>0}} \inf_{|\Sigma| \subset |mL|,\ \Sigma^G = \Sigma} c_K\Big(\frac{1}{m}|\Sigma|\Big).$$

When G is a finite group, one can pick for m large enough a G-invariant divisor $D_{\delta,m}$ associated with a G-invariant section σ, possibly after multiplying m by the order of G. One then gets the slightly simpler equality

$$\text{(13.31)} \qquad \alpha_{K,G}(L) := \inf_{\{h\ G\text{-equiv.},\, \Theta_{L,h} \geqslant 0\}} c_K(h) = \inf_{m \in \mathbb{Z}_{>0}} \inf_{D \in |mL|^G} c_K\Big(\frac{1}{m} D\Big).$$

In a similar manner, one can work on an orbifold X rather than on a non singular variety. The L^2 techniques work in this setting with almost no change (L^2 estimates are essentially insensitive to singularities, since one can just use an orbifold metric on the open set of regular points).

The main interest of Tian's invariant $\alpha_{X,G}$ (and of the related concept of log canonical threshold) is that it provides a neat criterion for the existence of Kähler-Einstein metrics for Fano manifolds (see [Tia87], [Siu88], [Nad89], [DK01]).

(13.32) Theorem. *Let X be a Fano manifold, i.e. a projective manifold with $-K_X$ ample. Assume that X admits a compact group of automorphisms G such that $\alpha_{X,G}(K_X) > n/(n+1)$. Then X possesses a G-invariant Kähler-Einstein metric.*

We will not give here the details of the proof, which rely on very delicate C^k-estimates (successively for $k = 0, 1, 2, \ldots$) for the Monge-Ampère operator. In fine, the required estimates can be shown to depend only on the boundedness of the integral $\int_X e^{-2\gamma\varphi}$ for a suitable constant $\gamma \in \,]\frac{n}{n+1}, 1]$, where φ is the potential of the Kähler metric $\omega \in c_1(X)$ (also viewed as the weight of a Hermitian metric on K_X). Now, one can restrict the estimate to G-invariant weights φ, and this translates into the sufficient condition of Theorem 13.32. The approach explained in [DK01] simplifies the analysis developed in earlier works by proving first a general semi-continuity theorem which implies the desired a priori bound under the assumption of Theorem 13.32. The semi-continuity theorem states as

(13.33) Theorem ([DK01]). *Let K be a compact set in a complex manifold X. Then the map $\varphi \mapsto c_K(\varphi)^{-1}$ is upper semi-continuous with respect to the weak ($= L^1_{\rm loc}$) topology on the space of plurisubharmonic functions. Moreover, if $\gamma < c_K(\varphi)$, then $\int_K |e^{-2\gamma\psi} - e^{-2\gamma\varphi}|$ converges to 0 when ψ converges to φ in the weak topology.*

Sketch of proof. We will content ourselves by explaining the main points. It is convenient to observe (by a quite easy integration argument suggested to us by J. McNeal) that $c_K(\varphi)$ can be calculated by estimating the Lebesgue volume $\mu_U(\{\varphi < \log r\}$ of tubular neighborhoods as $r \to 0$:

(13.34)

$$c_K(\varphi) = \sup\big\{c \geqslant 0\,;\, r^{-2c}\mu_U(\{\varphi < \log r\}) \text{ is bounded as } r \to 0, \text{ for some } U \supset K\big\}.$$

The first step is the following important monotonicity result, which is a straightforward consequence of the L^2 extension theorem.

(13.35) Proposition. *Let φ be a quasi-psh function on a complex manifold X, and let $Y \subset X$ be a complex submanifold such that $\varphi_{|Y} \not\equiv -\infty$ on every connected component of Y. Then, if K is a compact subset of Y, we have*

$$c_K(\varphi_{|Y}) \leqslant c_K(\varphi).$$

(Here, of course, $c_K(\varphi)$ is computed on X, i.e., by means of neighborhoods of K in X).

We need only proving monotonicity for $c_{z_0}(\varphi_{|Y})$ when z_0 is a point of Y. This is done by just extending the holomorphic function $f(z) = 1$ on $B(z_0, r) \cap Y$ with respect to the weight $e^{-2\gamma\varphi}$ whenever $\gamma < c_{z_0}(\varphi_{|Y})$.

(13.36) Proposition. *Let X, Y be complex manifolds of respective dimensions n, m, let $\mathcal{I} \subset \mathcal{O}_X$, $\mathcal{J} \subset \mathcal{O}_Y$ be coherent ideals, and let $K \subset X$, $L \subset Y$ be compact sets. Put $\mathcal{I} \oplus \mathcal{J} := \mathrm{pr}_1^\star\mathcal{I} + \mathrm{pr}_2^\star\mathcal{J} \subset \mathcal{O}_{X\times Y}$. Then*

$$c_{K\times L}(\mathcal{I} \oplus \mathcal{J}) = c_K(\mathcal{I}) + c_L(\mathcal{J}).$$

Proof. It is enough to show that $c_{(x,y)}(\mathcal{I} \oplus \mathcal{J}) = c_x(\mathcal{I}) + c_y(\mathcal{J})$ at every point $(x, y) \in X \times Y$. Without loss of generality, we may assume that $X \subset \mathbb{C}^n$, $Y \subset \mathbb{C}^m$ are open sets and $(x, y) = (0, 0)$. Let $g = (g_1, \ldots, g_p)$, resp. $h = (h_1, \ldots, h_q)$, be systems of generators of $\mathcal{I}$ (resp. $\mathcal{J}$) on a neighborhood of 0. Set

$$\varphi = \log \sum |g_j|, \qquad \psi = \log \sum |h_k|.$$

Then $\mathcal{I} \oplus \mathcal{J}$ is generated by the $p + q$-tuple of functions

$$g \oplus h = (g_1(x), \ldots, g_p(x), h_1(y), \ldots, h_q(y))$$

and the corresponding psh function $\Phi(x, y) = \log\big(\sum |g_j(x)| + \sum |h_k(y)|\big)$ has the same behavior along the poles as $\Phi'(x, y) = \max(\varphi(x), \psi(y))$ (up to a term $O(1) \leqslant \log 2$). Now, for sufficiently small neighborhoods U, V of 0, we have

$$\mu_{U\times V}\big(\{\max(\varphi(x), \psi(y)) < \log r\}\big) = \mu_U(\{\varphi < \log r\}) \times \mu_V(\{\psi < \log r\}),$$

and one can derive from this that

$$C_1 r^{2(c+c')} \leqslant \mu_{U\times V}\big(\{\max(\varphi(x), \psi(y)) < \log r\}\big) \leqslant C_2 r^{2(c+c')}\,|\log r|^{n-1+m-1}$$

with $c = c_0(\varphi) = c_0(\mathcal{I})$ and $c' = c_0(\psi) = c_0(\mathcal{J})$. We infer

$$c_{(0,0)}(\mathcal{I} \oplus \mathcal{J}) = c + c' = c_0(\mathcal{I}) + c_0(\mathcal{J}). \qquad \square$$

(13.37) Proposition. *Let f, g be holomorphic on a complex manifold X. Then, for every $x \in X$,*

$$c_x(f + g) \le c_x(f) + c_x(g).$$

More generally, if $\mathcal{I}$ and $\mathcal{J}$ are coherent ideals, then

$$c_x(\mathcal{I} + \mathcal{J}) \le c_x(\mathcal{I}) + c_x(\mathcal{J}).$$

Proof. Let Δ be the diagonal in $X \times X$. Then $\mathcal{I} + \mathcal{J}$ can be seen as the restriction of $\mathcal{I} \oplus \mathcal{J}$ to Δ. Hence Propositions 13.35 and 13.36 combined imply

$$c_x(\mathcal{I} + \mathcal{J}) = c_{(x,x)}((\mathcal{I} \oplus \mathcal{J})_{|\Delta}) \leqslant c_{(x,x)}(\mathcal{I} \oplus \mathcal{J}) = c_x(\mathcal{I}) + c_x(\mathcal{J}).$$

Since $(f + g) \subset (f) + (g)$, we get

$$c_x(f + g) \leqslant c_x((f) + (g)) \leqslant c_x(f) + c_x(g). \qquad \square$$

Now we can explain in rough terms the strategy of proof of Theorem 13.33. We start by approximating psh singularities with analytic singularities, using Theorem 13.21. By the argument of Corollary 13.23, we can even reduce ourselves to the case of invertible ideals (f) near $z_0 = 0$, and look at what happens when we have a uniformly convergent sequence $f_\nu \to f$. In this case, we use the Taylor expansion of f at 0 to write $f = p_N + s_N$ where p_N is a polynomial of degree N and $s_N(z) = O(|z|^{N+1})$. Clearly $c_0(s_N) \leqslant n/(N+1)$, and from this we infer $|c_0(f) - c_0(P_N)| \leqslant n/(N+1)$ by Proposition 13.37. Similarly, we get the uniform estimate $|c_0(f_\nu) - c_0(P_{\nu,N})| \leqslant n/(N+1)$ for all indices ν. This means that the proof of the semi-continuity theorem is reduced to handling the situation of a finite dimensional space of polynomials. This case is well-known — one can apply Hironaka's desingularization theorem, in a relative version involving the coefficients of our polynomials as parameters. The conclusion is obtained by putting together carefully all required uniform estimates (which involve a lot of L^2 estimates). □

13.E. Hodge Conjecture and approximation of (p,p)- currents

Let X be a complex n-dimensional manifold. We study here the approximation in the weak topology of a given closed (p,p)-current T by a sequence of real (or rational) analytic cycles, i.e. by locally finite sums of the form $\sum \lambda_j[Z_j]$ where $Z_j \subset X$ is a (closed) analytic set of pure codimension p, and λ_j are real or rational coefficients. The discussion of this section is based on ideas of [Dem82c] (although the main result of Section 7 of [Dem82c] suffers from an incorrect proof of Lemma 7.5 — fortunately all statements are entirely salvaged by the results previously explained in chapter 13).

We will concentrate ourselves on the case where X is projective, although the problem is interesting in other contexts, e.g. for Stein manifolds. We know that the map

$$T \in \mathscr{D}'^{p,p}_{\text{closed}}(X) \mapsto \{T\} \in H^{p,p}(X,\mathbb{C})$$

is continuous in the weak topology. Since the cohomology class $\{[Z]\}$ of an irreducible codimension p cycle lies in the set of integral (p,p) classes, i.e. in $H^{p,p}(X,\mathbb{R}) \cap H^{2p}(X,\mathbb{Z})/\operatorname{tors}$, the approximation is possible only when the cohomology class $\{T\}$ lies in the Hodge group $\operatorname{Hdg}^p_{\mathbb{R}}(X)$ defined by

$$(13.38)\quad \operatorname{Hdg}^p_{\mathbb{K}}(X) = \mathbb{K}\otimes_{\mathbb{Z}}(H^{p,p}(X,\mathbb{R})\cap H^{2p}(X,\mathbb{Z})/\operatorname{tors}), \qquad \mathbb{K} = \mathbb{R} \text{ or } \mathbb{K} = \mathbb{Q}.$$

(13.39) Notation. *We denote by $\mathscr{D}'^{p,p}_{\mathrm{Hdg}}(X)$ the set of closed real (p,p)-currents T whose cohomology class $\{T\}$ belongs to the Hodge group $\operatorname{Hdg}^p_{\mathbb{R}}(X)$: this is a closed subspace of $\mathscr{D}'^{p,p}(X)$ in the weak topology.*

The celebrated *Hodge conjecture* asserts that for every X projective algebraic and every $p = 0, 1, \ldots, n = \dim_{\mathbb{C}} X$, the group $\operatorname{Hdg}^p_{\mathbb{Q}}(X)$ is generated over $\mathbb{Q}$ by cohomology classes of algebraic codimension p cycles $[Z]$ of X (since we are working in finite dimensional vector spaces and since rationals are dense in the reals, the analogous statement over $\mathbb{R}$ is completely equivalent to the statement over $\mathbb{Q}$).

(13.40) Theorem. *Let X be a projective n-dimensional manifold. The following properties are equivalent :*

(a) *The Hodge conjecture holds true in codimension p, namely $\mathrm{Hdg}^p_{\mathbb{Q}}(X)$ is generated by codimension p algebraic cycles.*

(b) *Every closed current $T \in \mathscr{D}'^{p,p}_{\mathrm{Hdg}}(X)$ is a weak limit of algebraic cycles $\sum \lambda_j [Z_j]$ of codimension p with rational coefficients.*

Proof. It is clear that (b) implies (a), hence it is enough to show that (a) $\Rightarrow$ (b).

Fix a current $T \in \mathscr{D}'^{p,p}_{\mathrm{Hdg}}(X)$. Assumption (a) implies that there exists a codimension p cycle $T_0 = \sum \lambda^0_j [Z^0_j]$ with real coefficients such that the cohomology classes of T and T_0 coincide. By the $\partial\overline{\partial}$-lemma, we conclude that there exists a real $(p-1,p-1)$-current U such that

$$T - T_0 = dd^c U.$$

Now, we can approximate the coefficients of T_0 by rational numbers and U by a smooth $(p-1,p-1)$-form (just use a partition of unity with respect to ccordinate charts, and apply a convolution in each chart). It is therefore sufficient to prove the following lemma. □

(13.41) Lemma. *Let X be a projective n-dimensional manifold and $T = dd^c U$ be a closed (p,p)-current with zero cohomology class on X, with U and T smooth. Then T is a weak limit of algebraic codimension p cycles with rational coefficients and zero cohomology class. In that case, one can even take the approximating cycles to be of the form $\sum_{1\leqslant j\leqslant N} \lambda_j [Z_j]$ where $Z_j \subset X$ are non singular algebraic subvarieties and $N = d_n$ depends only on dimension.*

Proof. A standard polarization trick shows that the space of (k,k)-forms on $\mathbb{C}^n$ is generated by decomposable forms of the type

$$dd^c|z^\alpha_1|^2 \wedge \cdots \wedge dd^c|z^\alpha_k|^2$$

for a suitable family of linear coordinate systems $z^\alpha = (z^\alpha_1, \ldots, z^\alpha_n)$, $1 \leqslant \alpha \leqslant \binom{n}{k}^2$. In particular, in every coordinate patch of X, we can write U in a unique way

$$U = \sum_\alpha \varphi_\alpha \, dd^c|z^\alpha_1|^2 \wedge \cdots \wedge dd^c|z^\alpha_{p-1}|^2.$$

Now, by using a partition of unity, we see that it is enough to prove the result when U and T can be written under the form

$$U = \varphi_1 \wedge dd^c\varphi_2 \wedge \cdots \wedge dd^c\varphi_p, \qquad T = dd^c U = dd^c\varphi_1 \wedge dd^c\varphi_2 \wedge \cdots \wedge dd^c\varphi_p$$

with certain global smooth functions φ_j on X. The number of such terms needed to generate a given smooth $(p-1,p-1)$ form U depends only on dimension; in fact this follows by an easy argument based on the topological dimension of X, if we allow non connected coordinate open sets consisting of unions of disjoint balls. Fix a positive line bundle (L, h) on X and a multiple $m_j L$ such that $S_j = m_j\Theta_{L,h} + dd^c\varphi_j > 0$ for every j. Now we simply write $dd^c\varphi_j = S_j - \Theta_j$, where $\Theta_j = m_j\Theta_{L,h}$, and we use Corollary 13.23 to approximate both $(1,1)$ forms $S_j > 0$ and $\Theta_j > 0$ by divisors coming from sections of $\ell m_j L$, $\ell \gg 1$. As L is ample, we can even perturb these divisors a little bit to get them non singular. In this way, we show by induction on $k = 1, 2, \ldots, p$ that each product

$$dd^c\varphi_1 \wedge dd^c\varphi_2 \wedge \cdots \wedge dd^c\varphi_k$$

is a weak limit of rational cycles generated by smooth irreducible components $Z_j \subset X$, and more precisely that

$$(13.42) \qquad dd^c\varphi_1 \wedge \cdots \wedge dd^c\varphi_k = \lim_{\ell\to+\infty} \sum_{1\leqslant j\leqslant 2^k} \lambda_{j,\ell}[Z_{j,\ell}], \qquad Z_{j,\ell} \text{ smooth.}$$

This is true for $k = 1$ by what we have just explained. If the result holds true for $k - 1$, we write

$$\begin{aligned} dd^c\varphi_1 \wedge \cdots \wedge dd^c\varphi_k &= \lim_{\ell\to+\infty} \sum_{1\leqslant j\leqslant 2^{k-1}} \lambda_{j,\ell}[Z_{j,\ell}] \wedge dd^c\varphi_k \\ &= \lim_{\ell\to+\infty} \sum_{1\leqslant j\leqslant 2^{k-1}} \lambda_{j,\ell}[Z_{j,\ell}] \wedge (S_k - \Theta_k). \end{aligned}$$

By Corollary 13.23 applied to each algebraic submanifold $Z_{j,\ell} \subset X$ and to the restriction of $m_k L$ to $Z_{j,\ell}$, equipped respectively with the metrics $h^{m_k}e^{-\varphi_k}$ and h^{m_k}, we find non singular $\mathbb{Q}$-divisors $\frac{1}{q_\ell}D'_{j,\ell}$, $\frac{1}{q_\ell}D''_{j,\ell}$ on $Z_{j,\ell}$ which approximate respectively $[Z_{j,\ell}] \wedge S_k$ and $[Z_{j,\ell}] \wedge \Theta_k$. This implies

$$dd^c\varphi_1 \wedge \cdots \wedge dd^c\varphi_k = \lim_{\ell\to+\infty} \sum_{1\leqslant j\leqslant 2^{k-1}} \frac{\lambda_{j,\ell}}{q_\ell}([D'_{j,\ell}] - [D''_{j,\ell}]).$$

Assertion (13.42) follows by induction, and the lemma is proved. □

(13.43) Remark. The above proof gives absolutely no control on the sign of coefficients λ_j in the approximating cycles $\sum_j \lambda_j[Z_j]$. When the current T is strongly positive (in the sense that for $\|T\|$-almost every x the value $T(x)$ lies in the convex cone generated by positive decomposable (p,p)-forms), it would be interesting to know whether the cycles $\sum_j \lambda_j[Z_j]$ can be taken to be positive. This is true for $p = 1$ by Corollary 13.23, but seems to be a very hard problem in general, except for the trivial cases $p = 0$, $p = n$. The answer is not even known to be true locally, e.g. for closed strongly positive (p,p)-currents on the unit ball of $\mathbb{C}^n$ (and $p \neq 0, 1, n$). We however expect that one can always take $N = 2$ in 13.41 (with possibly mixed signs), assuming of course that the Hodge conjecture holds true. □

(13.44) Remark. It is well known that the Hodge conjecture holds true for (p,p) classes if and only if it holds for $(n-p,n-p)$ classes. In fact, if $\omega = \Theta_{A,h} > 0$ is the curvature form of a very ample divisor A, the Hard Lefschetz theorem shows that there are isomorphisms

$$\bullet \wedge \omega^{n-2p} : H^{p,p}(X,\mathbb{R}) \to H^{n-p,n-p}(X,\mathbb{R}), \qquad \mathrm{Hdg}^p_{\mathbb{Q}}(X) \to \mathrm{Hdg}^{n-p}_{\mathbb{Q}}(X),$$

where the right hand isomorphism comes from the fact that $\{\omega\}$ is an integral class. In particular, both statements 13.40 (a) and 13.40 (b) hold true for the border cases $p = 0,\ 1, n-1,\ n$. □

Chapter 14

Subadditivity of Multiplier Ideals and Fujita's Approximate Zariski Decomposition

The goal of this chapter is to compare the multiplier ideal sheaf $\mathcal{I}(\varphi + \psi)$ of a sum of subharmonic functions to each of the multiplier ideal sheaves $\mathcal{I}(\varphi)$, $\mathcal{I}(\psi)$. We first notice the following basic restriction formula for multiplier ideals, which is just a rephrasing of the Ohsawa-Takegoshi extension theorem.

(14.1) Restriction Formula. *Let φ be a plurisubharmonic function on a complex manifold X, and let $Y \subset X$ be a submanifold. Then*

$$\mathcal{I}(\varphi_{|Y}) \subset \mathcal{I}(\varphi)_{|Y}.$$

Thus, in some sense, the singularities of φ can only get worse if we restrict to a submanifold (if the restriction of φ to some connected component of Y is identically $-\infty$, we agree that the corresponding multiplier ideal sheaf is zero). The proof is straightforward and just amounts to extending locally a germ of function f on Y near a point $y_0 \in Y$ to a function $\tilde{f}$ on a small Stein neighborhood of y_0 in X, which is possible by the Ohsawa-Takegoshi extension theorem. As a direct consequence, we get:

(14.2) Subadditivity Theorem.

(a) *Let X_1, X_2 be complex manifolds, $\pi_i : X_1 \times X_2 \to X_i$, $i = 1, 2$ the projections, and let φ_i be a plurisubharmonic function on X_i. Then*

$$\mathcal{I}(\varphi_1 \circ \pi_1 + \varphi_2 \circ \pi_2) = \pi_1^* \mathcal{I}(\varphi_1) \cdot \pi_2^* \mathcal{I}(\varphi_2).$$

(b) *Let X be a complex manifold and let φ, ψ be plurisubharmonic functions on X. Then*

$$\mathcal{I}(\varphi + \psi) \subset \mathcal{I}(\varphi) \cdot \mathcal{I}(\psi)$$

Proof. (a) Let us fix two relatively compact Stein open subsets $U_1 \subset X_1$, $U_2 \subset X_2$. Then $\mathcal{H}^2(U_1 \times U_2, \varphi_1 \circ \pi_1 + \varphi_2 \circ \pi_2, \pi_1^* dV_1 \otimes \pi_2^* dV_2)$ is the Hilbert tensor product of $\mathcal{H}^2(U_1, \varphi_1, dV_1)$ and $\mathcal{H}^2(U_2, \varphi_2, dV_2)$, and admits $(f'_k \boxtimes f''_l)$ as a Hilbert basis, where (f'_k) and (f''_l) are respective Hilbert bases. Since $\mathcal{I}(\varphi_1 \circ \pi_1 + \varphi_2 \circ \pi_2)_{|U_1 \times U_2}$

is generated as an $\mathcal{O}_{U_1\times U_2}$ module by the $(f'_k \boxtimes f''_l)$ (Proposition 5.7), we conclude that (a) holds true.

(b) We apply (a) to $X_1 = X_2 = X$ and the restriction formula to $Y =$ diagonal of $X \times X$. Then

$$\begin{aligned}\mathcal{I}(\varphi+\psi) &= \mathcal{I}\big((\varphi\circ\pi_1+\psi\circ\pi_2)_{|Y}\big) \subset \mathcal{I}\big(\varphi\circ\pi_1+\psi\circ\pi_2\big)_{|Y}\\ &= \Big(\pi_1^*\mathcal{I}(\varphi)\otimes\pi_2^*\mathcal{I}(\psi)\Big)_{|Y} = \mathcal{I}(\varphi)\cdot\mathcal{I}(\psi).\end{aligned}$$ □

(14.3) Proposition. *Let $f : X \to Y$ be an arbirary holomorphic map, and let φ be a plurisubharmonic function on Y. Then $\mathcal{I}(\varphi\circ f) \subset f^*\mathcal{I}(\varphi)$.*

Proof. Let
$$\Gamma_f = \{(x, f(x)\,;\, x \in X\} \subset X \times Y$$
be the graph of f, and let $\pi_X : X\times Y \to X$, $\pi_Y : X\times Y \to Y$ be the natural projections. Then we can view $\varphi\circ f$ as the restriction of $\varphi\circ\pi_Y$ to Γ_f, as π_X is a biholomorphism from Γ_f to X. Hence the restriction formula implies

$$\mathcal{I}(\varphi\circ f) = \mathcal{I}\big((\varphi\circ\pi_Y)_{|\Gamma_f}\big) \subset \mathcal{I}(\varphi\circ\pi_Y)_{|\Gamma_f} = \big(\pi_Y^*\mathcal{I}(\varphi)\big)_{|\Gamma_f} = f^*\mathcal{I}(\varphi).$$ □

As an application of subadditivity, we now reprove a result of Fujita [Fuj93], relating the growth of sections of multiples of a line bundle to the Chern numbers of its "largest nef part". Fujita's original proof is by contradiction, using the Hodge index theorem and intersection inequalities. The present method arose in the course of joint work with R. Lazarsfeld [Laz99].

(14.4) Lemma. *The line bundle L is big if and only if there is a multiple m_0L such that $m_0L = E + A$, where E is an effective divisor and A an ample divisor.*

Proof. If the condition is satisfied, the decomposition $km_0L = kE + kA$ gives rise to an injection $H^0(X, kA) \hookrightarrow H^0(X, km_0L)$, thus $\mathrm{Vol}(L) \geqslant m_0^{-n}\,\mathrm{Vol}(A) > 0$. Conversely, assume that L is big, and take A to be a very ample nonsingular divisor in X. The exact sequence

$$0 \longrightarrow \mathcal{O}_X(kL-A) \longrightarrow \mathcal{O}_X(kL) \longrightarrow \mathcal{O}_A(kL_{|A}) \longrightarrow 0$$

gives rise to a cohomology exact sequence

$$0 \to H^0(X, kL-A) \longrightarrow H^0(X, kL) \longrightarrow H^0(A, kL_{|A}),$$

and $h^0(A, kL_{|A}) = O(k^{n-1})$ since $\dim A = n-1$. Now, the assumption that L is big implies that $h^0(X, kL) > ck^n$ for infinitely many k, hence $H^0(X, m_0L - A) \neq 0$ for some large integer m_0. If E is the divisor of a section in $H^0(X, m_0L - A)$, we find $m_0L - A = E$, as required. □

(14.5) Lemma. *Let G be an arbitrary line bundle. For every $\varepsilon > 0$, there exists a positive integer m and a sequence $\ell_\nu \uparrow +\infty$ such that*

$$h^0\big(X, \ell_\nu(mL - G)\big) \geqslant \frac{\ell_\nu^m m^n}{n!}\big(\operatorname{Vol}(L) - \varepsilon\big),$$

in other words, $\operatorname{Vol}(mL - G) \geqslant m^n(\operatorname{Vol}(L) - \varepsilon)$ for m large enough.

Proof. Clearly, $\operatorname{Vol}(mL - G) \geqslant \operatorname{Vol}(mL - (G + E))$ for every effective divisor E. We can take E so large that $G + E$ is very ample, and we are thus reduced to the case where G is very ample by replacing G with $G + E$. By definition of $\operatorname{Vol}(L)$, there exists a sequence $k_\nu \uparrow +\infty$ such that

$$h^0(X, k_\nu L) \geqslant \frac{k_\nu^n}{n!}\Big(\operatorname{Vol}(L) - \frac{\varepsilon}{2}\Big).$$

We take $m \gg 1$ (to be precisely chosen later), and $\ell_\nu = \big[\frac{k_\nu}{m}\big]$, so that $k_\nu = \ell_\nu m + r_\nu$, $0 \leqslant r_\nu < m$. Then

$$\ell_\nu(mL - G) = k_\nu L - (r_\nu L + \ell_\nu G).$$

Fix a constant $a \in \mathbb{N}$ such that $aG - L$ is an effective divisor. Then $r_\nu L \leqslant maG$ (with respect to the cone of effective divisors), hence

$$h^0\big(X, \ell_\nu(mL - G)\big) \geqslant h^0\big(X, k_\nu L - (\ell_\nu + am)G\big).$$

We select a smooth divisor D in the very ample linear system $|G|$. By looking at global sections associated with the exact sequences of sheaves

$$0 \to \mathcal{O}(-(j+1)D) \otimes \mathcal{O}(k_\nu L) \to \mathcal{O}(-jD) \otimes \mathcal{O}(k_\nu L) \to \mathcal{O}_D(k_\nu L - jD) \to 0,$$

$0 \leqslant j < s$, we infer inductively that

$$\begin{aligned} h^0(X, k_\nu L - sD) &\geqslant h^0(X, k_\nu L) - \sum_{0 \leqslant j < s} h^0\big(D, \mathcal{O}_D(k_\nu L - jD)\big) \\ &\geqslant h^0(X, k_\nu L) - s\, h^0\big(D, k_\nu L_{|D}\big) \\ &\geqslant \frac{k_\nu^n}{n!}\Big(\operatorname{Vol}(L) - \frac{\varepsilon}{2}\Big) - s\,C k_\nu^{n-1} \end{aligned}$$

where C depends only on L and G. Hence, by putting $s = \ell_\nu + am$, we get

$$\begin{aligned} h^0\big(X, \ell_\nu(mL - G)\big) &\geqslant \frac{k_\nu^n}{n!}\Big(\operatorname{Vol}(L) - \frac{\varepsilon}{2}\Big) - C(\ell_\nu + am)k_\nu^{n-1} \\ &\geqslant \frac{\ell_\nu^n m^n}{n!}\Big(\operatorname{Vol}(L) - \frac{\varepsilon}{2}\Big) - C(\ell_\nu + am)(\ell_\nu + 1)^{n-1} m^{n-1} \end{aligned}$$

and the desired conclusion follows by taking $\ell_\nu \gg m \gg 1$. □

We are now ready to prove Fujita's decomposition theorem, as reproved in [DEL00].

(14.6) Theorem (Fujita). *Let L be a big line bundle. Then for every $\varepsilon > 0$, there exists a modification $\mu : \widetilde{X} \to X$ and a decomposition $\mu^* L = E + A$, where E is an effective $\mathbb{Q}$-divisor and A an ample $\mathbb{Q}$-divisor, such that $A^n > \mathrm{Vol}(L) - \varepsilon$.*

(14.7) Remark. Of course, if $\mu^* L = E + A$ with E effective and A nef, we get an injection

$$H^0(\widetilde{X}, kA) \hookrightarrow H^0(\widetilde{X}, kE + kA) = H^0(\widetilde{X}, k\mu^* L) = H^0(X, kL)$$

for every integer k which is a multiple of the denominator of E, hence $A^n \leqslant \mathrm{Vol}(L)$. □

(14.8) Remark. Once Theorem 14.6 is proved, the same kind of argument easily shows that

$$\mathrm{Vol}(L) = \lim_{k \to +\infty} \frac{n!}{k^n} h^0(X, kL),$$

because the formula is true for every ample line bundle A.

Proof of Theorem 14.6. It is enough to prove the theorem with A being a big and nef divisor. In fact, Proposition 14.4 then shows that we can write $A = E' + A'$ where E' is an effective $\mathbb{Q}$-divisor and A' an ample $\mathbb{Q}$-divisor, hence

$$E + A = E + \varepsilon E' + (1 - \varepsilon)A + \varepsilon A'$$

where $A'' = (1 - \varepsilon)A + \varepsilon A'$ is ample and the intersection number A''^n approaches A^n as closely as we want. Let G be as in Theorem 6.22 (Siu's theorem on uniform global generation). Lemma 14.5 implies that $\mathrm{Vol}(mL - G) > m^n(\mathrm{Vol}(L) - \varepsilon)$ for m large. By Theorem 6.8 on the existence of analytic Zariski decomposition, there exists a Hermitian metric h_m of weight φ_m on $mL - G$ such that

$$H^0\big(X, \ell(mL - G)\big) = H^0\big(X, \ell(mL - G) \otimes \mathcal{I}(\ell\varphi_m)\big)$$

for every $\ell \geqslant 0$. We take a smooth modification $\mu : \widetilde{X} \to X$ such that

$$\mu^* \mathcal{I}(\varphi_m) = \mathcal{O}_{\widetilde{X}}(-E)$$

is an invertible ideal sheaf in $\mathcal{O}_{\widetilde{X}}$. This is possible by taking the blow-up of X with respect to the ideal $\mathcal{I}(\varphi_m)$ and by resolving singularities [Hir64]. Theorem 6.22 applied to $L' = mL - G$ implies that $\mathcal{O}(mL) \otimes \mathcal{I}(\varphi_m)$ is generated by its global sections, hence its pull-back $\mathcal{O}(m\,\mu^* L - E)$ is also generated. This implies

$$m\,\mu^* L = E + A$$

where E is an effective divisor and A is a nef (semi-ample) divisor in $\widetilde{X}$. We find

$$\begin{aligned} H^0(\widetilde{X}, \ell A) &= H^0\big(\widetilde{X}, \ell(m\,\mu^* L - E)\big) \\ &\supset H^0\big(\widetilde{X}, \mu^*\big(\mathcal{O}(\ell mL) \otimes \mathcal{I}(\varphi_m)^\ell\big)\big) \\ &\supset H^0\big(\widetilde{X}, \mu^*\big(\mathcal{O}(\ell mL) \otimes \mathcal{I}(\ell\varphi_m)\big)\big), \end{aligned}$$

thanks to the subadditivity property of multiplier ideals. Moreover, the direct image $\mu_*\mu^*\mathcal{I}(\ell\varphi_m)$ coincides with the integral closure of $\mathcal{I}(\ell\varphi_m)$, hence with $\mathcal{I}(\ell\varphi_m)$, because a multiplier ideal sheaf is always integrally closed. From this we infer

$$\begin{aligned} H^0(\widetilde{X},\ell A) &\supset H^0\big(X,\mathcal{O}(\ell mL)\otimes\mathcal{I}(\ell\varphi_m)\big) \\ &\supset H^0\big(X,\mathcal{O}(\ell(mL-G))\otimes\mathcal{I}(\ell\varphi_m)\big) \\ &= H^0\big(X,\mathcal{O}(\ell(mL-G))\big). \end{aligned}$$

By Lemma 14.5, we find

$$h^0(\widetilde{X},\ell A)\geqslant \frac{\ell^n}{n!}m^n\big(\mathrm{Vol}(L)-\varepsilon\big)$$

for infinitely many ℓ, therefore $\mathrm{Vol}(A)=A^n\geqslant m^n(\mathrm{Vol}(L)-\varepsilon)$. Theorem 14.6 is proved, up to a minor change of notation $E\mapsto\frac{1}{m}E$, $A\mapsto\frac{1}{m}A$. □

We conclude by using Fujita's theorem to establish a geometric interpretation of the volume $\mathrm{Vol}(L)$. Suppose as above that X is a smooth projective variety of dimension n, and that L is a big line bundle on X. Given a large integer $k\gg 0$, denote by $B_k\subset X$ the base-locus of the linear system $|kL|$. The *moving self-intersection number* $(kL)^{[n]}$ of $|kL|$ is defined by choosing n general divisors $D_1,\ldots,D_n\in|kL|$ and putting

$$(kL)^{[n]}=\#\Big(D_1\cap\ldots\cap D_n\cap(X-B_k)\Big).$$

In other words, we simply count the number of intersection points away from the base locus of n general divisors in the linear system $|kL|$. This notion arises for example in Matsusaka's proof of his "big theorem". We show that the volume $\mathrm{Vol}(L)$ of L measures the rate of growth with respect to k of these moving self-intersection numbers:

(14.9) Proposition. *One has*

$$\mathrm{Vol}(L)=\limsup_{k\to\infty}\frac{(kL)^{[n]}}{k^n}.$$

Proof. We start by interpreting $(kL)^{[n]}$ geometrically. Let $\mu_k:X_k\longrightarrow X$ be a modification of $|kL|$ such that $\mu_k^*|kL|=|V_k|+F_k$, where

$$P_k:=\mu_k^*(kL)-F_k$$

is generated by sections, and $H^0(X,\mathcal{O}_X(kL))=V_k=H^0(X_k,\mathcal{O}_{X_k}(P_k))$, so that $B_k=\mu_k(F_k)$. Then evidently $(kL)^{[n]}$ counts the number of intersection points of n general divisors in P_k, and consequently

$$(kL)^{[n]}=(P_k)^n.$$

Since then P_k is big (and nef) for $k \gg 0$, we have $\mathrm{Vol}(P_k) = (P_k)^n$. Also, $\mathrm{Vol}(kL) \geqslant \mathrm{Vol}(P_k)$ since P_k embeds in $\mu_k^*(kL)$. Hence

$$\mathrm{Vol}(kL) \geqslant (kL)^{[n]} \qquad \forall k \gg 0.$$

On the other hand, an easy argument in the spirit of Lemma 14.5 shows that $\mathrm{Vol}(kL) = k^n \cdot \mathrm{Vol}(L)$ (cf. also [ELN96], Lemma 3.4), and so we conclude that

$$\mathrm{Vol}(L) \geqslant \frac{(kL)^{[n]}}{k^n}. \tag{14.10}$$

for every $k \gg 0$.

For the reverse inequality we use Fujita's theorem. Fix $\varepsilon > 0$, and consider the decomposition $\mu^* L = A + E$ on $\mu : \widetilde{X} \longrightarrow X$ constructed in Fujita's theorem. Let k be any positive integer such that kA is integral and globally generated. By taking a common resolution we can assume that X_k dominates $\widetilde{X}$, and hence we can write

$$\mu_k^* kL \sim A_k + E_k$$

with A_k globally generated and

$$(A_k)^n \geqslant k^n \cdot (\mathrm{Vol}(L) - \varepsilon).$$

But then A_k embeds in P_k and both $\mathcal{O}(A_k)$ and $\mathcal{O}(P_k)$ are globally generated, consequently

$$(A_k)^n \leqslant (P_k)^n = (kL)^{[n]}.$$

Therefore

$$\frac{(kL)^{[n]}}{k^n} \geqslant \mathrm{Vol}(L) - \varepsilon. \tag{14.11}$$

But (14.11) holds for any sufficiently large and divisible k, and in view of (14.10) the proposition holds. □

Chapter 15

Hard Lefschetz Theorem with Multiplier Ideal Sheaves

15.A. A Bundle Valued Hard Lefschetz Theorem

The goal of this section is to prove the following surjectivity theorem, which can be seen as an extension of the hard Lefschetz theorem for sections of pseudo-effective line bundles. We closely follow the exposition of [DPS00].

(15.1) Theorem. *Let (L, h) be a pseudo-effective line bundle on a compact Kähler manifold (X, ω) of dimension n, let $\Theta_{L,h} \geqslant 0$ be its curvature current and $\mathcal{I}(h)$ the associated multiplier ideal sheaf. Then, the wedge multiplication operator $\omega^q \wedge \bullet$ induces a surjective morphism*

$$\Phi^q_{\omega,h} : H^0(X, \Omega^{n-q}_X \otimes L \otimes \mathcal{I}(h)) \longrightarrow H^q(X, \Omega^n_X \otimes L \otimes \mathcal{I}(h)).$$

The special case when L is nef is due to Takegoshi [Take97]. An even more special case is when L is semi-positive, i.e. possesses a smooth metric with semi-positive curvature. In that case the multiple ideal sheaf $\mathcal{I}(h)$ coincides with $\mathcal{O}_X$ and we get the following consequence already observed by Mourougane [Mou99].

(15.2) Corollary. *Let (L, h) be a semi-positive line bundle on a compact Kähler manifold (X, ω) of dimension n. Then, the wedge multiplication operator $\omega^q \wedge \bullet$ induces a surjective morphism*

$$\Phi^q_{\omega} : H^0(X, \Omega^{n-q}_X \otimes L) \longrightarrow H^q(X, \Omega^n_X \otimes L).$$

It should be observed that although all objects involved in Theorem 15.1 are algebraic when X is a projective manifold, there are no known algebraic proof of the statement; it is not even clear how to define algebraically $\mathcal{I}(h)$ for the case when $h = h_{min}$ is a metric with minimal singularity. However, even in the special circumstance when L is nef, the multiplier ideal sheaf is crucially needed (see Section 15.E for a counterexample).

The proof of Theorem 15.1 is based on the Bochner formula, combined with a use of harmonic forms with values in the Hermitian line bundle (L, h). The method can be applied only after h has been made smooth at least in the complement of an analytic set. However, we have to accept singularities even in the regularized metrics because only a very small incompressible loss of positivity is acceptable in the Bochner estimate (by the results of [Dem92], singularities can only be removed at the expense of a fixed loss of positivity). Also, we need the multiplier ideal sheaves to be preserved by the smoothing process. This is possible thanks to a suitable "equisingular" regularization process.

15.B. Equisingular Approximations of Quasi Plurisubharmonic Functions

Let φ be a quasi-psh function. We say that φ has logarithmic poles if φ is locally bounded outside an analytic set A and has singularities of the form

$$\varphi(z) = c \log \sum_k |g_k|^2 + O(1)$$

with $c > 0$ and g_k holomorphic, on a neighborhood of every point of A. Our goal is to show the following

(15.3) Theorem. *Let $T = \alpha + i\partial\overline{\partial}\varphi$ be a closed $(1,1)$-current on a compact Hermitian manifold (X, ω), where α is a smooth closed $(1,1)$-form and φ a quasi-psh function. Let γ be a continuous real $(1,1)$-form such that $T \geqslant \gamma$. Then one can write $\varphi = \lim_{\nu\to+\infty} \varphi_\nu$ where*

(a) *φ_ν is smooth in the complement $X \smallsetminus Z_\nu$ of an analytic set $Z_\nu \subset X$;*

(b) *$\{\varphi_\nu\}$ is a decreasing sequence, and $Z_\nu \subset Z_{\nu+1}$ for all ν;*

(c) *$\int_X (e^{-2\varphi} - e^{-2\varphi_\nu})dV_\omega$ is finite for every ν and converges to 0 as $\nu \to +\infty$;*

(d) *$\mathcal{I}(\varphi_\nu) = \mathcal{I}(\varphi)$ for all ν ("equisingularity");*

(e) *$T_\nu = \alpha + i\partial\overline{\partial}\varphi_\nu$ satisfies $T_\nu \geqslant \gamma - \varepsilon_\nu\omega$, where $\lim_{\nu\to+\infty} \varepsilon_\nu = 0$.*

(15.4) Remark. It would be interesting to know whether the φ_ν can be taken to have logarithmic poles along Z_ν. Unfortunately, the proof given below destroys this property in the last step. Getting it to hold true seems to be more or less equivalent to proving the semi-continuity property

$$\lim_{\varepsilon\to 0_+} \mathcal{I}((1+\varepsilon)\varphi) = \mathcal{I}(\varphi).$$

Actually, this can be checked in dimensions 1 and 2, but is unknown in higher dimensions (and probably quite hard to establish).

Proof of Theorem 15.3. Clearly, by replacing T with $T-\alpha$ and γ with $\gamma-\alpha$, we may assume that $\alpha=0$ and $T=\mathrm{i}\partial\overline{\partial}\varphi\geqslant\gamma$. We divide the proof in four steps.

Step 1. *Approximation by quasi-psh functions with logarithmic poles.*
By [Dem92], there is a decreasing sequence (ψ_ν) of quasi-psh functions with logarithmic poles such that $\varphi=\lim\psi_\nu$ and $\mathrm{i}\partial\overline{\partial}\psi_\nu\geqslant\gamma-\varepsilon_\nu\omega$. We need a little bit more information on those functions, hence we first recall the main techniques used for the construction of (ψ_ν). For $\varepsilon>0$ given, fix a covering of X by open balls $B_j=\{|z^{(j)}|<r_j\}$ with coordinates $z^{(j)}=(z_1^{(j)},\ldots,z_n^{(j)})$, such that

$$0\leqslant\gamma+c_j\,\mathrm{i}\partial\overline{\partial}|z^{(j)}|^2\leqslant\varepsilon\omega\qquad\text{on}\quad B_j,\tag{15.5}$$

for some real number c_j. This is possible by selecting coordinates in which γ is diagonalized at the center of the ball, and by taking the radii $r_j>0$ small enough (thanks to the fact that γ is continuous). We may assume that these coordinates come from a finite sample of coordinates patches covering X, on which we perform suitable linear coordinate changes (by invertible matrices lying in some compact subset of the complex linear group). By taking additional balls, we may also assume that $X=\bigcup B_j''$ where

$$B_j''\subset\subset B_j'\subset\subset B_j$$

are concentric balls $B_j'=\{|z^{(j)}|<r_j'=r_j/2\}$, $B_j''=\{|z^{(j)}|<r_j''=r_j/4\}$. We define

$$\psi_{\varepsilon,\nu,j}=\frac{1}{2\nu}\log\sum_{k\in\mathbb{N}}|f_{\nu,j,k}|^2-c_j|z^{(j)}|^2\qquad\text{on}\quad B_j,\tag{15.6}$$

where $(f_{\nu,j,k})_{k\in\mathbb{N}}$ is an orthonormal basis of the Hilbert space $\mathcal{H}_{\nu,j}$ of holomorphic functions on B_j with finite L^2 norm

$$\|u\|^2=\int_{B_j}|u|^2e^{-2\nu(\varphi+c_j|z^{(j)}|^2)}d\lambda(z^{(j)}).$$

(The dependence of $\psi_{\varepsilon,\nu,j}$ on ε is through the choice of the open covering (B_j)). Observe that the choice of c_j in (15.5) guarantees that $\varphi+c_j|z^{(j)}|^2$ is plurisubharmonic on B_j, and notice also that

$$\sum_{k\in\mathbb{N}}|f_{\nu,j,k}(z)|^2=\sup_{f\in\mathcal{H}_{\nu,j},\,\|f\|\leqslant1}|f(z)|^2\tag{15.7}$$

is the square of the norm of the continuous linear form $\mathcal{H}_{\nu,j}\to\mathbb{C}$, $f\mapsto f(z)$. We claim that there exist constants C_i, $i=1,2,\ldots$ depending only on X and γ (thus independent of ε and ν), such that the following uniform estimates hold:

$$\mathrm{i}\partial\overline{\partial}\psi_{\varepsilon,\nu,j}\geqslant-c_j\,\mathrm{i}\partial\overline{\partial}|z^{(j)}|^2\geqslant\gamma-\varepsilon\omega\qquad\text{on}\quad B_j'\ \ (B_j'\subset\subset B_j),\tag{15.8}$$

$$\varphi(z)\leqslant\psi_{\varepsilon,\nu,j}(z)\leqslant\sup_{|\zeta-z|\leqslant r}\varphi(\zeta)+\frac{n}{\nu}\log\frac{C_1}{r}+C_2r^2\qquad\forall z\in B_j',\ \ r<r_j-r_j',\tag{15.9}$$

$$|\psi_{\varepsilon,\nu,j}-\psi_{\varepsilon,\nu,k}|\leqslant\frac{C_3}{\nu}+C_4\varepsilon\big(\min(r_j,r_k)\big)^2\qquad\text{on}\quad B_j'\cap B_k'.\tag{15.10}$$

Actually, the Hessian estimate (15.8) is obvious from (15.5) and (15.6). As in the proof of ([Dem92], Prop. 3.1), (15.9) results from the Ohsawa-Takegoshi L^2 extension theorem (left hand inequality) and from the mean value inequality (right hand inequality). Finally, as in ([Dem92], Lemma 3.6 and Lemma 4.6), (15.10) is a consequence of Hörmander's L^2 estimates. We briefly sketch the idea. Assume that the balls B_j are small enough, so that the coordinates $z^{(j)}$ are still defined on a neighborhood of all balls $\overline{B}_k$ which intersect B_j (these coordinates can be taken to be linear transforms of coordinates belonging to a fixed finite set of coordinate patches covering X, selected once for all). Fix a point $z_0 \in B'_j \cap B'_k$. By (15.6) and (15.7), we have

$$\psi_{\varepsilon,\nu,j}(z_0) = \frac{1}{\nu}\log|f(z_0)| - c_j|z^{(j)}|^2$$

for some holomorphic function f on B_j with $\|f\| = 1$. We consider the weight function

$$\Phi(z) = 2\nu(\varphi(z) + c_k|z^{(k)}|^2) + 2n\log|z^{(k)} - z_0^{(k)}|,$$

on both B_j and B_k. The trouble is that a priori we have to deal with different weights, hence a comparison of weights is needed. By the Taylor formula applied at z_0, we get

$$\Big|c_k|z^{(k)} - z_0^{(k)}|^2 - c_j|z^{(j)} - z_0^{(j)}|^2\Big| \leqslant C\varepsilon\big(\min(r_j, r_k)\big)^2 \qquad \text{on } B_j \cap B_k$$

[the only nonzero term of degree 2 has type $(1,1)$ and its Hessian satisfies

$$-\varepsilon\omega \leqslant i\partial\overline{\partial}(c_k|z^{(k)}|^2 - c_j|z^{(j)}|^2) \leqslant \varepsilon\omega$$

by (15.5); we may suppose $r_j \ll \varepsilon$ so that the terms of order 3 and more are negligible]. By writing $|z^{(j)}|^2 = |z^{(j)} - z_0^{(j)}|^2 + |z_0^{(j)}|^2 + 2\,\mathrm{Re}\langle z^{(j)} - z_0^{(j)}, z_0^{(j)}\rangle$, we obtain

$$\begin{aligned} c_k|z^{(k)}|^2 - c_j|z^{(j)}|^2 &= 2c_k\,\mathrm{Re}\langle z^{(k)} - z_0^{(k)}, z_0^{(k)}\rangle - 2c_j\,\mathrm{Re}\langle z^{(j)} - z_0^{(j)}, z_0^{(j)}\rangle \\ &\quad + c_k|z_0^{(k)}|^2 - c_j|z_0^{(j)}|^2 \pm C\varepsilon(\min(r_j, r_k))^2. \end{aligned}$$

We use a cut-off function θ equal to 1 in a neighborhood of z_0 and with support in $B_j \cap B_k$; as $z_0 \in B'_j \cap B'_k$, the function θ can be taken to have its derivatives uniformly bounded when z_0 varies. We solve the equation $\overline{\partial}u = \overline{\partial}(\theta f e^{\nu g})$ on B_k, where g is the holomorphic function

$$g(z) = c_k\langle z^{(k)} - z_0^{(k)}, z_0^{(k)}\rangle - c_j\langle z^{(j)} - z_0^{(j)}, z_0^{(j)}\rangle.$$

Thanks to Hörmander's L^2 estimates [Hör66], the L^2 solution for the weight Φ yields a holomorphic function $f' = \theta f e^{\nu g} - u$ on B_k such that $f'(z_0) = f(z_0)$ and

$$\begin{aligned} \int_{B_k} |f'|^2 e^{-2\nu(\varphi + c_k|z^{(k)}|^2)} d\lambda(z^{(k)}) &\leqslant C' \int_{B_j\cap B_k} |f|^2|e^{\nu g}|^2 e^{-2\nu(\varphi + c_k|z^{(k)}|^2)} d\lambda(z^{(k)}) \\ &\leqslant C'\exp\Big(2\nu\big(c_k|z_0^{(k)}|^2 - c_j|z_0^{(j)}|^2 + C\varepsilon(\min(r_j, r_k))^2\big)\Big) \\ &\qquad \int_{B_j} |f|^2 e^{-2\nu(\varphi + c_j|z^{(j)}|^2)} d\lambda(z^{(j)}). \end{aligned}$$

Let us take the supremum of $\frac{1}{\nu}\log|f(z_0)| = \frac{1}{\nu}\log|f'(z_0)|$ over all f with $\|f\| \leqslant 1$. By the definition of $\psi_{\varepsilon,\nu,k}$ ((15.6) and (15.7)) and the bound on $\|f'\|$, we find

$$\psi_{\varepsilon,\nu,k}(z_0) \leqslant \psi_{\nu,j}(z_0) + \frac{\log C'}{2\nu} + C\varepsilon(\min(r_j, r_k))^2,$$

whence (15.10) by symmetry. Assume that ν is so large that $C_3/\nu < C_4\varepsilon(\inf_j r_j)^2$. We "glue" all functions $\psi_{\varepsilon,\nu,j}$ into a function $\psi_{\varepsilon,\nu}$ globally defined on X, and for this we set

$$\psi_{\varepsilon,\nu}(z) = \sup_{j,\ B'_j \ni z} \Big(\psi_{\varepsilon,\nu,j}(z) + 12\, C_4\varepsilon(r_j'^2 - |z^{(j)}|^2)\Big) \qquad \text{on } X.$$

Every point of X belongs to some ball B''_k, and for such a point we get

$$12\, C_4\varepsilon(r_k'^2 - |z^{(k)}|^2) \geqslant 12\, C_4\varepsilon(r_k'^2 - r_k''^2) > 2C_4 r_k^2 > \frac{C_3}{\nu} + C_4\varepsilon(\min(r_j, r_k))^2.$$

This, together with (15.10), implies that in $\psi_{\varepsilon,\nu}(z)$ the supremum is never reached for indices j such that $z \in \partial B'_j$, hence $\psi_{\varepsilon,\nu}$ is well defined and continuous, and by standard properties of upper envelopes of (quasi)-plurisubharmonic functions we get

$$\mathrm{i}\partial\overline{\partial}\psi_{\varepsilon,\nu} \geqslant \gamma - C_5\varepsilon\omega \tag{15.11}$$

for $\nu \geqslant \nu_0(\varepsilon)$ large enough. By inequality (15.9) applied with $r = e^{-\sqrt{\nu}}$, we see that $\lim_{\nu\to+\infty}\psi_{\varepsilon,\nu}(z) = \varphi(z)$. At this point, the difficulty is to show that $\psi_{\varepsilon,\nu}$ is decreasing with ν – this may not be formally true, but we will see at Step 3 that this is essentially true. Another difficulty is that we must simultaneously let ε go to 0, forcing us to change the covering as we want the error to get smaller and smaller in (15.11).

Step 2. A comparison of integrals.
We claim that

$$I := \int_X \big(e^{-2\varphi} - e^{-2\max(\varphi, \frac{\ell}{\ell-1}\psi_{\nu,\varepsilon})+a}\big)dV_\omega < +\infty \tag{15.12}$$

for every $\ell \in\,]1,\nu]$ and $a \in \mathbb{R}$. In fact

$$\begin{aligned} I &\leqslant \int_{\{\varphi < \frac{\ell}{\ell-1}\psi_{\varepsilon,\nu}+a\}} e^{-2\varphi}dV_\omega = \int_{\{\varphi<\frac{\ell}{\ell-1}\psi_{\varepsilon,\nu}\}+a} e^{2(\ell-1)\varphi-2\ell\varphi}dV_\omega \\ &\leqslant e^{2(\ell-1)a}\int_X e^{2\ell(\psi_{\varepsilon,\nu}-\varphi)}dV_\omega \leqslant C\Big(\int_X e^{2\nu(\psi_{\varepsilon,\nu}-\varphi)}dV_\omega\Big)^{\frac{\ell}{\nu}} \end{aligned}$$

by Hölder's inequality. In order to show that these integrals are finite, it is enough, by the definition and properties of the functions $\psi_{\varepsilon,\nu}$ and $\psi_{\varepsilon,\nu,j}$, to prove that

$$\int_{B'_j} e^{2\nu\psi_{\varepsilon,\nu,j}-2\nu\varphi}d\lambda = \int_{B'_j}\Big(\sum_{k=0}^{+\infty}|f_{\nu,j,k}|^2\Big)e^{-2\nu\varphi}d\lambda < +\infty.$$

By the strong Noetherian property of coherent ideal sheaves (see e.g. [GR84]), we know that the sequence of ideal sheaves generated by the holomorphic functions $(f_{\nu,j,k}(z)\overline{f_{\nu,j,k}(\overline{w})})_{k\leqslant k_0}$ on $B_j \times B_j$ is locally stationary as k_0 increases, hence independant of k_0 on $B'_j \times B'_j \subset\subset B_j \times B_j$ for k_0 large enough. As the sum of the series $\sum_k f_{\nu,j,k}(z)\overline{f_{\nu,j,k}(\overline{w})}$ is bounded by

$$\Big(\sum_k |f_{\nu,j,k}(z)|^2 \sum_k |f_{\nu,j,k}(\overline{w})|^2\Big)^{1/2}$$

and thus uniformly covergent on every compact subset of $B_j \times B_j$, and as the space of sections of a coherent ideal sheaf is closed under the topology of uniform convergence on compact subsets, we infer from the Noetherian property that the holomorphic function $\sum_{k=0}^{+\infty} f_{\nu,j,k}(z)\overline{f_{\nu,j,k}(\overline{w})}$ is a section of the coherent ideal sheaf generated by $\{f_{\nu,j,k}(z)\overline{f_{\nu,j,k}(\overline{w})}\}_{k\leqslant k_0}$ over $B'_j \times B'_j$, for k_0 large enough. Hence, by restricting to the conjugate diagonal $w = \overline{z}$, we get

$$\sum_{k=0}^{+\infty} |f_{\nu,j,k}(z)|^2 \leqslant C \sum_{k=0}^{k_0} |f_{\nu,j,k}(z)|^2 \qquad \text{on } B'_j.$$

This implies

$$\int_{B'_j} \Big(\sum_{k=0}^{+\infty} |f_{\nu,j,k}|^2\Big) e^{-2\varphi} d\lambda \leqslant C \int_{B'_j} \Big(\sum_{k=0}^{k_0} |f_{\nu,j,k}|^2\Big) e^{-2\varphi} d\lambda = C(k_0+1).$$

(15.12) is proved.

Step 3. Subadditivity of the approximating sequence $\psi_{\varepsilon,\nu}$.
We want to compare $\psi_{\varepsilon,\nu_1+\nu_2}$ and ψ_{ε,ν_1}, ψ_{ε,ν_2} for every pair of indices ν_1, ν_2, first when the functions are associated with the same covering $X = \bigcup B_j$. Consider a function $f \in \mathcal{H}_{\nu_1+\nu_2,j}$ with

$$\int_{B_j} |f(z)|^2 e^{-2(\nu_1+\nu_2)\varphi_j(z)} d\lambda(z) \leqslant 1, \qquad \varphi_j(z) = \varphi(z) + c_j|z^{(j)}|^2.$$

We may view f as a function $\hat{f}(z,z)$ defined on the diagonal Δ of $B_j \times B_j$. Consider the Hilbert space of holomorphic functions u on $B_j \times B_j$ such that

$$\int_{B_j\times B_j} |u(z,w)|^2 e^{-2\nu_1\varphi_j(z)-2\nu_2\varphi_j(w)} d\lambda(z) d\lambda(w) < +\infty.$$

By the Ohsawa-Takegoshi L^2 extension theorem [OT87], there exists a function $\widetilde{f}(z,w)$ on $B_j \times B_j$ such that $\widetilde{f}(z,z) = f(z)$ and

$$\begin{aligned}\int_{B_j\times B_j} &|\widetilde{f}(z,w)|^2 e^{-2\nu_1\varphi_j(z)-2\nu_2\varphi_j(w)} d\lambda(z) d\lambda(w)\\ &\leqslant C_7 \int_{B_j} |f(z)|^2 e^{-2(\nu_1+\nu_2)\varphi_j(z)} d\lambda(z) = C_7,\end{aligned}$$

where the constant C_7 only depends on the dimension n (it is actually independent of the radius r_j if say $0 < r_j \leqslant 1$). As the Hilbert space under consideration on $B_j \times B_j$ is the completed tensor product $\mathcal{H}_{\nu_1,j} \widehat{\otimes} \mathcal{H}_{\nu_2,j}$, we infer that

$$\widetilde{f}(z,w) = \sum_{k_1,k_2} c_{k_1,k_2} f_{\nu_1,j,k_1}(z) f_{\nu_2,j,k_2}(w)$$

with $\sum_{k_1,k_2} |c_{k_1,k_2}|^2 \leqslant C_7$. By restricting to the diagonal, we obtain

$$|f(z)|^2 = |\widetilde{f}(z,z)|^2 \leqslant \sum_{k_1,k_2} |c_{k_1,k_2}|^2 \sum_{k_1} |f_{\nu_1,j,k_1}(z)|^2 \sum_{k_2} |f_{\nu_2,j,k_2}(z)|^2.$$

From (15.5) and (15.6), we get

$$\psi_{\varepsilon,\nu_1+\nu_2,j} \leqslant \frac{\log C_7}{\nu_1+\nu_2} + \frac{\nu_1}{\nu_1+\nu_2}\psi_{\varepsilon,\nu_1,j} + \frac{\nu_2}{\nu_1+\nu_2}\psi_{\varepsilon,\nu_2,j},$$

in particular

$$\psi_{\varepsilon,2^\nu,j} \leqslant \psi_{\varepsilon,2^{\nu-1},j} + \frac{C_8}{2^\nu},$$

and we see that $\psi_{\varepsilon,2^\nu} + C_8 2^{-\nu}$ is a decreasing sequence. By Step 2 and Lebesgue's monotone convergence theorem, we infer that for every $\varepsilon, \delta > 0$ and $a \leqslant a_0 \ll 0$ fixed, the integral

$$I_{\varepsilon,\delta,\nu} = \int_X \left(e^{-2\varphi} - e^{-2\max(\varphi,(1+\delta)(\psi_{2^\nu,\varepsilon}+a))}\right) dV_\omega$$

converges to 0 as ν tends to $+\infty$ (take $\ell = \frac{1}{\delta} + 1$ and $2^\nu > \ell$ and a_0 such that $\delta \sup_X \varphi + a_0 \leqslant 0$; we do not have monotonicity strictly speaking but need only replace a by $a + C_8 2^{-\nu}$ to get it, thereby slightly enlarging the integral).

Step 4. Selection of a suitable upper envelope.

For the simplicity of notation, we assume here that $\sup_X \varphi = 0$ (possibly after subtracting a constant), hence we can take $a_0 = 0$ in the above. We may even further assume that all our functions $\psi_{\varepsilon,\nu}$ are nonpositive. By Step 3, for each $\delta = \varepsilon = 2^{-k}$, we can select an index $\nu = p(k)$ such that

$$\text{(15.13)} \qquad I_{2^{-k},2^{-k},p(k)} = \int_X \left(e^{-2\varphi} - e^{-2\max(\varphi,(1+2^{-k})\psi_{2^{-k},2^{p(k)}})}\right) dV_\omega \leqslant 2^{-k}.$$

By construction, we have an estimate $i\partial\overline{\partial}\psi_{2^{-k},2^{p(k)}} \geqslant \gamma - C_5 2^{-k}\omega$, and the functions $\psi_{2^{-k},2^{p(k)}}$ are quasi-psh with logarithmic poles. Our estimates (especially (15.9)) imply that $\lim_{k\to+\infty} \psi_{2^{-k},2^{p(k)}}(z) = \varphi(z)$ as soon as $2^{-p(k)} \log(1/\inf_j r_j(k)) \to 0$ (notice that the r_j's now depend on $\varepsilon = 2^{-k}$). We set

$$\text{(15.14)} \qquad \varphi_\nu(z) = \sup_{k \geqslant \nu} (1+2^{-k})\psi_{2^{-k},2^{p(k)}}(z).$$

By construction $\{\varphi_\nu\}$ is a decreasing sequence and satisfies the estimates

$$\varphi_\nu \geqslant \max\big(\varphi, (1+2^{-\nu})\psi_{2^{-\nu},2^{p(\nu)}}\big), \qquad i\partial\overline{\partial}\varphi_\nu \geqslant \gamma - C_5 2^{-\nu}\omega.$$

Inequality (15.13) implies that

$$\int_X (e^{-2\varphi} - e^{-2\varphi_\nu})dV_\omega \leqslant \sum_{k=\nu}^{+\infty} 2^{-k} = 2^{1-\nu}.$$

Finally, if Z_ν is the set of poles of $\psi_{2^{-\nu},2^{p(\nu)}}$, then $Z_\nu \subset Z_{\nu+1}$ and φ_ν is continuous on $X \smallsetminus Z_\nu$. The reason is that in a neighborhood of every point $z_0 \in X \smallsetminus Z_\nu$, the term $(1+2^{-k})\psi_{2^{-k},2^{p(k)}}$ contributes to φ_ν only when it is larger than $(1+2^{-\nu})\psi_{2^{-\nu},2^{p(\nu)}}$. Hence, by the almost-monotonicity, the relevant terms of the sup in (15.14) are squeezed between $(1+2^{-\nu})\psi_{2^{-\nu},2^{p(\nu)}}$ and $(1+2^{-k})(\psi_{2^{-\nu},2^{p(\nu)}}+C_8 2^{-\nu})$, and therefore there is uniform convergence in a neighborhood of z_0. Finally, condition (c) implies that

$$\int_U |f|^2(e^{-2\varphi} - e^{-2\varphi_\nu})dV_\omega < +\infty$$

for every germ of holomorphic function $f \in \mathcal{O}(U)$ at a point $x \in X$. Therefore both integrals $\int_U |f|^2 e^{-2\varphi}dV_\omega$ and $\int_U |f|^2 e^{-2\varphi_\nu}dV_\omega$ are simultaneously convergent or divergent, i.e. $\mathcal{I}(\varphi) = \mathcal{I}(\varphi_\nu)$. Theorem 15.3 is proved, except that φ_ν is possibly just continuous instead of being smooth. This can be arranged by Richberg's regularization theorem [Ri68], at the expense of an arbitrary small loss in the Hessian form. □

(15.15) Remark. By a very slight variation of the proof, we can strengthen condition (c) and obtain that for every $t > 0$

$$\int_X (e^{-2t\varphi} - e^{-2t\varphi_\nu})dV_\omega$$

is finite for ν large enough and converges to 0 as $\nu \to +\infty$. This implies that the sequence of multiplier ideals $\mathcal{I}(t\varphi_\nu)$ is a stationary decreasing sequence, with $\mathcal{I}(t\varphi_\nu) = \mathcal{I}(t\varphi)$ for ν large.

15.C. A Bochner Type Inequality

Let (L, h) be a smooth Hermitian line bundle on a (non necessarily compact) Kähler manifold (Y, ω). We denote by $|\ \ | = |\ \ |_{\omega,h}$ the pointwise Hermitian norm on $\Lambda^{p,q}T_Y^* \otimes L$ associated with ω and h, and by $\|\ \ \| = \|\ \ \|_{\omega,h}$ the global L^2 norm

$$\|u\|^2 = \int_Y |u|^2 dV_\omega \qquad \text{where} \quad dV_\omega = \frac{\omega^n}{n!}$$

We consider the $\overline{\partial}$ operator acting on (p,q)-forms with values in L, its adjoint $\overline{\partial}_h^*$ with respect to h and the complex Laplace-Beltrami operator $\Delta_h'' = \overline{\partial}\overline{\partial}_h^* + \overline{\partial}_h^*\overline{\partial}$. Let v be a smooth $(n-q,0)$-form with compact support in Y. Then $u = \omega^q \wedge v$ satisfies

$$\|\overline{\partial}u\|^2 + \|\overline{\partial}_h^* u\|^2 = \|\overline{\partial}v\|^2 + \int_Y \sum_{I,J} \Big(\sum_{j\in J} \lambda_j\Big)|u_{IJ}|^2 \tag{15.16}$$

where $\lambda_1 \leqslant \cdots \leqslant \lambda_n$ are the curvature eigenvalues of $\Theta_{L,h}$ expressed in an orthonormal frame $(\partial/\partial z_1, \ldots, \partial/\partial z_n)$ (at some fixed point $x_0 \in Y$), in such a way that

$$\omega_{x_0} = \mathrm{i} \sum_{1\leqslant j\leqslant n} dz_j \wedge d\overline{z}_j, \qquad (\Theta_{L,h})_{x_0} = \mathrm{i}\partial\overline{\partial}\varphi_{x_0} = \mathrm{i} \sum_{1\leqslant j\leqslant n} \lambda_j dz_j \wedge d\overline{z}_j.$$

The proof of (15.16) proceeds by checking that

$$(\overline{\partial}_\varphi^*\overline{\partial} + \overline{\partial}\,\overline{\partial}_\varphi^*)(v\wedge\omega^q) - (\overline{\partial}_\varphi^*\overline{\partial}v)\wedge\omega^q = q\,\mathrm{i}\partial\overline{\partial}\varphi\wedge\omega^{q-1}\wedge v, \tag{15.17}$$

taking the inner product with $u = \omega^q \wedge v$ and integrating by parts in the left hand side. In order to check (15.16), we use the identity $\overline{\partial}_\varphi^* = e^{\varphi}\overline{\partial}^*(e^{-\varphi}\bullet) = \overline{\partial}^* + \nabla^{0,1}\varphi \,\lrcorner\, \bullet$. Let us work in a local trivialization of L such that $\varphi(x_0) = 0$ and $\nabla\varphi(x_0) = 0$. At x_0 we then find

$$\begin{aligned}&(\overline{\partial}_\varphi^*\overline{\partial} + \overline{\partial}\,\overline{\partial}_\varphi^*)(\omega^q\wedge v) - \omega^q\wedge(\overline{\partial}_\varphi^*\overline{\partial}v)\\ &\qquad = \big[(\overline{\partial}^*\overline{\partial} + \overline{\partial}\,\overline{\partial}^*)(\omega^q\wedge v) - \omega^q\wedge(\overline{\partial}^*\overline{\partial}v)\big] + \overline{\partial}(\nabla^{0,1}\varphi \,\lrcorner\, (\omega^q\wedge v)).\end{aligned}$$

However, the term $[\cdots]$ corresponds to the case of a trivial vector bundle and it is well known in that case that $[\Delta'', \omega^q\wedge\bullet] = 0$, hence $[\cdots] = 0$. On the other hand

$$\nabla^{0,1}\varphi \,\lrcorner\, (\omega^q\wedge v) = q(\nabla^{0,1}\varphi \,\lrcorner\, \omega)\wedge\omega^{q-1}\wedge v = -q\,\mathrm{i}\partial\varphi\wedge\omega^{q-1}\wedge v,$$

and so

$$(\overline{\partial}_\varphi^*\overline{\partial} + \overline{\partial}\,\overline{\partial}_\varphi^*)(\omega^q\wedge v) - \omega^q\wedge(\overline{\partial}_\varphi^*\overline{\partial}v) = q\,\mathrm{i}\partial\overline{\partial}\varphi\wedge\omega^{q-1}\wedge v.$$

Our formula is thus proved when v is smooth and compactly supported. In general, we have:

(15.18) Proposition. *Let (Y,ω) be a* complete *Kähler manifold and (L,h) a smooth Hermitian line bundle such that the curvature possesses a uniform lower bound $\Theta_{L,h} \geqslant -C\omega$. For every measurable $(n-q,0)$-form v with L^2 coefficients and values in L such that $u = \omega^q\wedge v$ has differentials $\overline{\partial}u$, $\overline{\partial}^*u$ also in L^2, we have*

$$\|\overline{\partial}u\|^2 + \|\overline{\partial}_h^* u\|^2 = \|\overline{\partial}v\|^2 + \int_Y \sum_{I,J} \Big(\sum_{j\in J} \lambda_j\Big)|u_{IJ}|^2$$

(*here, all differentials are computed in the sense of distributions*).

Proof. Since (Y, ω) is assumed to be complete, there exists a sequence of smooth forms v_ν with compact support in Y (obtained by truncating v and taking the convolution with a regularizing kernel) such that $v_\nu \to v$ in L^2 and such that $u_\nu = \omega^q \wedge v_\nu$ satisfies $u_\nu \to u$, $\overline{\partial} u_\nu \to \overline{\partial} u$, $\overline{\partial}^* u_\nu \to \overline{\partial}^* u$ in L^2. By the curvature assumption, the final integral in the right hand side of (15.16) must be under control (i.e. the integrand becomes nonnegative if we add a term $C\|u\|^2$ on both sides, $C \gg 0$). We thus get the equality by passing to the limit and using Lebesgue's monotone convergence theorem. □

15.D. Proof of Theorem 15.1

To fix the ideas, we first indicate the proof in the much simpler case when (L, h) is Hermitian semi-positive, and then treat the general case.

(15.19) Special Case. (L, h) is (smooth) Hermitian semi-positive.

Let $\{\beta\} \in H^q(X, \Omega_X^n \otimes L)$ be an arbitrary cohomology class. By standard L^2 Hodge theory, $\{\beta\}$ can be represented by a smooth harmonic $(0, q)$-form β with values in $\Omega_X^n \otimes L$. We can also view β as a (n, q)-form with values in L. The pointwise Lefschetz isomorphism produces a unique $(n - q, 0)$-form α such that $\beta = \omega^q \wedge \alpha$. Proposition 15.18 then yields

$$\|\overline{\partial}\alpha\|^2 + \int_Y \sum_{I,J} \Big(\sum_{j\in J} \lambda_j\Big) |\alpha_{IJ}|^2 = \|\overline{\partial}\beta\|^2 + \|\overline{\partial}_h^*\beta\|^2 = 0,$$

and the curvature eigenvalues λ_j are nonnegative by our assumption. Hence $\overline{\partial}\alpha = 0$ and $\{\alpha\} \in H^0(X, \Omega_X^{n-q} \otimes L)$ is mapped to $\{\beta\}$ by $\Phi_{\omega,h}^q = \omega^q \wedge \bullet$.

(15.20) General Case.

There are several difficulties. The first difficulty is that the metric h is no longer smooth and we cannot directly represent cohomology classes by harmonic forms. We circumvent this problem by smoothing the metric on an (analytic) Zariski open subset and by avoiding the remaining poles on the complement. However, some careful estimates have to be made in order to take the error terms into account.

Fix $\varepsilon = \varepsilon_\nu$ and let $h_\varepsilon = h_{\varepsilon_\nu}$ be an approximation of h, such that h_ε is smooth on $X \smallsetminus Z_\varepsilon$ (Z_ε being an analytic subset of X), $\Theta_{L,h_\varepsilon} \geqslant -\varepsilon\omega$, $h_\varepsilon \leqslant h$ and $\mathcal{I}(h_\varepsilon) = \mathcal{I}(h)$. This is possible by Theorem 15.3. Now, we can find a family

$$\omega_{\varepsilon,\delta} = \omega + \delta(\mathrm{i}\partial\overline{\partial}\psi_\varepsilon + \omega), \qquad \delta > 0$$

of *complete Kähler* metrics on $X \smallsetminus Z_\varepsilon$, where ψ_ε is a quasi-psh function on X with $\psi_\varepsilon = -\infty$ on Z_ε, ψ_ε on $X \smallsetminus Z_\varepsilon$ and $\mathrm{i}\partial\overline{\partial}\psi_\varepsilon + \omega \geqslant 0$ (see e.g. [Dem82b], Théorème 1.5). By construction, $\omega_{\varepsilon,\delta} \geqslant \omega$ and $\lim_{\delta\to 0} \omega_{\varepsilon,\delta} = \omega$. We look at the L^2 Dolbeault complex $K^\bullet_{\varepsilon,\delta}$ of $(n,\bullet)$-forms on $X \smallsetminus Z_\varepsilon$, where the L^2 norms are induced by $\omega_{\varepsilon,\delta}$ on differential forms and by h_ε on elements in L. Specifically

$$K^q_{\varepsilon,\delta} = \Big\{u{:}X\smallsetminus Z_\varepsilon\to\Lambda^{n,q}T^*_X\otimes L;\ \int_{X\smallsetminus Z_\varepsilon}(|u|^2_{\Lambda^{n,q}\omega_{\varepsilon,\delta}\otimes h_\varepsilon}+|\overline{\partial}u|^2_{\Lambda^{n,q+1}\omega_{\varepsilon,\delta}\otimes h_\varepsilon})dV_{\omega_{\varepsilon,\delta}}<\infty\Big\}.$$

Let $\mathcal{K}^q_{\varepsilon,\delta}$ be the corresponding sheaf of germs of locally L^2 sections on X (the local L^2 condition should hold on X, not only on $X \smallsetminus Z_\varepsilon$!). Then, for all $\varepsilon > 0$ and $\delta \geqslant 0$, $(\mathcal{K}^q_{\varepsilon,\delta}, \overline{\partial})$ is a resolution of the sheaf $\Omega^n_X \otimes L \otimes \mathcal{I}(h_\varepsilon) = \Omega^n_X \otimes L \otimes \mathcal{I}(h)$. This is because L^2 estimates hold locally on small Stein open sets, and the L^2 condition on $X \smallsetminus Z_\varepsilon$ forces holomorphic sections to extend across Z_ε ([Dem82b], Lemma 6.9).

Let $\{\beta\} \in H^q(X, \Omega^n_X \otimes L \otimes \mathcal{I}(h))$ be a cohomology class represented by a smooth form with values in $\Omega^n_X \otimes L \otimes \mathcal{I}(h)$ (one can use a Čech cocycle and convert it to an element in the $\mathcal{C}^\infty$ Dolbeault complex by means of a partition of unity, thanks to the usual De Rham-Weil isomorphism). Then

$$\|\beta\|^2_{\varepsilon,\delta} \leqslant \|\beta\|^2 = \int_X |\beta|^2_{\Lambda^{n,q}\omega\otimes h} dV_\omega < +\infty.$$

The reason is that $|\beta|^2_{\Lambda^{n,q}\omega\otimes h} dV_\omega$ decreases as ω increases. This is just an easy calculation, shown by comparing two metrics ω, ω' which are expressed in diagonal form in suitable coordinates; the norm $|\beta|^2_{\Lambda^{n,q}\omega\otimes h}$ turns out to decrease faster than the volume dV_ω increases; see e.g. [Dem82b], Lemma 3.2; a special case is $q = 0$, then $|\beta|^2_{\Lambda^{n,q}\omega\otimes h} dV_\omega = i^{n^2}\beta \wedge \overline{\beta}$ with the identification $L \otimes \overline{L} \simeq \mathbb{C}$ given by the metric h, hence the integrand is even independent of ω in that case.

By the proof of the De Rham-Weil isomorphism, the map $\alpha \mapsto \{\alpha\}$ from the cocycle space $Z^q(\mathcal{K}^\bullet_{\varepsilon,\delta})$ equipped with its L^2 topology, into $H^q(X, \Omega^n_X \otimes L \otimes \mathcal{I}(h))$ equipped with its finite vector space topology, is continuous. Also, Banach's open mapping theorem implies that the coboundary space $B^q(\mathcal{K}^\bullet_{\varepsilon,\delta})$ is closed in $Z^q(\mathcal{K}^\bullet_{\varepsilon,\delta})$. This is true for all $\delta \geqslant 0$ (the limit case $\delta = 0$ yields the strongest L^2 topology in bidegree (n, q)). Now, β is a $\overline{\partial}$-closed form in the Hilbert space defined by $\omega_{\varepsilon,\delta}$ on $X \smallsetminus Z_\varepsilon$, so there is a $\omega_{\varepsilon,\delta}$-harmonic form $u_{\varepsilon,\delta}$ in the same cohomology class as β, such that

$$\|u_{\varepsilon,\delta}\|_{\varepsilon,\delta} \leqslant \|\beta\|_{\varepsilon,\delta}.$$

(15.21) Remark. The existence of a harmonic representative holds true only for $\delta > 0$, because we need to have a complete Kähler metric on $X \smallsetminus Z_\varepsilon$. The trick of employing $\omega_{\varepsilon,\delta}$ instead of a fixed metric ω, however, is not needed when Z_ε is (or can be taken to be) empty. This is the case if (L, h) is such that $\mathcal{I}(h) = \mathcal{O}_X$ and L is nef. Indeed, in that case, from the very definition of nefness, it is easy to prove that we can take the φ_ν's to be everywhere smooth in Theorem 15.3. However, we will see in Section 15.E that multiplier ideal sheaves are needed even in case L is nef, when $\mathcal{I}(h) \neq \mathcal{O}_X$.

Let $v_{\varepsilon,\delta}$ be the unique $(n-q,0)$-form such that $u_{\varepsilon,\delta} = v_{\varepsilon,\delta} \wedge \omega^q_{\varepsilon,\delta}$ ($v_{\varepsilon,\delta}$ exists by the pointwise Lefschetz isomorphism). Then

$$\|v_{\varepsilon,\delta}\|_{\varepsilon,\delta} = \|u_{\varepsilon,\delta}\|_{\varepsilon,\delta} \leqslant \|\beta\|_{\varepsilon,\delta} \leqslant \|\beta\|.$$

As $\sum_{j\in J}\lambda_j \geqslant -q\varepsilon$ by the assumption on Θ_{L,h_ε}, the Bochner formula yields

$$\|\overline{\partial} v_{\varepsilon,\delta}\|^2_{\varepsilon,\delta} \leqslant q\varepsilon\|u_{\varepsilon,\delta}\|^2_{\varepsilon,\delta} \leqslant q\varepsilon\|\beta\|^2.$$

These uniform bounds imply that there are subsequences $u_{\varepsilon,\delta_\nu}$ and $v_{\varepsilon,\delta_\nu}$ with $\delta_\nu \to 0$, possessing weak-L^2 limits $u_\varepsilon = \lim_{\nu\to+\infty} u_{\varepsilon,\delta_\nu}$ and $v_\varepsilon = \lim_{\nu\to+\infty} v_{\varepsilon,\delta_\nu}$. The limit $u_\varepsilon = \lim_{\nu\to+\infty} u_{\varepsilon,\delta_\nu}$ is with respect to $L^2(\omega) = L^2(\omega_{\varepsilon,0})$. To check this, notice that in bidegree $(n-q,0)$, the space $L^2(\omega)$ has the weakest topology of all spaces $L^2(\omega_{\varepsilon,\delta})$; indeed, an easy calculation as in ([Dem82b], Lemma 3.2) yields

$$|f|^2_{\Lambda^{n-q,0}\omega\otimes h} dV_\omega \leqslant |f|^2_{\Lambda^{n-q,0}\omega_{\varepsilon,\delta}\otimes h} dV_{\omega_{\varepsilon,\delta}} \qquad \text{if } f \text{ is of type } (n-q,0).$$

On the other hand, the limit $v_\varepsilon = \lim_{\nu\to+\infty} v_{\varepsilon,\delta_\nu}$ takes place in all spaces $L^2(\omega_{\varepsilon,\delta})$, $\delta > 0$, since the topology gets stronger and stronger as $\delta \downarrow 0$ [possibly not in $L^2(\omega)$, though, because in bidegree (n,q) the topology of $L^2(\omega)$ might be strictly stronger than that of all spaces $L^2(\omega_{\varepsilon,\delta})$]. The above estimates yield

$$\|v_\varepsilon\|^2_{\varepsilon,0} = \int_X |v_\varepsilon|^2_{\Lambda^{n-q,0}\omega\otimes h_\varepsilon} dV_\omega \leqslant \|\beta\|^2,$$
$$\|\overline{\partial} v_\varepsilon\|^2_{\varepsilon,0} \leqslant q\varepsilon\|\beta\|^2_{\varepsilon,0},$$
$$u_\varepsilon = \omega^q \wedge v_\varepsilon \equiv \beta \qquad \text{in } H^q(X, \Omega^n_X \otimes L \otimes \mathcal{I}(h_\varepsilon)).$$

Again, by arguing in a given Hilbert space $L^2(h_{\varepsilon_0})$, we find L^2 convergent subsequences $u_\varepsilon \to u$, $v_\varepsilon \to v$ as $\varepsilon \to 0$, and in this way get $\overline{\partial} v = 0$ and

$$\|v\|^2 \leqslant \|\beta\|^2,$$
$$u = \omega^q \wedge v \equiv \beta \qquad \text{in } H^q(X, \Omega^n_X \otimes L \otimes \mathcal{I}(h)).$$

Theorem 15.1 is proved. Notice that the equisingularity property $\mathcal{I}(h_\varepsilon) = \mathcal{I}(h)$ is crucial in the above proof, otherwise we could not infer that $u \equiv \beta$ from the fact that $u_\varepsilon \equiv \beta$. This is true only because all cohomology classes $\{u_\varepsilon\}$ lie in the same fixed cohomology group $H^q(X, \Omega^n_X \otimes L \otimes \mathcal{I}(h))$, whose topology is induced by the topology of $L^2(\omega)$ on $\overline{\partial}$-closed forms (e.g. through the De Rham-Weil isomorphism). □

15.E. A Counterexample

In view of Corollary 15.2, one might wonder whether the morphism Φ^q_ω would not still be surjective when L is a nef vector bundle. We will show that this is unfortunately not so, even in the case of algebraic surfaces.

Let B be an elliptic curve and let V be the rank 2 vector bundle over B which is defined as the (unique) non split extension

$$0 \to \mathcal{O}_B \to V \to \mathcal{O}_B \to 0.$$

In particular, the bundle V is numerically flat, i.e. $c_1(V) = 0$, $c_2(V) = 0$. We consider the ruled surface $X = \mathbb{P}(V)$. On that surface there is a unique section $C = \mathbb{P}(\mathcal{O}_B) \subset X$ with $C^2 = 0$ and

$$\mathcal{O}_X(C) = \mathcal{O}_{\mathbb{P}(V)}(1)$$

is a nef line bundle. It is easy to see that

$$h^0(X, \mathcal{O}_{\mathbb{P}(V)}(m)) = h^0(B, S^m V) = 1$$

for all $m \in \mathbb{N}$ (otherwise we would have $mC = aC + M$ where aC is the fixed part of the linear system $|mC|$ and $M \neq 0$ the moving part, thus $M^2 \geqslant 0$ and $C \cdot M > 0$, contradiction). We claim that

$$h^0(X, \Omega^1_X(kC)) = 2$$

for all $k \geqslant 2$. This follows by tensoring the exact sequence

$$0 \to \Omega^1_{X|C} \to \Omega^1_X \to \pi^*\Omega^1_C \simeq \mathcal{O}_C \to 0$$

by $\mathcal{O}_X(kC)$ and observing that

$$\Omega^1_{X|C} = K_X = \mathcal{O}_X(-2C).$$

From this, we get

$$0 \to H^0(X, \mathcal{O}_X((k-2)C)) \to H^0(X, \Omega^1_X \mathcal{O}(kC)) \to H^0(X, \mathcal{O}_X(kC))$$

where $h^0(X, \mathcal{O}_X((k-2)C)) = h^0(X, \mathcal{O}_X(kC)) = 1$ for all $k \geqslant 2$. Moreover, the last arrow is surjective because we can multiply a section of $H^0(X, \mathcal{O}_X(kC))$ by a nonzero section in $H^0(X, \pi^*\Omega^1_B)$ to get a preimage. Our claim follows. We now consider the diagram

$$\begin{array}{ccc} H^0(X, \Omega^1_X(2C)) & \xrightarrow{\wedge\omega} & H^1(X, K_X(2C)) \\ \simeq \downarrow & & \downarrow \varphi \\ H^0(X, \Omega^1_X(3C)) & \xrightarrow[\psi]{\wedge\omega} & H^1(X, K_X(3C)). \end{array}$$

Since $K_X(2C) \simeq \mathcal{O}_X$ and $K_X(3C) \simeq \mathcal{O}_X(C)$, the cohomology sequence of

$$0 \to K_X(2C) \to K_X(3C) \to K_X(3C)|C \simeq \mathcal{O}_C \to 0$$

immediately implies $\varphi = 0$ (notice that $h^1(X, K_X(2C)) = h^1(X, K_X(3C)) = 1$, since $h^1(B, \mathcal{O}_B) = h^1(B, V) = 1$), and $h^2(X, K_X(2C)) = h^2(B, \mathcal{O}_B) = 0$). Therefore the diagram implies $\psi = 0$, and we get:

(15.22) Proposition. *$L = \mathcal{O}_{\mathbb{P}(V)}(3)$ is a counterample to (15.2) in the nef case.*

By Corollary 15.2, we infer that $\mathcal{O}_X(3)$ cannot be Hermitian semi-positive and we thus again obtain – by a quite different method — the result of [DPS94], example 1.7.

(15.23) Corollary. *Let B be an elliptic curve, V the vector bundle given by the unique non-split extension*

$$0 \to \mathcal{O}_B \to V \to \mathcal{O}_B \to 0.$$

Let $X = \mathbb{P}(V)$. Then $L = \mathcal{O}_X(1)$ is nef but not Hermitian semi-positive (nor does any multiple, e.g. the anticanonical line bundle $-K_X = \mathcal{O}_X(-2)$ is nef but not semi-positive).

Chapter 16

Invariance of Plurigenera of Projective Varieties

The goal of this chapter is to give a proof of the following fundamental result on the invariance of plurigenera, which has been proved by Y.T. Siu [Siu98] in the case of varieties of general type (in which case the proof has been translated in a purely algebraic form by Y. Kawamata [Kaw99]), and by [Siu00] in general. Let us recall that X is said to be of general type if $\kappa(K_X) = n = \dim X$.

(16.1) Theorem (Siu). *Let $X \to S$ be a proper holomorphic map defining a family of smooth projective varieties of general type on an irreducible base S. Then the plurigenus $p_m(X_t) = h^0(X_t, mK_{X_t})$ of fibers is independent of t for all $m \geqslant 0$.*

The proof somehow involves taking "limits" of divisors as $m \to +\infty$, and therefore transcendental methods are a strong contender in this circle of ideas, because currents provide a natural compactification of the space of divisors. Quite recently, M. Păun obtained a very short and elegant proof of Theorem 16.1 based merely on the Ohwawa-Takegoshi extension theorem, and we are going to sketch his arguments below (see also M. Păun [Pau07], B. Claudon [Cla07] and S. Takayama [Taka07]). In fact, following Păun, one can prove more general results valid for cohomology with twisted coefficients. Remarkably enough, no algebraic proof of these results are known at this point, in the case of varieties of nonnegative Kodaira dimension which are not of general type.

Notice that by connecting any two points of S by a chain of analytic disks, it is enough to consider the case where $S = \Delta$ is a disk.

(16.2) Theorem (generalized version of Păun's theorem). *Let $\pi : \mathscr{X} \to \Delta$ be a smooth projective family over the unit disk, and let $(L_j, h_j)_{0 \leqslant j \leqslant m-1}$ be (singular) Hermitian line bundles with semi-positive curvature currents $i\Theta_{L_j,h_j} \geqslant 0$ on $\mathscr{X}$. Assume that*

(a) *the restriction of h_j to the central fiber X_0 is well defined (i.e. not identically $+\infty$).*

(b) *the multiplier ideal sheaf $\mathscr{I}(h_{j|X_0})$ is trivial for $1 \leqslant j \leqslant m-1$.*

Then any section σ of $\mathcal{O}(mK_{\mathcal{X}} + \sum L_j)_{|X_0} \otimes \mathcal{I}(h_{0|X_0})$ over the central fiber X_0 extends to $\mathcal{X}$.

The invariance of plurigenera is just the case when all line bundles L_j and their metrics h_j are trivial. Since the dimension $t \mapsto h^0(X_t, mK_{X_t})$ is always upper semicontinuous and since Theorem 16.2 implies the lower semicontinuity, we conclude that the dimension is constant along analytic disks (hence along any irreducible base S, by joining any two points through a chain of analytic disks).

In order to prove Theorem 16.2, we first state the technical version of the Ohsawa-Takegoshi L^2 extension theorem needed for the proof, which is a special case of the Ohsawa-Takegoshi Theorem — the reader is invited to check that the statement indeed follows from Theorem 12.6.

(16.3) Lemma. *Let $\pi : \mathcal{X} \to \Delta$ be as before and let (L, h) be a (singular) Hermitian line bundle with semi-positive curvature current $i\Theta_{L,h} \geqslant 0$ on $\mathcal{X}$. Let ω be a global Kähler metric on $\mathcal{X}$, and $dV_{\mathcal{X}}$, dV_{X_0} the respective induced volume elements on X_0 and $\mathcal{X}$. Assume that h_{X_0} is well defined. Then any holomorphic section u of $\mathcal{O}(K_{\mathcal{X}} + L) \otimes \mathcal{I}(h_{|X_0})$ extends into a section $\widetilde{u}$ over $\mathcal{X}$ satisfying an L^2 estimate*

$$\int_{\mathcal{X}} \|\widetilde{u}\|^2_{\omega\otimes h} dV_{\mathcal{X}} \leqslant C_0 \int_{X_0} \|u\|^2_{\omega\otimes h} dV_{X_0},$$

where $C_0 \geqslant 0$ is some universal constant (independent of $\mathcal{X}$, L, ...).

Proof of Theorem (16.2). We write $h_j = e^{-\varphi_j}$ in terms of local plurisubharmonic weights. Fix an auxiliary line bundle A (which will later be taken to be sufficiently ample), and define inductively a sequence of line bundles F_p by putting $F_0 = A$ and

$$F_p = F_{p-1} + K_{\mathcal{X}} + L_r \qquad \text{if } p = mq + r,\ 0 \leqslant r \leqslant m-1.$$

By construction we have $F_{p+m} = F_p + mK_{\mathcal{X}} + \sum_j L_j$ and

$$F_0 = A,\ F_1 = A + K_{\mathcal{X}} + L_1,\ \ldots, F_p = A + pK_{\mathcal{X}} + L_1 + \cdots + L_p,\ 1 \leqslant p \leqslant m-1.$$

The game is to construct inductively families of sections, say $\{\widetilde{u}_j^{(p)}\}_{j=1,\ldots,N_p}$, of F_p over $\mathcal{X}$, together with ad hoc L^2 estimates, in such a way that

(a) for $p = 0, \ldots, m-1$, F_p is generated by its sections $\{\widetilde{u}_j^{(p)}\}_{j=1,\ldots,N_p}$;

(b) we have the m-periodicity relations $N_{p+m} = N_p$ and $\widetilde{u}_j^{(p)}$ is an extension of $u_j^{(p)} := \sigma^q u_j^{(r)}$ over $\mathcal{X}$ for $p = mq + r$, where $u_j^{(r)} := \widetilde{u}^{(r)}_{j|X_0}$, $0 \leqslant r \leqslant m-1$.

Property (a) can certainly be achieved by taking A ample enough so that $F_0, \ldots, F_{m-1}$ are generated by their sections, and by choosing the $\widetilde{u}_j^{(p)}$ appropriately for the first m indices $p = 0, \ldots, m-1$. Now, by induction, we equip F_{p-1} with the tautological metric $|\xi|^2 / \sum |\widetilde{u}_j^{(p-1)}(x)|^2$, and $F_p - K_{\mathcal{X}} = F_{p-1} + L_r$ with that metric multiplied by $h_r = e^{-\varphi_r}$; it is clear that these metrics have semi-positive curvature currents (the metric on F_p itself if obtained by using a smooth Kähler metric ω on $\mathscr{X}$). In this setting, we apply the Ohsawa-Takegoshi theorem to the line bundle $F_{p-1}+L_r$ to extend $u_j^{(p)}$ into a section $\widetilde{u}_j^{(p)}$ over $\mathscr{X}$. By construction the pointwise norm of that section in $F_{p|X_0}$ in a local trivialization of the bundles involved is the ratio

$$\frac{|u_j^{(p)}|^2}{\sum_\ell |u_\ell^{(p-1)}|^2} e^{-\varphi_r},$$

up to some fixed smooth positive factor depending only on the metric induced by ω on $K_{\mathscr{X}}$. However, by the induction relations, we have

$$\frac{\sum_j |u_j^{(p)}|^2}{\sum_\ell |u_\ell^{(p-1)}|^2} e^{-\varphi_r} = \begin{cases} \dfrac{\sum_j |u_j^{(r)}|^2}{\sum_\ell |u_\ell^{(r-1)}|^2} e^{-\varphi_r} & \text{for } p = mq + r,\ 0 < r \leqslant m-1, \\ \dfrac{\sum_j |u_j^{(0)}|^2}{\sum_\ell |u_\ell^{(m-1)}|^2} |\sigma|^2 e^{-\varphi_0} & \text{for } p \equiv 0 \bmod m. \end{cases}$$

Since the sections $\{u_j^{(r)}\}$ generate their line bundle, the ratios involved are positive functions without zeroes and poles, hence smooth and bounded [possibly after shrinking the base disc Δ, as is permitted]. On the other hand, assumption (b) and the fact that σ has coefficients in the multiplier ideal sheaf $\mathscr{I}(h_{0|X_0})$ tell us that $e^{-\varphi_r}$, $1 \leqslant r < m$ and $|\sigma|^2 e^{-\varphi_0}$ are locally integrable on X_0. It follows that there is a constant $C_1 \geqslant 0$ such that

$$\int_{X_0} \frac{\sum_j |u_j^{(p)}|^2}{\sum_\ell |u_\ell^{(p-1)}|^2} e^{-\varphi_r} dV_\omega \leqslant C_1$$

for all $p \geqslant 1$ (of course, the integral certainly involves finitely many trivializations of the bundles involved, whereas the integrand expression is just local in each chart). Inductively, the L^2 extension theorem produces sections $\widetilde{u}_j^{(p)}$ of F_p over $\mathscr{X}$ such that

$$\int_{\mathscr{X}} \frac{\sum_j |\widetilde{u}_j^{(p)}|^2}{\sum_\ell |\widetilde{u}_\ell^{(p-1)}|^2} e^{-\varphi_r} dV_\omega \leqslant C_2 = C_0 C_1.$$

The next idea is to extract the limits of p-th roots of these sections to get a singular Hermitian metric on $mK_{\mathscr{X}} + \sum L_j$. As the functions $e^{-\varphi_r}$ are locally bounded below (φ_r being psh), the Hölder inequality implies that

$$\int_{\mathscr{X}} \Big(\sum_j |\widetilde{u}_j^{(p)}|^2 \Big)^{1/p} dV_\omega \leqslant C_3.$$

The mean value inequality for plurisubharmonic functions shows a fortiori that the sequence of psh functions $\frac{1}{p} \log \sum_j |\widetilde{u}_j^{(p)}|^2$ is locally uniformly bounded from above. These functions should be thought of as weights on the $\mathbb{Q}$-line bundles

$$\frac{1}{p}(A + q(mK_{\mathscr{X}} + \sum L_j) + L_1 + \cdots + L_r) \text{ converging to } K_{\mathscr{X}} + \frac{1}{m} \sum L_j \quad \text{as } p \to +\infty,$$

and thus they are potentials of currents in a bounded subset of the Kähler cone. Moreover, the sections $\widetilde{u}_j^{(p)}$ extend $\sigma^q u_j^r$ on X_0, and so we have in particular

$$\lim_{p\to+\infty}\frac{1}{p}\log\sum_j|u_j^{(p)}|^2=\lim_{p\to+\infty}\frac{1}{p}\log\big(|\sigma|^{2q}\sum_j|u_j^{(0)}|^2\big)=\frac{1}{m}\log|\sigma|^2\not\equiv-\infty\ \text{ on } X_0.$$

Therefore, by well known facts of potential theory, the sequence $\frac{1}{p}\log\sum_j|u_j^{(p)}|^2$ must have some subsequence which converges in L^1_{loc} topology to the potential ψ of a current in the first Chern class of $K_{\mathcal{X}}+\frac{1}{m}\sum L_j$, in the form of an upper regularized limit

$$\psi(z)=\limsup_{\zeta\to z}\lim_{\nu\to+\infty}\frac{1}{p_\nu}\log\sum_j|\widetilde{u}_j^{(p_\nu)}(\zeta)|^2,$$

which is such that $\psi(z)\geqslant\frac{1}{m}\log|\sigma|^2$ on X_0. Hence $mK_{\mathcal{X}}+\sum L_j$ possesses a Hermitian metric $H=e^{-m\psi}$, and we have by construction $\|\sigma\|_H\leqslant 1$ and $\Theta_H\geqslant 0$. In order to conclude, we equip the bundle

$$G=(m-1)K_{\mathcal{X}}+\sum L_j$$

with the metric $\gamma=H^{1-1/m}\prod h_j^{1/m}$, and $mK_{\mathcal{X}}+\sum L_j=K_{\mathcal{X}}+G$ with the metric $\omega\otimes\gamma$. Clearly γ has a semi-positive curvature current on $\mathcal{X}$ and in a local trivialization we have

$$\|\sigma\|^2_{\omega\otimes\gamma}\leqslant C|\sigma|^2\exp\Big(-(m-1)\psi+\frac{1}{m}\sum\varphi_j\Big)\leqslant C\Big(|\sigma|^2\prod e^{-\varphi_j}\Big)^{1/m}$$

on X_0. Since $|\sigma|^2e^{-\varphi_0}$ and $e^{-\varphi_r}$, $r>0$ are all locally integrable, we see that $\|\sigma\|^2_{\omega\otimes\gamma}$ is also locally integrable on X_0 by the Hölder inequality. A new (and final) application of the L^2 extension theorem to the Hermitian line bundle (G,γ) implies that σ can be extended to $\mathcal{X}$. Theorem 16.2 is proved. □

Chapter 17

Numerical Characterization of the Kähler Cone

The main goal of this Chapter is to describe a structure theorem for the Kähler cone of any compact Kähler manifold, first obtained in [DP04]. The result can be seen as the Kähler generalization of the Nakai-Moishezon criterion for ample line bundles on projective varieties.

17.A. Positive Classes in Intermediate (p,p)-bidegrees

We first discuss some general positivity concepts for cohomology classes of type (p,p), although we will not be able to say much about these. Recall that we have a Serre duality pairing

$$(17.1)\qquad H^{p,q}(X,\mathbb{C})\times H^{n-p,n-q}(X,\mathbb{C})\longrightarrow\mathbb{C},\qquad (\alpha,\beta)\longmapsto\int_X\alpha\wedge\beta\in\mathbb{C}.$$

In particular, if we restrict to real classes, this yields a duality pairing

$$(17.2)\qquad H^{p,p}(X,\mathbb{R})\times H^{n-p,n-p}(X,\mathbb{R})\longrightarrow\mathbb{R},\qquad (\alpha,\beta)\longmapsto\int_X\alpha\wedge\beta\in\mathbb{R}.$$

Now, one can define $H^{p,p}_{\mathrm{SP}}(X,\mathbb{R})$ to be the closure of the cone of classes of d-closed strongly positive smooth (p,p)-forms (a (p,p)-form in $\Lambda^{p,p}T^*_X$ is by definition strongly positive if it is in the convex cone generated by decomposable (p,p) forms $iu_1\wedge\overline{u}_1\wedge\cdots\wedge iu_p\wedge\overline{u}_p$ where the u_j are $(1,0)$-forms). Clearly, $H^{1,1}_{\mathrm{SP}}(X,\mathbb{R})=\overline{\mathcal{K}}$ and the cup product defines a multilinear map

$$(17.3)\qquad \overline{\mathcal{K}}\times\cdots\times\overline{\mathcal{K}}\longrightarrow H^{p,p}_{\mathrm{SP}}(X,\mathbb{R})$$

on the p-fold product of the Kähler cone and its closure. We also have $H^{p,p}_{\mathrm{SP}}(X,\mathbb{R})\subset H^{p,p}_{\geqslant 0}(X,\mathbb{R})$ where $H^{p,p}_{\geqslant 0}(X,\mathbb{R})$ is the cone of classes of d-closed weakly positive currents of type (p,p), and the Serre duality pairing induces a positive intersection product

$$(17.4)\qquad H^{p,p}_{\mathrm{SP}}(X,\mathbb{R})\times H^{n-p,n-p}_{\geqslant 0}(X,\mathbb{R})\longrightarrow\mathbb{R}_+,\qquad (\alpha,T)\longmapsto\int_X\alpha\wedge T\in\mathbb{R}_+$$

(notice that if α is strongly positive and $T \geqslant 0$, then $\alpha \wedge T$ is a positive measure).

If $\mathscr{C}$ is a convex cone in a finite dimensional vector space E, we denote by $\mathscr{C}^\vee$ the dual cone, i.e. the set of linear forms $u \in E^*$ which take nonnegative values on all elements of $\mathscr{C}$. By the Hahn-Banach theorem, we always have $\mathscr{C}^{\vee\vee} = \overline{\mathscr{C}}$. A basic problem would be to investigate whether $H^{p,p}_{\mathrm{SP}}(X,\mathbb{R})$ and $H^{n-p,n-p}_{\geqslant 0}(X,\mathbb{R})$ are always dual cones, and another even harder question, which somehow encompasses the Hodge conjecture, would be to relate these cones to the cones generated by cohomology classes of effective analytic cycles. We are essentially unable to address these extremely difficult questions, except in the special cases $p = 1$ or $p = n-1$ which are much better understood and are the main target of the following sections.

17.B. Numerically Positive Classes of Type (1,1)

We describe here the main results obtained in [DP04]. The upshot is that the Kähler cone depends only on the intersection product of the cohomology ring, the Hodge structure and the homology classes of analytic cycles. More precisely, we have:

(17.5) Theorem. *Let X be a compact Kähler manifold. Let $\mathscr{P}$ be the set of real $(1,1)$ cohomology classes $\{\alpha\}$ which are* numerically positive *on analytic cycles, i.e. such that $\int_Y \alpha^p > 0$ for every irreducible analytic set Y in X, $p = \dim Y$. Then the Kähler cone $\mathscr{K}$ of X is one of the connected components of $\mathscr{P}$.*

(17.6) Special Case. *If X is projective algebraic, then $\mathscr{K} = \mathscr{P}$.*

These results (which are new even in the projective case) can be seen as a generalization of the well-known Nakai-Moishezon criterion. Recall that the Nakai-Moishezon criterion provides a necessary and sufficient criterion for a line bundle to be ample: *a line bundle $L \to X$ on a projective algebraic manifold X is ample if and only if*

$$L^p \cdot Y = \int_Y c_1(L)^p > 0,$$

for every algebraic subset $Y \subset X$, $p = \dim Y$.

It turns out that the numerical conditions $\int_Y \alpha^p > 0$ also characterize arbitrary transcendental Kähler classes when X is projective: this is precisely the meaning of the Special Case 17.6.

(17.7) Example. The following example shows that the cone $\mathscr{P}$ need not be connected (and also that the components of $\mathscr{P}$ need not be convex, either). Let us consider for instance a complex torus $X = \mathbb{C}^n/\Lambda$. It is well-known that a generic torus X does not possess any analytic subset except finite subsets and X itself. In that case, the numerical positivity is expressed by the single condition $\int_X \alpha^n > 0$. However, on a torus, $(1,1)$-classes are in one-to-one correspondence with constant Hermitian forms α on $\mathbb{C}^n$. Thus, for X generic, $\mathscr{P}$ is the set of Hermitian forms on $\mathbb{C}^n$ such that $\det(\alpha) > 0$, and Theorem 17.5 just expresses the elementary result of linear algebra saying that the set $\mathcal{K}$ of positive definite forms is one of the connected components of the open set $\mathcal{P} = \{\det(\alpha) > 0\}$ of Hermitian forms of positive determinant (the other components, of course, are the sets of forms of signature (p, q), $p + q = n$, q even. They are not convex when $p > 0$ and $q > 0$).

Sketch of proof of Theorems 17.5 *and* 17.6. By Definition 13.15, a *Kähler current* is a closed positive current T of type $(1, 1)$ such that $T \geqslant \varepsilon\omega$ for some smooth Kähler metric ω and $\varepsilon > 0$ small enough. The crucial steps of the proof of Theorem 17.5 are contained in the following statements.

(17.8) Proposition (Păun [Pau98a, 98b]). *Let X be a compact complex manifold (or more generally a compact complex space). Then*

(a) *The cohomology class of a closed positive $(1,1)$-current $\{T\}$ is nef if and only if the restriction $\{T\}_{|Z}$ is nef for every irreducible component Z in any of the Lelong sublevel sets $E_c(T)$.*

(b) *The cohomology class of a Kähler current $\{T\}$ is a Kähler class (i.e. the class of a* smooth *Kähler form) if and only if the restriction $\{T\}_{|Z}$ is a Kähler class for every irreducible component Z in any of the Lelong sublevel sets $E_c(T)$.*

The proof of Proposition 17.8 is not extremely hard if we take for granted the fact that Kähler currents can be approximated by Kähler currents with logarithmic poles, a fact which was first proved in Section 13.B (see also [Dem92]). Thus in (b), we may assume that $T = \alpha + i\partial\overline{\partial}\varphi$ is a current with analytic singularities, where φ is a quasi-psh function with logarithmic poles on some analytic set Z, and φ smooth on $X \smallsetminus Z$. Now, we proceed by an induction on dimension (to do this, we have to consider analytic spaces rather than with complex manifolds, but it turns out that this makes no difference for the proof). Hence, by the induction hypothesis, there exists a smooth potential ψ on Z such that $\alpha_{|Z} + i\partial\overline{\partial}\psi > 0$ along Z. It is well known that one can then find a potential $\widetilde{\psi}$ on X such that $\alpha + i\partial\overline{\partial}\widetilde{\psi} > 0$ in a neighborhood V of Z (but possibly non positive elsewhere). Essentially, it is enough to take an arbitrary extension of ψ to X and to add a large multiple of the square of the distance to Z, at least near smooth points; otherwise, we stratify Z by its successive singularity loci, and proceed again by induction on the dimension of these loci. Finally, we use a a standard gluing procedure : the current $T = \alpha + i\max_\varepsilon(\varphi, \widetilde{\psi} - C)$, $C \gg 1$, will be equal to $\alpha + i\partial\overline{\partial}\varphi > 0$ on $X \smallsetminus V$, and to a smooth Kähler form on V. □

The next (and more substantial step) consists of the following result which is reminiscent of the Grauert-Riemenschneider conjecture ([Siu84; Dem85b], cf. Corollary 8.3).

(17.9) Theorem ([DP04]). *Let X be a compact Kähler manifold and let $\{\alpha\}$ be a nef class (i.e. $\{\alpha\} \in \overline{\mathcal{K}}$). Assume that $\int_X \alpha^n > 0$. Then $\{\alpha\}$ contains a Kähler current T, in other words $\{\alpha\} \in \mathcal{E}^\circ$.*

Step 1. The basic argument is to prove that for every irreducible analytic set $Y \subset X$ of codimension p, the class $\{\alpha\}^p$ contains a closed positive (p,p)-current Θ such that $\Theta \geqslant \delta[Y]$ for some $\delta > 0$. For this, we use in an essential way the Calabi-Yau theorem [Yau78] on solutions of Monge-Ampère equations, which yields the following result as a special case:

(17.10) Lemma ([Yau78]). *Let (X,ω) be a compact Kähler manifold and $n = \dim X$. Then for any smooth volume form $f > 0$ such that $\int_X f = \int_X \omega^n$, there exists a Kähler metric $\widetilde{\omega} = \omega + i\partial\overline{\partial}\varphi$ in the same Kähler class as ω, such that $\widetilde{\omega}^n = f$.* □

We exploit this by observing that $\alpha + \varepsilon\omega$ is a Kähler class. Hence we can solve the Monge-Ampère equation

(17.10 a) $$(\alpha + \varepsilon\omega + i\partial\overline{\partial}\varphi_\varepsilon)^n = C_\varepsilon \omega_\varepsilon^n$$

where $\{\omega_\varepsilon\}$ is a family of Kähler metrics contained in the Kähler class $\{\omega\}$ chosen such that a fixed fraction of their volume is concentrated in an ε-tubular neighborhood V_ε of Y; these metrics ω_ε can be easily constructed by adding to ω the $\partial\overline{\partial}$ of a potential of the form of Y — in X and $V_\varepsilon := \{\sum \theta_\alpha |g_{j,\alpha}|^2 < \varepsilon^2\}$. Here we take.

$$C_\varepsilon = \frac{\int_X (\alpha + \varepsilon\omega)^n}{\int_X \omega^n} \geqslant C_0 = \frac{\int_X \alpha^n}{\int_X \omega^n} > 0.$$

Let us put $\alpha_\varepsilon := \alpha + \varepsilon\omega + i\partial\partial\varphi_\varepsilon$ and denote by

$$\lambda_1(z) \leqslant \cdots \leqslant \lambda_n(z)$$

the eigenvalues of $\alpha_\varepsilon(z)$ with respect to $\omega_\varepsilon(z)$, at every point $z \in X$ (these functions are continuous with respect to z, and of course depend also on ε). The equation (17.10 a) is equivalent to the fact that

(17.10 b) $$\lambda_1(z) \cdots \lambda_n(z) = C_\varepsilon$$

is constant, and the most important observation for us is that the constant C_ε is bounded away from 0, thanks to our assumption $\int_X \alpha^n > 0$.

Fix a regular point $x_0 \in Y$ and a small neighborhood U (meeting only the irreducible component of x_0 in Y). By the choice of ω_ε, we have (exercise!) a uniform lower bound

(17.10 c) $$\int_{U \cap V_\varepsilon} \omega_\varepsilon^p \wedge \omega^{n-p} \geqslant \delta_p(U) > 0.$$

Now, by looking at the p smallest (resp. $(n-p)$ largest) eigenvalues λ_j of α_ε with respect to ω_ε, we find

(17.10 d) $$\alpha_\varepsilon^p \geqslant \lambda_1 \cdots \lambda_p\, \omega_\varepsilon^p,$$

(17.10 e) $$\alpha_\varepsilon^{n-p} \wedge \omega_\varepsilon^p \geqslant \frac{1}{n!} \lambda_{p+1} \cdots \lambda_n\, \omega_\varepsilon^n,$$

The last inequality (17.10 e) implies

$$\int_X \lambda_{p+1}\cdots\lambda_n\,\omega_\varepsilon^n \leqslant n!\int_X \alpha_\varepsilon^{n-p}\wedge\omega_\varepsilon^p = n!\int_X (\alpha+\varepsilon\omega)^{n-p}\wedge\omega^p \leqslant M$$

for some constant $M > 0$ (we assume $\varepsilon \leqslant 1$, say). In particular, for every $\delta > 0$, the subset $E_\delta \subset X$ of points z such that $\lambda_{p+1}(z)\cdots\lambda_n(z) > M/\delta$ satisfies $\int_{E_\delta}\omega_\varepsilon^n \leqslant \delta$, hence

(17.10 f) $$\int_{E_\delta}\omega_\varepsilon^p\wedge\omega^{n-p} \leqslant 2^{n-p}\int_{E_\delta}\omega_\varepsilon^n \leqslant 2^{n-p}\delta.$$

The combination of (17.10 c) and (17.10 f) yields

$$\int_{(U\cap V_\varepsilon)\smallsetminus E_\delta}\omega_\varepsilon^p\wedge\omega^{n-p} \geqslant \delta_p(U) - 2^{n-p}\delta.$$

On the other hand (17.10 b) and (17.10 d) imply

$$\alpha_\varepsilon^p \geqslant \frac{C_\varepsilon}{\lambda_{p+1}\cdots\lambda_n}\omega_\varepsilon^p \geqslant \frac{C_\varepsilon}{M/\delta}\omega_\varepsilon^p \qquad \text{on } (U\cap V_\varepsilon)\smallsetminus E_\delta.$$

From this we infer

(17.10 g) $$\int_{U\cap V_\varepsilon}\alpha_\varepsilon^p\wedge\omega^{n-p} \geqslant \frac{C_\varepsilon}{M/\delta}\int_{(U\cap V_\varepsilon)\smallsetminus E_\delta}\omega_\varepsilon^p\wedge\omega^{n-p} \geqslant \frac{C_\varepsilon}{M/\delta}(\delta_p(U) - 2^{n-p}\delta) > 0$$

provided that δ is taken small enough, e.g. $\delta = 2^{-(n-p+1)}\delta_p(U)$. The family of (p,p)-forms α_ε^p is uniformly bounded in mass since

$$\int_X \alpha_\varepsilon^p\wedge\omega^{n-p} = \int_X (\alpha+\varepsilon\omega)^p\wedge\omega^{n-p} \leqslant \text{Const.}$$

Inequality (17.10 g) implies that any weak limit Θ of (α_ε^p) carries a positive mass on $U\cap Y$. By Skoda's extension theorem [Sko82], $\mathbf{1}_Y\Theta$ is a closed positive current with support in Y, hence $\mathbf{1}_Y\Theta = \sum c_j[Y_j]$ is a combination of the various components Y_j of Y with coefficients $c_j > 0$. Our construction shows that Θ belongs to the cohomology class $\{\alpha\}^p$. Step 1 of Theorem 17.9 is proved.

Step 2. The second and final step consists in using a "diagonal trick": for this, we apply Step 1 to

$$\widetilde{X} = X\times X, \qquad \widetilde{Y} = \text{diagonal}\,\Delta\subset\widetilde{X}, \qquad \widetilde{\alpha} = \mathrm{pr}_1^*\alpha + \mathrm{pr}_2^*\alpha.$$

It is then clear that $\widetilde{\alpha}$ is nef on $\widetilde{X}$ and that

$$\int_{\widetilde{X}}(\widetilde{\alpha})^{2n} = \binom{2n}{n}\Big(\int_X \alpha^n\Big)^2 > 0.$$

It follows by Step 1 that the class $\{\widetilde{\alpha}\}^n$ contains a Kähler current Θ of bidegree (n,n) such that $\Theta \geqslant \delta[\Delta]$ for some $\delta > 0$. Therefore the push-forward

$$T := (\mathrm{pr}_1)_*(\Theta \wedge \mathrm{pr}_2^*\omega)$$

is a positive $(1,1)$-current such that

$$T \geqslant \delta(\mathrm{pr}_1)_*([\Delta] \wedge \mathrm{pr}_2^*\omega) = \delta\omega.$$

It follows that T is a Kähler current. On the other hand, T is numerically equivalent to $(\mathrm{pr}_1)_*(\widetilde{\alpha}^n \wedge \mathrm{pr}_2^*\omega)$, which is the form given in coordinates by

$$x \mapsto \int_{y\in X} \big(\alpha(x) + \alpha(y)\big)^n \wedge \omega(y) = C\alpha(x)$$

where $C = n\int_X \alpha(y)^{n-1} \wedge \omega(y)$. Hence $T \equiv C\alpha$, which implies that $\{\alpha\}$ contains a Kähler current. Theorem 17.9 is proved. □

End of Proof of Theorems 17.5 *and* 17.6. Clearly the open cone $\mathcal{K}$ is contained in $\mathcal{P}$, hence in order to show that $\mathcal{K}$ is one of the connected components of $\mathcal{P}$, we need only show that $\mathcal{K}$ is closed in $\mathcal{P}$, i.e. that $\overline{\mathcal{K}} \cap \mathcal{P} \subset \mathcal{K}$. Pick a class $\{\alpha\} \in \overline{\mathcal{K}} \cap \mathcal{P}$. In particular $\{\alpha\}$ is nef and satisfies $\int_X \alpha^n > 0$. By Theorem 17.9 we conclude that $\{\alpha\}$ contains a Kähler current T. However, an induction on dimension using the assumption $\int_Y \alpha^p$ for all analytic subsets Y (we also use resolution of singularities for Y at this step) shows that the restriction $\{\alpha\}_{|Y}$ is the class of a Kähler current on Y. We conclude that $\{\alpha\}$ is a Kähler class by Proposition 17.8 (b), therefore $\{\alpha\} \in \mathcal{K}$, as desired. □

The Projective Case 17.6 is a consequence of the following variant of Theorem 17.5.

(17.11) Corollary. *Let X be a compact Kähler manifold. A $(1,1)$ cohomology class $\{\alpha\}$ on X is Kähler if and only if there exists a Kähler metric ω on X such that $\int_Y \alpha^k \wedge \omega^{p-k} > 0$ for all irreducible analytic sets Y and all $k = 1, 2, \ldots, p = \dim Y$.*

Proof. The assumption clearly implies that

$$\int_Y (\alpha + t\omega)^p > 0$$

for all $t \in \mathbb{R}_+$, hence the half-line $\alpha + (\mathbb{R}_+)\omega$ is entirely contained in the cone $\mathcal{P}$ of numerically positive classes. Since $\alpha + t_0\omega$ is Kähler for t_0 large, we conclude that the half-line in entirely contained in the connected component $\mathcal{K}$, and therefore $\alpha \in \mathcal{K}$. □

In the projective case, we can take $\omega = c_1(H)$ for a given very ample divisor H, and the condition $\int_Y \alpha^k \wedge \omega^{p-k} > 0$ is equivalent to

$$\int_{Y\cap H_1\cap\cdots\cap H_{p-k}} \alpha^k > 0$$

for a suitable complete intersection $Y \cap H_1 \cap \cdots \cap H_{p-k}$, $H_j \in |H|$. This shows that algebraic cycles are sufficient to test the Kähler property, and the special case 17.6 follows. On the other hand, we can pass to the limit in Corollary 17.11 by replacing α by $\alpha + \varepsilon\omega$, and in this way we get also a characterization of nef classes.

(17.12) Corollary. *Let X be a compact Kähler manifold. A $(1,1)$ cohomology class $\{\alpha\}$ on X is nef if and only if there exists a Kähler metric ω on X such that $\int_Y \alpha^k \wedge \omega^{p-k} \geqslant 0$ for all irreducible analytic sets Y and all $k = 1, 2, \ldots, p = \dim Y$.*

By a formal convexity argument, one can derive from Corollary 17.11 or 17.12 the following interesting consequence about the dual of the cone $\mathcal{K}$.

(17.13) Theorem. *Let X be a compact Kähler manifold.*

(a) *A $(1,1)$ cohomology class $\{\alpha\}$ on X is nef if and only for every irreducible analytic set Y in X, $p = \dim X$ and every Kähler metric ω on X we have $\int_Y \alpha \wedge \omega^{p-1} \geqslant 0$. (Actually this numerical condition is needed only for Kähler classes $\{\omega\}$ which belong to a 2-dimensional space $\mathbb{R}\{\alpha\} + \mathbb{R}\{\omega_0\}$, where $\{\omega_0\}$ is a given Kähler class).*

(b) *The dual of the nef cone $\overline{\mathcal{K}}$ is the closed convex cone in $H^{n-1,n-1}(X, \mathbb{R})$ generated by cohomology classes of currents of the form $[Y] \wedge \omega^{p-1}$ in $H^{n-1,n-1}(X, \mathbb{R})$, where Y runs over the collection of irreducible analytic subsets of X and $\{\omega\}$ over the set of Kähler classes of X. This dual cone coincides with $H^{n-1,n-1}_{\geqslant 0}(X, \mathbb{R})$.*

Proof. (a) Clearly a nef class $\{\alpha\}$ satisfies the given numerical condition. The proof of the converse is more tricky. First, observe that for every integer $p \geqslant 1$, there exists a polynomial identity of the form

$$(17.14) \qquad (y - \delta x)^p - (1 - \delta)^p x^p = (y - x) \int_0^1 A_p(t, \delta)\big((1 - t)x + ty\big)^{p-1} dt$$

where $A_p(t, \delta) = \sum_{0 \leqslant m \leqslant p} a_m(t)\delta^m \in \mathbb{Q}[t, \delta]$ is a polynomial of degree $\leqslant p - 1$ in t (moreover, the polynomial A_p is unique under this limitation for the degree). To see this, we observe that $(y - \delta x)^p - (1 - \delta)^p x^p$ vanishes identically for $x = y$, so it is divisible by $y - x$. By homogeneity in (x, y), we have an expansion of the form

$$(y - \delta x)^p - (1 - \delta)^p x^p = (y - x) \sum_{0 \leqslant \ell \leqslant p-1,\, 0 \leqslant m \leqslant p} b_{\ell,m} x^\ell y^{p-1-\ell} \delta^m$$

in the ring $\mathbb{Z}[x, y, \delta]$. Formula (17.14) is then equivalent to

$$(17.15) \qquad b_{\ell,m} = \int_0^1 a_m(t) \binom{p-1}{\ell} (1 - t)^\ell t^{p-1-\ell} \, dt.$$

Since $(U, V) \mapsto \int_0^1 U(t)V(t)dt$ is a non degenerate linear pairing on the space of polynomials of degree $\leqslant p - 1$ and since $\big(\binom{p-1}{\ell}(1 - t)^\ell t^{p-1-\ell}\big)_{0 \leqslant \ell \leqslant p-1}$ is a basis of this space, (17.15) can be achieved for a unique choice of the polynomials $a_m(t)$. A straightforward calculation shows that $A_p(t,0) = p$ identically. We can therefore choose $\delta_0 \in [0,1[$ so small that $A_p(t,\delta) > 0$ for all $t \in [0,1]$, $\delta \in [0,\delta_0]$ and $p = 1, 2, \ldots, n$.

Now, fix a Kähler metric ω such that $\omega' = \alpha + \omega$ yields a Kähler class $\{\omega'\}$ (just take a large multiple $\omega = k\omega_0$, $k \gg 1$, of the given Kähler metric ω_0 to initialize the process). A substitution $x = \omega$ and $y = \omega'$ in our polynomial identity yields

$$(\alpha + (1-\delta)\omega)^p - (1-\delta)^p\omega^p = \int_0^1 A_p(t,\delta)\,\alpha \wedge \big((1-t)\omega + t\omega'\big)^{p-1} dt.$$

For every irreducible analytic subset $Y \subset X$ of dimension p we find

$$\int_Y (\alpha + (1-\delta)\omega)^p - (1-\delta)^p \int_Y \omega^p = \int_0^1 A_p(t,\delta)dt \Big(\int_Y \alpha \wedge ((1-t)\omega + t\omega')^{p-1} \Big).$$

However, $(1-t)\omega + t\omega'$ is a Kähler class (contained in $\mathbb{R}\{\alpha\} + \mathbb{R}\{\omega_0\}$) and therefore $\int_Y \alpha \wedge \big((1-t)\omega + t\omega'\big)^{p-1} \geqslant 0$ by the numerical condition. This implies $\int_Y (\alpha + (1-\delta)\omega)^p > 0$ for all $\delta \in [0,\delta_0]$. We have produced a segment entirely contained in $\mathcal{P}$ such that one extremity $\{\alpha + \omega\}$ is in $\mathcal{K}$, so the other extremity $\{\alpha + (1-\delta_0)\omega\}$ is also in $\mathcal{K}$. By repeating the argument inductively after replacing ω with $(1-\delta_0)\omega$, we see that $\{\alpha + (1-\delta_0)^\nu \omega\} \in \mathcal{K}$ for every integer $\nu \geqslant 0$. From this we infer that $\{\alpha\}$ is nef, as desired.

(b) Part (a) can be reformulated by saying that the dual cone $\overline{\mathcal{K}}^\vee$ is the closure of the convex cone generated by $(n-1, n-1)$ cohomology classes of the form $[Y] \wedge \omega^{p-1}$. Since these classes are contained in $H^{n-1,n-1}_{\geqslant 0}(X,\mathbb{R})$ which is also contained in $\overline{\mathcal{K}}^\vee$ by (17.6), we infer that

$$\overline{\mathcal{K}}^\vee = H^{n-1,n-1}_{\geqslant 0}(X,\mathbb{R}) = \overline{\mathrm{Cone}(\{[Y] \wedge \omega^{p-1}\})}. \qquad \square$$

17.C. Deformations of Compact Kähler Manifolds

Our main Theorem 17.5 also has an important application to the deformation theory of compact Kähler manifolds.

(17.16) Theorem. *Let $\pi : \mathcal{X} \to S$ be a deformation of compact Kähler manifolds over an irreducible base S. Then there exists a countable union $S' = \bigcup S_\nu$ of analytic subsets $S_\nu \subsetneq S$, such that the Kähler cones $\mathcal{K}_t \subset H^{1,1}(X_t,\mathbb{C})$ of the fibers $X_t = \pi^{-1}(t)$ are invariant over $S \smallsetminus S'$ under parallel transport with respect to the $(1,1)$-projection $\nabla^{1,1}$ of the Gauss-Manin connection ∇ in the decomposition of*

$$\nabla = \begin{pmatrix} \nabla^{2,0} & * & 0 \\ * & \nabla^{1,1} & * \\ 0 & * & \nabla^{0,2} \end{pmatrix}$$

on the Hodge bundle $H^2 = H^{2,0} \oplus H^{1,1} \oplus H^{0,2}$.

We moreover conjecture that for an arbitrary deformation $\mathcal{X} \to S$ of compact complex manifolds, the Kähler property is open with respect to the countable Zariski topology on the base S of the deformation.

Let us recall the general fact that all fibers X_t of a deformation over a connected base S are diffeomorphic, since $\mathcal{X} \to S$ is a locally trivial differentiable bundle. This implies that the cohomology bundle

$$S \ni t \mapsto H^k(X_t, \mathbb{C})$$

is locally constant over the base S. The corresponding (flat) connection of this bundle is called the Gauss-Manin connection, and will be denoted here by ∇. As is well known, the Hodge filtration

$$F^p(H^k(X_t, \mathbb{C})) = \bigoplus_{r+s=k, r \geqslant p} H^{r,s}(X_t, \mathbb{C})$$

defines a *holomorphic* subbundle of $H^k(X_t, \mathbb{C})$ (with respect to its locally constant structure). On the other hand, the Dolbeault groups are given by

$$H^{p,q}(X_t, \mathbb{C}) = F^p(H^k(X_t, \mathbb{C})) \cap \overline{F^{k-p}(H^k(X_t, \mathbb{C}))}, \qquad k = p + q,$$

and they form *real analytic* subbundles of $H^k(X_t, \mathbb{C})$. We are interested especially in the decomposition

$$H^2(X_t, \mathbb{C}) = H^{2,0}(X_t, \mathbb{C}) \oplus H^{1,1}(X_t, \mathbb{C}) \oplus H^{0,2}(X_t, \mathbb{C})$$

and the induced decomposition of the Gauss-Manin connection acting on H^2

$$\nabla = \begin{pmatrix} \nabla^{2,0} & * & * \\ * & \nabla^{1,1} & * \\ * & * & \nabla^{0,2} \end{pmatrix}.$$

Here the stars indicate suitable bundle morphisms – actually with the lower left and upper right stars being zero by Griffiths' transversality property, but we do not really care here. The notation $\nabla^{p,q}$ stands for the induced (real analytic, not necessarily flat) connection on the subbundle $t \mapsto H^{p,q}(X_t, \mathbb{C})$.

Sketch of Proof of Theorem 17.16. The result is local on the base, hence we may assume that S is contractible. Then the family is differentiably trivial, the Hodge bundle $t \mapsto H^2(X_t, \mathbb{C})$ is the trivial bundle and $t \mapsto H^2(X_t, \mathbb{Z})$ is a trivial lattice. We use the existence of a relative cycle space $C^p(\mathcal{X}/S) \subset C^p(\mathcal{X})$ which consists of all cycles contained in the fibres of $\pi : X \to S$. It is equipped with a canonical holomorphic projection

$$\pi_p : C^p(\mathcal{X}/S) \to S.$$

We then define the S_ν's to be the images in S of those connected components of $C^p(\mathcal{X}/S)$ which do not project onto S. By the fact that the projection is proper on each component, we infer that S_ν is an analytic subset of S. The definition of the S_ν's imply that the cohomology classes induced by the analytic cycles $\{[Z]\}$, $Z \subset X_t$, remain exactly the same for all $t \in S \smallsetminus S'$. This result implies in its turn that the conditions defining the numerically positive cones $\mathcal{P}_t$ remain the same, except for the fact that the spaces $H^{1,1}(X_t, \mathbb{R}) \subset H^2(X_t, \mathbb{R})$ vary along with the Hodge decomposition. At this point, a standard calculation implies that the $\mathcal{P}_t$ are invariant by parallel transport under $\nabla^{1,1}$. This is done as follows.

Since S is irreducible and S' is a countable union of analytic sets, it follows that $S \smallsetminus S'$ is arcwise connected by piecewise smooth analytic arcs. Let

$$\gamma : [0,1] \to S \smallsetminus S', \qquad u \mapsto t = \gamma(u)$$

be such a smooth arc, and let $\alpha(u) \in H^{1,1}(X_{\gamma(u)}, \mathbb{R})$ be a family of real $(1,1)$-cohomology classes which are constant by parallel transport under $\nabla^{1,1}$. This is equivalent to assuming that

$$\nabla(\alpha(u)) \in H^{2,0}(X_{\gamma(u)}, \mathbb{C}) \oplus H^{0,2}(X_{\gamma(u)}, \mathbb{C})$$

for all u. Suppose that $\alpha(0)$ is a numerically positive class in $X_{\gamma(0)}$. We then have

$$\alpha(0)^p \cdot \{[Z]\} = \int_Z \alpha(0)^p > 0$$

for all p-dimensional analytic cycles Z in $X_{\gamma(0)}$. Let us denote by

$$\zeta_Z(t) \in H^{2q}(X_t, \mathbb{Z}), \qquad q = \dim X_t - p,$$

the family of cohomology classes equal to $\{[Z]\}$ at $t = \gamma(0)$, such that $\nabla \zeta_Z(t) = 0$ (i.e. constant with respect to the Gauss-Manin connection). By the above discussion, $\zeta_Z(t)$ is of type (q,q) for all $t \in S$, and when $Z \subset X_{\gamma(0)}$ varies, $\zeta_Z(t)$ generates all classes of analytic cycles in X_t if $t \in S \smallsetminus S'$. Since ζ_Z is ∇-parallel and $\nabla\alpha(u)$ has no component of type $(1,1)$, we find

$$\frac{d}{du}(\alpha(u)^p \cdot \zeta_Z(\gamma(u)) = p\alpha(u)^{p-1} \cdot \nabla\alpha(u) \cdot \zeta_Z(\gamma(u)) = 0.$$

We infer from this that $\alpha(u)$ is a numerically positive class for all $u \in [0,1]$. This argument shows that the set $\mathcal{P}_t$ of numerically positive classes in $H^{1,1}(X_t, \mathbb{R})$ is invariant by parallel transport under $\nabla^{1,1}$ over $S \smallsetminus S'$.

By a standard result of Kodaira-Spencer [KS60] relying on elliptic PDE theory, every Kähler class in X_{t_0} can be deformed to a nearby Kähler class in nearby fibres X_t. This implies that the connected component of $\mathcal{P}_t$ which corresponds to the Kähler cone $\mathcal{K}_t$ must remain the same. The theorem is proved. □

As a by-product of our techniques, especially the regularization theorem for currents, we also get the following result for which we refer to [DP04].

(17.17) Theorem. *A compact complex manifold carries a Kähler current if and only if it is bimeromorphic to a Kähler manifold (or equivalently, dominated by a Kähler manifold).*

This class of manifolds is called the *Fujiki class* $\mathscr{C}$. If we compare this result with the solution of the Grauert-Riemenschneider conjecture 8.3, we are led to the following statement, which is a weaker form of Conjecture 8.21.

(17.18) Conjecture. *Let X be a compact complex manifold of dimension n. Assume that X possesses a cohomology class $\{\alpha\}$ of type (1,1) such that $\int_{X(u,\leqslant 1)} u^n > 0$ for some smooth representative $u \in \alpha$. Then $\{\alpha\}$ contains a Kähler current and X is in the Fujiki class $\mathscr{C}$.*

In the case where α is nef and the assumption is replaced by $\int_X \alpha^n > 0$, Conjecture 17.18 has been recently confirmed by D. Popovici [Pop08] (with a highly technical proof). We want also to mention that most of the above results were already known in the cases of complex surfaces (i.e. in dimension 2), thanks to the work of N. Buchdahl [Buc99, 00] and of A. Lamari [Lam99a, 99b].

Shortly after the original [DP04] manuscript appeared in April 2001, Daniel Huybrechts [Huy01] informed us that Theorem 17.5 can be used to calculate the Kähler cone of a very general hyperkähler manifold: the Kähler cone is then equal to a suitable connected component of the positive cone defined by the Beauville-Bogomolov quadratic form. In the case of an arbitrary hyperkähler manifold, S.Boucksom [Bou02] later showed that a $(1, 1)$ class $\{\alpha\}$ is Kähler if and only if it lies in the positive part of the Beauville-Bogomolov quadratic cone and moreover $\int_C \alpha > 0$ for all *rational curves* $C \subset X$ (see also [Huy99]).

Chapter 18

Structure of the Pseudo-effective Cone and Mobile Intersection Theory

18.A. Classes of Mobile Curves and of Mobile $(n-1, n-1)$-currents

We introduce various positive cones in $H^{n-1,n-1}(X,\mathbb{R})$, some of which exhibit certain "mobility" properties, in the sense that they can be more or less freely deformed. Ampleness is clearly such a property, since a very ample divisor A can be moved in its linear system $|A|$ so as to cover the whole ambient variety. By extension, a Kähler class $\{\omega\} \in H^{1,1}(X,\mathbb{R})$ is also considered to be mobile, as illustrated alternatively by the fact that the Monge-Ampère volume form $(\omega + i\partial\overline{\partial}\varphi)^n$ of a Kähler metric in the same cohomology class can be taken to be equal to an arbitrary volume form $f > 0$ with $\int_X f = \int_X \omega^n$ (thanks to Yau's theorem [Yau78]).

(18.1) Definition. *Let X be a smooth projective variety.*

(a) *One defines* NE(X) *to be the convex cone generated by cohomology classes of all effective curves in $H^{n-1,n-1}(X,\mathbb{R})$*

(b) *We say that C is a* mobile curve *if $C = C_{t_0}$ is a member of an analytic family $\{C_t\}_{t\in S}$ such that $\bigcup_{t\in S} C_t = X$ and, as such, is a reduced irreducible 1-cycle. We define the mobile cone* ME(X)*, to be the convex cone generated by all mobile curves.*

(c) *If X is projective, we say that an effective 1-cycle C is a* strongly mobile *if we have*

$$C = \mu_*(\widetilde{A}_1 \cap \cdots \cap \widetilde{A}_{n-1})$$

for suitable very ample divisors $\widetilde{A}_j$ on $\widetilde{X}$, where $\mu : \widetilde{X} \to X$ is a modification. We let $\mathrm{ME}^s(X)$ *be the convex cone generated by all strongly mobile effective 1-cycles (notice that by taking $\widetilde{A}_j$ general enough these classes can be represented by reduced irreducible curves; also, by Hironaka, one could just restrict oneself to compositions of blow-ups with smooth centers).*

Clearly, we have

$$\mathrm{ME}^s(X) \subset \mathrm{ME}(X) \subset \mathrm{NE}(X).$$

The cone $\mathrm{NE}(X)$ is contained in the analogue of the Neron-Severi group for $(n-1,n-1)$-classes, namely

$$\mathrm{NS}_{\mathbb{R}}^{n-1}(X) := (H^{n-1,n-1}(X,\mathbb{R}) \cap H^{2n-2}(X,\mathbb{Z})/\mathrm{tors}) \otimes_{\mathbb{Z}} \mathbb{R}$$

(sometimes also denoted $N_1(X)$ in the literature). We wish to introduce similar concepts for cones of non necessarily integral classes, on arbitrary compact Kähler manifolds. The relevant definition is as follows.

(18.2) Definition. *Let X be a compact Kähler manifold.*

(a) *We define $\mathcal{N} = H_{\geqslant 0}^{n-1,n-1}(X,\mathbb{R})$ to be the (closed) convex cone in $H^{n-1,n-1}(X,\mathbb{R})$ generated by classes of positive currents T of type $(n-1,n-1)$, i.e., of bidimension $(1,1)$.*

(b) *We define the cone $\mathcal{M}^s \subset H^{n-1,n-1}(X,\mathbb{R})$ of* strongly mobile classes *to be the closure of the convex cone generated by classes of currents of the form*

$$\mu_*(\widetilde{\omega}_1 \wedge \cdots \wedge \widetilde{\omega}_{n-1})$$

where $\mu : \widetilde{X} \to X$ is an arbitrary modification, and the $\widetilde{\omega}_j$ are Kähler forms on $\widetilde{X}$.

(c) *We define the cone $\mathcal{M} \subset H^{n-1,n-1}(X,\mathbb{R})$ of* mobile classes *to be the closure of the convex cone generated by classes of currents of the form*

$$\mu_*([\widetilde{Y}_{t_0}] \wedge \widetilde{\omega}_1 \wedge \cdots \wedge \widetilde{\omega}_{p-1})$$

where $\mu : \widetilde{X} \to X$ is an arbitrary modification, the $\widetilde{\omega}_j$ are Kähler forms on $\widetilde{X}$ and $(\widetilde{Y}_t)_{t\in S}$ is an analytic family of effective p-dimensional analytic cycles covering $\widetilde{X}$ such that $\widetilde{Y}_{t_0}$ is reduced and irreducible, with p running over all $\{1,2,\ldots,n\}$.

Clearly, we have

$$\mathcal{M}^s \subset \mathcal{M} \subset \mathcal{N}.$$

For X projective, it is also immediately clear from the definitions that

$$(18.3)\qquad \begin{cases} \overline{\mathrm{NE}(X)} \subset \mathcal{N}_{\mathrm{NS}} := \mathcal{N} \cap \mathrm{NS}_{\mathbb{R}}^{n-1}(X), \\ \overline{\mathrm{ME}(X)} \subset \mathcal{M}_{\mathrm{NS}} := \mathcal{M} \cap \mathrm{NS}_{\mathbb{R}}^{n-1}(X), \\ \overline{\mathrm{ME}^s(X)} \subset \mathcal{M}^s_{\mathrm{NS}} := \mathcal{M}^s \cap \mathrm{NS}_{\mathbb{R}}^{n-1}(X). \end{cases}$$

The upshot of these definitions lie in the following easy observation.

(18.4) Proposition. *Let X be a compact Kähler manifold. The Serre duality pairing*

$$H^{1,1}(X,\mathbb{R}) \times H^{n-1,n-1}(X,\mathbb{R}) \longrightarrow \mathbb{R}, \qquad (\alpha,\beta) \longmapsto \int_X \alpha \wedge \beta$$

takes nonnegative values

(a) *for all pairs* $(\alpha,\beta)\in\overline{\mathcal{K}}\times\mathcal{N}$;

(b) *for all pairs* $(\alpha,\beta)\in\mathcal{E}\times\mathcal{M}$.

Proof. (a) is obvious. In order to prove (b), we may assume that $\beta=\mu_*([Y_{t_0}]\wedge\widetilde{\omega}_1\wedge\cdots\wedge\widetilde{\omega}_{p-1})$ for some modification $\mu:\widetilde{X}\to X$, where $\{\alpha\}=\{T\}$ is the class of a positive $(1,1)$-current on X and $\widetilde{\omega}_j$ are Kähler forms on $\widetilde{X}$. Then for $t\in S$ generic

$$\begin{aligned}(18.5)\qquad \int_X \alpha\wedge\beta &= \int_X T\wedge\mu_*([\widetilde{Y}_t]\wedge\widetilde{\omega}_1\wedge\cdots\wedge\widetilde{\omega}_{p-1})\\ &= \int_{\widetilde{X}} \mu^*T\wedge[\widetilde{Y}_t]\wedge\widetilde{\omega}_1\wedge\cdots\wedge\widetilde{\omega}_{p-1}\\ &= \int_{\widetilde{Y}_t}(\mu^*T)_{\restriction\widetilde{Y}_t}\wedge\widetilde{\omega}_1\wedge\cdots\wedge\widetilde{\omega}_{p-1}\geqslant 0\end{aligned}$$

provided that we show that the final integral is well defined and that the formal calculations involved in (18.5) are correct. Here, we have used the fact that a closed positive $(1,1)$-current T always has a pull-back μ^*T, which follows from the observation that if $T=\alpha+\mathrm{i}\partial\overline{\partial}\varphi$ with α smooth and φ quasi-psh, we may always set $\mu^*T=\mu^*\alpha+\mathrm{i}\partial\overline{\partial}(\varphi\circ\mu)$, with $\varphi\circ\mu$ quasi-psh and not identically $-\infty$ on $\widetilde{X}$. Similarly, we see that the restriction $(\mu^*T)_{\restriction\widetilde{Y}_t}$ is a well defined positive $(1,1)$-current for t generic, by putting

$$(\mu^*T)_{\restriction\widetilde{Y}_t}=(\mu^*\alpha)_{\restriction\widetilde{Y}_t}+\mathrm{i}\partial\overline{\partial}((\varphi\circ\mu)_{\restriction\widetilde{Y}_t})$$

and choosing t such that $\widetilde{Y}_t$ is not contained in the pluripolar set of $-\infty$ poles of $\varphi\circ\mu$ (this is possible thanks to the assumption that $\widetilde{Y}_t$ covers $\widetilde{X}$; locally near any given point we can modify α so that $\alpha=0$ on a small neighborhood V, and then φ is psh on V). Finally, in order to justify the formal calculations we can use a regularization argument for T, writing $T=\lim T_k$ with $T_k=\alpha+\mathrm{i}\partial\overline{\partial}\varphi_k$ and a decreasing sequence of smooth almost plurisubharmonic potentials $\varphi_k\downarrow\varphi$ such that the Levi forms have a uniform lower bound $\mathrm{i}\partial\overline{\partial}\varphi_k\geqslant -C\omega$ (such a sequence exists by [Dem92]). Then $(\mu^*T_k)_{\restriction\widetilde{Y}_t}\to(\mu^*T)_{\restriction\widetilde{Y}_t}$ in the weak topology of currents. □

Proposition 18.4 leads to the natural question whether the cones $(\overline{\mathcal{K}},\mathcal{N})$ and $(\mathcal{E},\mathcal{M})$ are dual under Serre duality, The second part of the question is addressed in the next section. The results proved in Chapter 17 yield a complete answer to the first part – even in the general Kähler setting.

(18.6) Theorem. *Let X be a compact Kähler manifold. Then*

(a) $\overline{\mathcal{K}}$ *and* $\mathcal{N}$ *are dual cones.*

(b) *If X is projective algebraic, then* $\overline{\mathcal{K}}_{\mathrm{NS}}=\mathrm{Nef}(X)$ *and* $\mathcal{N}_{\mathrm{NS}}=\overline{\mathrm{NE}(X)}$ *and these cones are dual.*

Proof. (a) is a weaker version of Theorem 17.13 (b).

(b) The equality $\overline{\mathcal{K}}_{\rm NS} = \mathrm{Nef}(X)$ has already been discussed and is a consequence of the Kodaira embedding theorem. Now, we know that

$$\overline{\mathrm{NE}(X)} \subset \mathcal{N}_{\rm NS} \subset \overline{\mathcal{K}}_{\rm NS}^{\vee} = \mathrm{Nef}(X)^{\vee},$$

where the second inclusion is a consequence of Proposition 18.4(a). However, it is already wellknown that $\overline{\mathrm{NE}(X)}$ and $\overline{\mathrm{NE}(X)}$ are dual cones (see [Har70]), hence the inclusions are equalities (we could also obtain a self-contained proof by reconsidering the arguments used for Theorem 17.13 (a) when α and ω_0 are rational classes; one sees by the density of the rationals that the numerical condition for α is needed only for elements of the form $[Y] \wedge \omega^{p-1}$ with $\omega \in \mathbb{Q}\{\alpha\} + \mathbb{Q}\{\omega_0\}$ a rational class, so $[Y] \wedge \omega^{p-1}$ is then a $\mathbb{Q}$-effective curve). □

18.B. Zariski Decomposition and Mobile Intersections

Let X be compact Kähler and let $\alpha \in \mathcal{E}^{\circ}$ be in the *interior* of the pseudo–effective cone. In analogy with the algebraic context such a class α is called "big", and it can then be represented by a *Kähler current* T, i.e. a closed positive $(1,1)$-current T such that $T \geqslant \delta\omega$ for some smooth Hermitian metric ω and a constant $\delta \ll 1$. We first need a variant of the approximation theorem proved in Section 13.B.

(18.7) Regularization Theorem for Currents. *Let X be a compact complex manifold equipped with a Hermitian metric ω. Let $T = \alpha + i\partial\overline{\partial}\varphi$ be a closed $(1,1)$-current on X, where α is smooth and φ is a quasi-plurisubharmonic function. Assume that $T \geqslant \gamma$ for some real $(1,1)$-form γ on X with real coefficients. Then there exists a sequence $T_m = \alpha + i\partial\overline{\partial}\varphi_m$ of closed $(1,1)$-currents such that*

(a) *φ_m (and thus T_m) is smooth on the complement $X \smallsetminus Z_m$ of an analytic set Z_m, and the Z_m's form an increasing sequence*

$$Z_0 \subset Z_1 \subset \cdots \subset Z_m \subset \cdots \subset X.$$

(b) *There is a uniform estimate $T_m \geqslant \gamma - \delta_m\omega$ with $\lim \downarrow \delta_m = 0$ as m tends to $+\infty$.*

(c) *The sequence (φ_m) is non increasing, and we have $\lim \downarrow \varphi_m = \varphi$. As a consequence, T_m converges weakly to T as m tends to $+\infty$.*

(d) *Near Z_m, the potential φ_m has logarithmic poles, namely, for every $x_0 \in Z_m$, there is a neighborhood U of x_0 such that $\varphi_m(z) = \lambda_m \log \sum_\ell |g_{m,\ell}|^2 + O(1)$ for suitable holomorphic functions $(g_{m,\ell})$ on U and $\lambda_m > 0$. Moreover, there is a (global) proper modification $\mu_m : \widetilde{X}_m \to X$ of X, obtained as a sequence of blow-ups with smooth centers, such that $\varphi_m \circ \mu_m$ can be written locally on $\widetilde{X}_m$ as*

$$\varphi_m \circ \mu_m(w) = \lambda_m\Big(\sum n_\ell \log |\widetilde{g}_\ell|^2 + f(w)\Big)$$

where $(\widetilde{g}_\ell = 0)$ *are local generators of suitable (global) divisors* D_ℓ *on* $\widetilde{X}_m$ *such that* $\sum D_\ell$ *has normal crossings,* n_ℓ *are positive integers, and the* f *'s are smooth functions on* $\widetilde{X}_m$.

Sketch of proof. We essentially repeat the proofs of Theorems 13.2 and 13.12 with additional considerations. One fact that does not follow readily from these proofs is the monotonicity of the sequence φ_m (which we will not really need anyway). For this, we can take $m = 2^\nu$ and use the subadditivity technique already explained in Step 3 of the proof of Theorem 15.3 (b). The map μ_m is obtained by blowing-up the (global) ideals $\mathcal{J}_m$ defined by the holomorphic functions $(g_{j,m})$ in the local approximations $\varphi_m \sim \frac{1}{2m}\log\sum_j |g_{j,m}|^2$. By Hironaka [Hir64], we can achieve that $\mu_m^*\mathcal{J}_m$ is an invertible ideal sheaf associated with a normal crossing divisor. □

(18.8) Corollary. *If* T *is a Kähler current, then one can write* $T = \lim T_m$ *for a sequence of Kähler currents* T_m *which have logarithmic poles with coefficients in* $\frac{1}{m}\mathbb{Z}$, *i.e. there are modifications* $\mu_m : X_m \to X$ *such that*

$$\mu_m^* T_m = [E_m] + \beta_m$$

where E_m *is an effective* $\mathbb{Q}$*-divisor on* X_m *with coefficients in* $\frac{1}{m}\mathbb{Z}$ *(the "fixed part") and* β_m *is a closed semi-positive form (the "mobile part").*

Proof. We apply Theorem 18.7 with $\gamma = \varepsilon\omega$ and m so large that $\delta_m \leqslant \varepsilon/2$. Then T_m has analytic singularities and $T_m \geqslant \frac{\varepsilon}{2}\omega$, so we get a composition of blow-ups $\mu_m : X_m \to X$ such

$$\mu_m^* T_m = [E_m] + \beta_m,$$

where E_m is an effective $\mathbb{Q}$-divisor and $\beta_m \geqslant \frac{\varepsilon}{2}\mu_m^*\omega$. In particular, β_m is strictly positive outside the exceptional divisors, by playing with the multiplicities of the components of the exceptional divisors in E_m, we could even achieve that β_m is a Kähler class on X_m. Notice also that by construction, μ_m is obtained by blowing-up the multiplier ideal sheaves $\mathcal{I}(mT) = \mathcal{I}(m\varphi)$ associated to a potential φ of T. □

The more familiar algebraic analogue would be to take $\alpha = c_1(L)$ with a big line bundle L and to blow-up the base locus of $|mL|$, $m \gg 1$, to get a $\mathbb{Q}$-divisor decomposition

$$\mu_m^* L \sim E_m + D_m, \qquad E_m \text{ effective}, \quad D_m \text{ free}.$$

Such a blow-up is usually referred to as a "log resolution" of the linear system $|mL|$, and we say that $E_m + D_m$ is an approximate Zariski decomposition of L. We will also use this terminology for Kähler currents with logarithmic poles.

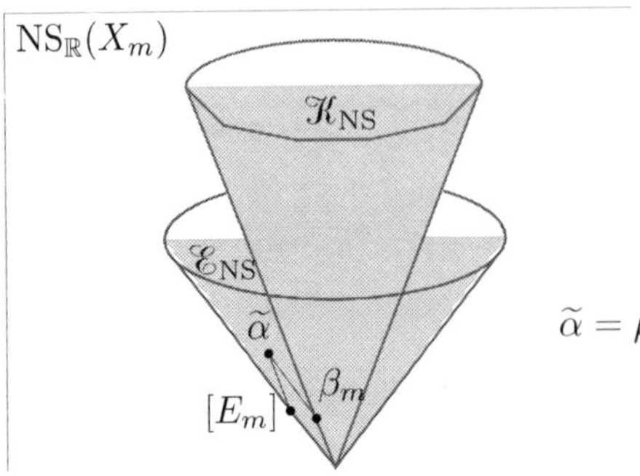

In analogy with the concept of volume of a line bundle, we introduce the following more general definition, which was already briefly mentioned in the statement of Conjecture 8.21.

(18.9) Definition. *We define the* volume, *or* mobile self-intersection *of a class* $\alpha \in H^{1,1}X, \mathbb{R})$ *to be*

$$\mathrm{Vol}(\alpha) = \sup_{T\in\alpha} \int_{X\smallsetminus \mathrm{Sing}(T)} T^n = \sup_{T\in\alpha} \int_{\widetilde{X}} \beta^n > 0,$$

where the supremum is taken over all Kähler currents $T \in \alpha$ *with logarithmic poles, and* $\mu^*T = [E] + \beta$ *with respect to some modification* $\mu : \widetilde{X} \to X$*. Correspondingly, we set*

$$\mathrm{Vol}(\alpha) = 0 \qquad \textit{if } \alpha \notin \mathcal{E}^\circ.$$

By Theorem 14.6, if L is a big line bundle, we have

$$\mathrm{Vol}(c_1(L)) = \lim_{m\to+\infty} D_m^n = \lim_{m\to+\infty} \frac{n!}{m^n} h^0(X, mL),$$

and in these terms, we get the following statement.

(18.10) Proposition. *Let* L *be a big line bundle on the projective manifold* X*. Let* $\varepsilon > 0$*. Then there exists a modification* $\mu : X_\epsilon \to X$ *and a decomposition* $\mu^*(L) = E + \beta$ *with* E *an effective* $\mathbb{Q}$*-divisor and* β *a big and nef* $\mathbb{Q}$*-divisor such that*

$$\mathrm{Vol}(L) - \varepsilon \leqslant \mathrm{Vol}(\beta) \leqslant \mathrm{Vol}(L).$$

It is very useful to observe that the supremum in Definition 18.9 is actually achieved by a collection of currents whose singularities satisfy a filtering property. Namely, if $T_1 = \alpha + i\partial\overline{\partial}\varphi_1$ and $T_2 = \alpha + i\partial\overline{\partial}\varphi_2$ are two Kähler currents with logarithmic poles in the class of α, then

$$T = \alpha + i\partial\overline{\partial}\varphi, \qquad \varphi = \max(\varphi_1, \varphi_2) \tag{18.11}$$

is again a Kähler current with weaker singularities than T_1 and T_2. One could define as well

$$(18.11') \qquad T = \alpha + i\partial\overline{\partial}\varphi, \qquad \varphi = \frac{1}{2m}\log(e^{2m\varphi_1} + e^{2m\varphi_2}),$$

where $m = \mathrm{lcm}(m_1, m_2)$ is the lowest common multiple of the denominators occuring in T_1, T_2. Now, take a simultaneous log-resolution $\mu_m : X_m \to X$ for which the singularities of T_1 and T_2 are resolved as $\mathbb{Q}$-divisors E_1 and E_2. Then clearly the associated divisor in the decomposition $\mu_m^* T = [E] + \beta$ is given by $E = \min(E_1, E_2)$. By doing so, the volume $\int_{X_m} \beta^n$ gets increased, as we shall see in the proof of Theorem 18.12 below.

(18.12) Theorem (Boucksom [Bou02]). *Let X be a compact Kähler manifold. We denote here by $H^{k,k}_{\geqslant 0}(X)$ the cone of cohomology classes of type (k,k) which have non-negative intersection with all closed semi-positive smooth forms of bidegree $(n-k, n-k)$.*

(a) *For each integer $k = 1, 2, \ldots, n$, there exists a canonical "mobile intersection product"*

$$\mathcal{E} \times \cdots \times \mathcal{E} \to H^{k,k}_{\geqslant 0}(X), \quad (\alpha_1, \ldots, \alpha_k) \mapsto \langle \alpha_1 \cdot \alpha_2 \cdot \cdots \cdot \alpha_{k-1} \cdot \alpha_k \rangle$$

such that $\mathrm{Vol}(\alpha) = \langle \alpha^n \rangle$ *whenever α is a big class.*

(b) *The product is increasing, homogeneous of degree 1 and superadditive in each argument, i.e.*

$$\langle \alpha_1 \cdots (\alpha_j' + \alpha_j'') \cdots \alpha_k \rangle \geqslant \langle \alpha_1 \cdots \alpha_j' \cdots \alpha_k \rangle + \langle \alpha_1 \cdots \alpha_j'' \cdots \alpha_k \rangle.$$

It coincides with the ordinary intersection product when the $\alpha_j \in \overline{\mathcal{K}}$ are nef classes.

(c) *The mobile intersection product satisfies the Teissier-Hovanskii inequalities*

$$\langle \alpha_1 \cdot \alpha_2 \cdot \cdots \cdot \alpha_n \rangle \geqslant (\langle \alpha_1^n \rangle)^{1/n} \cdots (\langle \alpha_n^n \rangle)^{1/n} \qquad (\text{with } \langle \alpha_j^n \rangle = \mathrm{Vol}(\alpha_j)).$$

(d) *For $k = 1$, the above "product" reduces to a (non linear) projection operator*

$$\mathcal{E} \to \mathcal{E}_1, \qquad \alpha \to \langle \alpha \rangle$$

onto a certain convex subcone $\mathcal{E}_1$ of $\mathcal{E}$ such that $\overline{\mathcal{K}} \subset \mathcal{E}_1 \subset \mathcal{E}$. Moreover, there is a "divisorial Zariski decomposition"

$$\alpha = \{N(\alpha)\} + \langle \alpha \rangle$$

where $N(\alpha)$ is a uniquely defined effective divisor which is called the "negative divisorial part" of α. The map $\alpha \mapsto N(\alpha)$ is homogeneous and subadditive, and $N(\alpha) = 0$ if and only if $\alpha \in \mathcal{E}_1$.

(e) *The components of* $N(\alpha)$ *always consist of divisors whose cohomology classes are linearly independent, especially* $N(\alpha)$ *has at most* $\rho = \mathrm{rank}_{\mathbb{Z}}\,\mathrm{NS}(X)$ *components.*

Proof. We essentially repeat the arguments developped in [Bou02], with some simplifications arising from the fact that X is supposed to be Kähler from the start.

(a) First assume that all classes α_j are big, i.e. $\alpha_j \in \mathscr{E}^\circ$. Fix a smooth closed $(n-k, n-k)$ *semi-positive* form u on X. We select Kähler currents $T_j \in \alpha_j$ with logarithmic poles, and a simultaneous log-resolution $\mu : \widetilde{X} \to X$ such that

$$\mu^* T_j = [E_j] + \beta_j.$$

We consider the direct image current $\mu_*(\beta_1 \wedge \cdots \wedge \beta_k)$ (which is a closed positive current of bidegree (k,k) on X) and the corresponding integrals

$$\int_{\widetilde{X}} \beta_1 \wedge \cdots \wedge \beta_k \wedge \mu^* u \geqslant 0.$$

If we change the representative T_j with another current T'_j, we may always take a simultaneous log-resolution such that $\mu^* T'_j = [E'_j] + \beta'_j$, and by using (18.11′) we can always assume that $E'_j \leqslant E_j$. Then $D_j = E_j - E'_j$ is an effective divisor and we find $[E_j] + \beta_j \equiv [E'_j] + \beta'_j$, hence $\beta'_j \equiv \beta_j + [D_j]$. A substitution in the integral implies

$$\begin{aligned}
&\int_{\widetilde{X}} \beta'_1 \wedge \beta_2 \wedge \cdots \wedge \beta_k \wedge \mu^* u \\
&\qquad = \int_{\widetilde{X}} \beta_1 \wedge \beta_2 \wedge \cdots \wedge \beta_k \wedge \mu^* u + \int_{\widetilde{X}} [D_1] \wedge \beta_2 \wedge \cdots \wedge \beta_k \wedge \mu^* u \\
&\qquad \geqslant \int_{\widetilde{X}} \beta_1 \wedge \beta_2 \wedge \cdots \wedge \beta_k \wedge \mu^* u.
\end{aligned}$$

Similarly, we can replace successively all forms β_j by the β'_j, and by doing so, we find

$$\int_{\widetilde{X}} \beta'_1 \wedge \beta'_2 \wedge \cdots \wedge \beta'_k \wedge \mu^* u \geqslant \int_{\widetilde{X}} \beta_1 \wedge \beta_2 \wedge \cdots \wedge \beta_k \wedge \mu^* u.$$

We claim that the closed positive currents $\mu_*(\beta_1 \wedge \cdots \wedge \beta_k)$ are uniformly bounded in mass. In fact, if ω is a Kähler metric in X, there exists a constant $C_j \geqslant 0$ such that $C_j\{\omega\} - \alpha_j$ is a Kähler class. Hence $C_j \omega - T_j \equiv \gamma_j$ for some Kähler form γ_j on X. By pulling back with μ, we find $C_j \mu^* \omega - ([E_j] + \beta_j) \equiv \mu^* \gamma_j$, hence

$$\beta_j \equiv C_j \mu^* \omega - ([E_j] + \mu^* \gamma_j).$$

By performing again a substitution in the integrals, we find

$$\int_{\widetilde{X}} \beta_1 \wedge \cdots \wedge \beta_k \wedge \mu^* u \leqslant C_1 \cdots C_k \int_{\widetilde{X}} \mu^* \omega^k \wedge \mu^* u = C_1 \cdots C_k \int_X \omega^k \wedge u$$

and this is true especially for $u = \omega^{n-k}$. We can now arrange that for each of the integrals associated with a countable dense family of forms u, the supremum is achieved by a sequence of currents $(\mu_m)_*(\beta_{1,m} \wedge \cdots \wedge \beta_{k,m})$ obtained as direct images by a suitable sequence of modifications $\mu_m : \widetilde{X}_m \to X$. By extracting a subsequence, we can achieve that this sequence is weakly convergent and we set

$$\langle \alpha_1 \cdot \alpha_2 \cdot \cdots \cdot \alpha_k \rangle = \lim_{m \to +\infty} \uparrow \{(\mu_m)_*(\beta_{1,m} \wedge \beta_{2,m} \wedge \cdots \wedge \beta_{k,m})\}$$

(the monotonicity is not in terms of the currents themselves, but in terms of the integrals obtained when we evaluate against a smooth closed semi-positive form u). By evaluating against a basis of positive classes $\{u\} \in H^{n-k,n-k}(X)$, we infer by Serre duality that the class of $\langle \alpha_1 \cdot \alpha_2 \cdot \cdots \cdot \alpha_k \rangle$ is uniquely defined (although, in general, the representing current is not unique).

(b) It is indeed clear from the definition that the mobile intersection product is homogeneous, increasing and superadditive in each argument, at least when the α_j's are in $\mathcal{E}^\circ$. However, we can extend the product to the closed cone $\mathcal{E}$ by monotonicity, by setting

$$\langle \alpha_1 \cdot \alpha_2 \cdot \cdots \cdot \alpha_k \rangle = \lim_{\delta \downarrow 0} \downarrow \langle (\alpha_1 + \delta\omega) \cdot (\alpha_2 + \delta\omega) \cdot \cdots \cdot (\alpha_k + \delta\omega) \rangle$$

for arbitrary classes $\alpha_j \in \mathcal{E}$ (again, monotonicity occurs only where we evaluate against closed semi-positive forms u). By weak compactness, the mobile intersection product can always be represented by a closed positive current of bidegree (k, k).

(c) The Teissier-Hovanskii inequalities are a direct consequence of the fact that they hold true for nef classes, so we just have to apply them to the classes $\beta_{j,m}$ on $\widetilde{X}_m$ and pass to the limit.

(d) When $k = 1$ and $\alpha \in \mathcal{E}^0$, we have

$$\alpha = \lim_{m \to +\infty} \{(\mu_m)_* T_m\} = \lim_{m \to +\infty} (\mu_m)_*[E_m] + \{(\mu_m)_* \beta_m\}$$

and $\langle \alpha \rangle = \lim_{m \to +\infty} \{(\mu_m)_* \beta_m\}$ by definition. However, the images $F_m = (\mu_m)_* F_m$ are effective $\mathbb{Q}$-divisors in X, and the filtering property implies that F_m is a decreasing sequence. It must therefore converge to a (uniquely defined) limit $F = \lim F_m := N(\alpha)$ which is an effective $\mathbb{R}$-divisor, and we get the asserted decomposition in the limit.

Since $N(\alpha) = \alpha - \langle \alpha \rangle$ we easily see that $N(\alpha)$ is subadditive and that $N(\alpha) = 0$ if α is the class of a smooth semi-positive form. When α is no longer a big class, we define

$$\langle \alpha \rangle = \lim_{\delta \downarrow 0} \downarrow \langle \alpha + \delta\omega \rangle, \qquad N(\alpha) = \lim_{\delta \downarrow 0} \uparrow N(\alpha + \delta\omega)$$

(the subadditivity of N implies $N(\alpha + (\delta + \varepsilon)\omega) \leqslant N(\alpha + \delta\omega)$). The divisorial Zariski decomposition follows except maybe for the fact that $N(\alpha)$ might be a convergent countable sum of divisors. However, this will be ruled out when (e) is proved. As $N(\bullet)$ is subadditive and homogeneous, the set $\mathcal{E}_1 = \{\alpha \in \mathcal{E}\ ;\ N(\alpha) = 0\}$ is a closed convex conne, and we find that $\alpha \mapsto \langle\alpha\rangle$ is a projection of $\mathcal{E}$ onto $\mathcal{E}_1$ (according to [Bou02], $\mathcal{E}_1$ consists of those pseudo-effective classes which are "nef in codimension 1").

(e) Let $\alpha \in \mathcal{E}^\circ$, and assume that $N(\alpha)$ contains linearly dependent components F_j. Then already all currents $T \in \alpha$ should be such that $\mu^* T = [E] + \beta$ where $F = \mu_* E$ contains those linearly dependent components. Write $F = \sum \lambda_j F_j$, $\lambda_j > 0$ and assume that

$$\sum_{j \in J} c_j F_j \equiv 0$$

for a certain non trivial linear combination. Then some of the coefficients c_j must be negative (and some other positive). Then E is numerically equivalent to

$$E' \equiv E + t\mu^*\Big(\sum \lambda_j F_j\Big),$$

and by choosing $t > 0$ appropriate, we obtain an effective divisor E' which has a zero coefficient on one of the components $\mu^* F_{j_0}$. By replacing E with $\min(E, E')$ via (18.11′), we eliminate the component $\mu^* F_{j_0}$. This is a contradiction since $N(\alpha)$ was supposed to contain F_{j_0}. □

(18.13) Definition. *For a class $\alpha \in H^{1,1}(X, \mathbb{R})$, we define the* numerical dimension $\mathrm{nd}(\alpha)$ *to be* $\mathrm{nd}(\alpha) = -\infty$ *if α is not pseudo-effective, and*

$$\mathrm{nd}(\alpha) = \max\{p \in \mathbb{N}\ ;\ \langle\alpha^p\rangle \neq 0\}, \qquad \mathrm{nd}(\alpha) \in \{0, 1, \ldots, n\}$$

if α is pseudo-effective.

By the results of [DP04], a class is big ($\alpha \in \mathcal{E}^\circ$) if and only if $\mathrm{nd}(\alpha) = n$. Classes of numerical dimension 0 can be described much more precisely, again following Boucksom [Bou02].

(18.14) Theorem. *Let X be a compact Kähler manifold. Then the subset $\mathcal{D}_0$ of irreducible divisors D in X such that $\mathrm{nd}(D) = 0$ is countable, and these divisors are rigid as well as their multiples. If $\alpha \in \mathcal{E}$ is a pseudo-effective class of numerical dimension 0, then α is numerically equivalent to an effective $\mathbb{R}$-divisor $D = \sum_{j \in J} \lambda_j D_j$, for some finite subset $(D_j)_{j \in J} \subset \mathcal{D}_0$ such that the cohomology classes $\{D_j\}$ are linearly independent and some $\lambda_j > 0$. If such a linear combination is of numerical dimension 0, then so is any other linear combination of the same divisors.*

Proof. It is immediate from the definition that a pseudo-effective class is of numerical dimension 0 if and only if $\langle\alpha\rangle = 0$, in other words if $\alpha = N(\alpha)$. Thus $\alpha \equiv \sum \lambda_j D_j$ as described in 18.14, and since $\lambda_j \langle D_j\rangle \leqslant \langle\alpha\rangle$, the divisors D_j must themselves have numerical dimension 0. There is at most one such divisor D in any given cohomology class in $NS(X) \cap \mathcal{E} \subset H^2(X, \mathbb{Z})$, otherwise two such divisors $D \equiv D'$ would yield a blow-up $\mu : \widetilde{X} \to X$ resolving the intersection, and by taking $\min(\mu^* D, \mu^* D')$ via (18.11′), we would find $\mu^* D \equiv E + \beta$, $\beta \neq 0$, so that $\{D\}$ would not be of numerical dimension 0. This implies that there are at most countably many divisors of numerical dimension 0, and that these divisors are rigid as well as their multiples. □

(18.15) Remark. If L is an arbitrary holomorphic line bundle, we define its numerical dimension to be $\mathrm{nd}(L) = \mathrm{nd}(c_1(L))$. Using the canonical maps $\Phi_{|mL|}$ and pulling-back the Fubini-Study metric it is immediate to see that $\mathrm{nd}(L) \geqslant \kappa(L)$ (which generalizes the analogue inequality already seen for nef line bundles, see (6.18)).

The above general concept of numerical dimension leads to a very natural formulation of the abundance conjecture for Kähler varieties.

(18.16) Generalized Abundance Conjecture. *Let X be an arbitrary compact Kähler manifold X.*

(a) *The Kodaira dimension of X should be equal to its numerical dimension:* $\kappa(K_X) = \mathrm{nd}(K_X)$.

(b) *More generally, let Δ be a $\mathbb{Q}$-divisor which is klt (Kawamata log terminal, i.e. such that $c_X(\Delta) > 1$). Then $\kappa(K_X + \Delta) = \mathrm{nd}(K_X + \Delta)$.*

This appears to be a fairly strong statement. In fact, already in the case $\Delta = 0$, it is not difficult to show that the generalized abundance conjecture would contain the $C_{n,m}$ conjectures.

(18.17) Remark. It is obvious that abundance holds in the case $\mathrm{nd}(K_X) = -\infty$ (if L is not pseudo-effective, no multiple of L can have sections), or in the case $\mathrm{nd}(K_X) = n$ which implies K_X big (the latter property follows e.g. from the solution of the Grauert-Riemenschneider conjecture in the form proven in [Dem85b], see also [DP04]).

In the remaining cases, the most tractable situation is the case when $\mathrm{nd}(K_X) = 0$. In fact Theorem 18.14 then gives $K_X \equiv \sum \lambda_j D_j$ for some effective divisor with numerically independent components, $\mathrm{nd}(D_j) = 0$. It follows that the λ_j are rational and therefore

$$(*) \qquad K_X \sim \sum \lambda_j D_j + F \qquad \text{where } \lambda_j \in \mathbb{Q}^+,\ \mathrm{nd}(D_j) = 0 \text{ and } F \in \mathrm{Pic}^0(X).$$

If we assume additionally that $q(X) = h^{0,1}(X)$ is zero, then mK_X is linearly equivalent to an integral divisor for some multiple m, and it follows immediately that $\kappa(X) = 0$. The case of a general projective manifold with $\mathrm{nd}(K_X) = 0$ and positive irregularity $q(X) > 0$ has been solved by Campana-Peternell [CP04], Proposition 3.7. It would be interesting to understand the Kähler case as well.

18.C. The Orthogonality Estimate

The goal of this section is to show that, in an appropriate sense, approximate Zariski decompositions are almost orthogonal.

(18.18) Theorem. *Let X be a projective manifold, and let $\alpha = \{T\} \in \mathcal{E}^{\circ}_{\mathrm{NS}}$ be a big class represented by a Kähler current T. Consider an approximate Zariski decomposition*

$$\mu_m^* T_m = [E_m] + [D_m]$$

Then

$$(D_m^{n-1} \cdot E_m)^2 \leqslant 20\,(C\omega)^n \big(\mathrm{Vol}(\alpha) - D_m^n\big)$$

where $\omega = c_1(H)$ is a Kähler form and $C \geqslant 0$ is a constant such that $\pm\alpha$ is dominated by $C\omega$ (i.e., $C\omega \pm \alpha$ is nef). In other words, E_m and D_m become "more and more orthogonal" as D_m^n approaches the volume.

Proof. For every $t \in [0,1]$, we have

$$\mathrm{Vol}(\alpha) = \mathrm{Vol}(E_m + D_m) \geqslant \mathrm{Vol}(tE_m + D_m).$$

Now, by our choice of C, we can write E_m as a difference of two nef divisors

$$E_m = \mu^*\alpha - D_m = \mu_m^*(\alpha + C\omega) - (D_m + C\mu_m^*\omega). \qquad \square$$

(18.19) Lemma. *For all nef $\mathbb{R}$-divisors A, B we have*

$$\mathrm{Vol}(A - B) \geqslant A^n - nA^{n-1} \cdot B$$

as soon as the right hand side is positive.

Proof. In case A and B are integral (Cartier) divisors, this is a consequence of the holomorphic Morse inequalities 7.4 (see [Dem01]); one can also argue by an elementary estimate of to $H^0(X, mA - B_1 - \ldots - B_m)$ via the Riemann-Roch formula (assuming A and B very ample, $B_1, \ldots, B_m \in |B|$ generic). If A and B are $\mathbb{Q}$-Cartier, we conclude by the homogeneity of the volume. The general case of $\mathbb{R}$-divisors follows by approximation using the upper semi-continuity of the volume [Bou02, 3.1.26]. $\square$

(18.20) Remark. We hope that Lemma 18.19 also holds true on an arbitrary Kähler manifold for arbitrary nef (non necessarily integral) classes. This would follow from Conjecture 8.21 generalizing holomorphic Morse inequalities to non integral classes, exactly by the same proof as Theorem 8.5.

(18.21) Lemma. *Let $\beta_1, \ldots, \beta_n$ and $\beta'_1, \ldots, \beta'_n$ be nef classes on a compact Kähler manifold $\widetilde{X}$ such that each difference $\beta'_j - \beta_j$ is pseudo-effective. Then the n-th intersection products satisfy*

$$\beta_1 \cdots \beta_n \leqslant \beta'_1 \cdots \beta'_n.$$

Proof. We can proceed step by step and replace just one β_j by $\beta' j \equiv \beta_j + T_j$ where T_j is a closed positive $(1,1)$-current and the other classes $\beta'_k = \beta_k$, $k \neq j$ are limits of Kähler forms. The inequality is then obvious. □

End of proof of Theorem 18.18. In order to exploit the lower bound of the volume, we write

$$tE_m + D_m = A - B, \qquad A = D_m + t\mu_m^*(\alpha + C\omega), \quad B = t(D_m + C\mu_m^*\omega).$$

By our choice of the constant C, both A and B are nef. Lemma 18.19 and the binomial formula imply

$$\begin{aligned}\mathrm{Vol}(tE_m{+}D_m) &\geqslant A^n - nA^{n-1}\cdot B\\ &= D_m^n + nt\,D_m^{n-1}\cdot\mu_m^*(\alpha+C\omega) + \sum_{k=2}^{n} t^k\binom{n}{k} D_m^{n-k}\cdot\mu_m^*(\alpha+C\omega)^k\\ &\quad - nt\,D_m^{n-1}\cdot(D_m + C\mu_m^*\omega)\\ &\quad - nt^2\sum_{k=1}^{n-1} t^{k-1}\binom{n-1}{k} D_m^{n-1-k}\cdot\mu_m^*(\alpha+C\omega)^k\cdot(D_m + C\mu_m^*\omega).\end{aligned}$$

Now, we use the obvious inequalities

$$D_m \leqslant \mu_m^*(C\omega), \qquad \mu_m^*(\alpha + C\omega) \leqslant 2\mu_m^*(C\omega), \qquad D_m + C\mu_m^*\omega \leqslant 2\mu_m^*(C\omega)$$

in which all members are nef (and where the inequality $\leqslant$ means that the difference of classes is pseudo-effective). We use Lemma 18.21 to bound the last summation in the estimate of the volume, and in this way we get

$$\mathrm{Vol}(tE_m + D_m) \geqslant D_m^n + ntD_m^{n-1}\cdot E_m - nt^2\sum_{k=1}^{n-1} 2^{k+1}t^{k-1}\binom{n-1}{k}(C\omega)^n.$$

We will always take t smaller than $1/10n$ so that the last summation is bounded by $4(n-1)(1+1/5n)^{n-2} < 4ne^{1/5} < 5n$. This implies

$$\mathrm{Vol}(tE_m + D_m) \geqslant D_m^n + nt\,D_m^{n-1}\cdot E_m - 5n^2t^2(C\omega)^n.$$

Now, the choice $t = \frac{1}{10n}(D_m^{n-1}\cdot E_m)((C\omega)^n)^{-1}$ gives by substituting

$$\frac{1}{20}\frac{(D_m^{n-1}\cdot E_m)^2}{(C\omega)^n} \leqslant \mathrm{Vol}(E_m + D_m) - D_m^n \leqslant \mathrm{Vol}(\alpha) - D_m^n$$

(and we have indeed $t \leqslant \frac{1}{10n}$ by Lemma 18.21), whence Theorem 18.18. Of course, the constant 20 is certainly not optimal. □

(18.22) Corollary. *If $\alpha \in \mathscr{E}_{\mathrm{NS}}$, then the divisorial Zariski decomposition $\alpha = N(\alpha) + \langle\alpha\rangle$ is such that*

$$\langle\alpha^{n-1}\rangle\cdot N(\alpha) = 0.$$

Proof. By replacing α with $\alpha+\delta c_1(H)$, one sees that it is sufficient to consider the case where α is big. Then the orthogonality estimate implies

$$\begin{aligned}(\mu_m)_*(D_m^{n-1})\cdot(\mu_m)_*E_m &= D_m^{n-1}\cdot(\mu_m)^*(\mu_m)_*E_m\\ &\leqslant D_m^{n-1}\cdot E_m \leqslant C(\mathrm{Vol}(\alpha)-D_m^n)^{1/2}.\end{aligned}$$

Since $\langle\alpha^{n-1}\rangle=\lim(\mu_m)_*(D_m^{n-1})$, $N(\alpha)=\lim(\mu_m)_*E_m$ and $\lim D_m^n=\mathrm{Vol}(\alpha)$, we get the desired conclusion in the limit. □

18.D. Dual of the Pseudo-effective Cone

The following statement was first proved in [BDPP04].

(18.23) Theorem. *If X is projective, the cones $\mathcal{E}_{\mathrm{NS}}=\overline{\mathrm{Eff}(X)}$ and $\overline{\mathrm{ME}^s(X)}$ are dual.*

In other words, a line bundle L is pseudo-effective if (and only if) $L\cdot C\geqslant 0$ for all *mobile curves*, i.e., $L\cdot C\geqslant 0$ for every very generic curve C (not contained in a countable union of algebraic subvarieties). In fact, by definition of $\mathrm{ME}^s(X)$, it is enough to consider only those curves C which are images of generic complete intersection of very ample divisors on some variety $\widetilde{X}$, under a modification $\mu:\widetilde{X}\to X$. By a standard blowing-up argument, it also follows that a line bundle L on a normal Moishezon variety is pseudo-effective if and only if $L\cdot C\geqslant 0$ for every mobile curve C.

Proof. By Proposition 18.4 (b) we have in any case

$$\mathcal{E}_{\mathrm{NS}}\subset(\mathrm{ME}^s(X))^\vee.$$

If the inclusion is strict, there is an element $\alpha\in\partial\mathcal{E}_{\mathrm{NS}}$ on the boundary of $\mathcal{E}_{\mathrm{NS}}$ which is in the interior of $\mathrm{ME}^s(X)^\vee$.

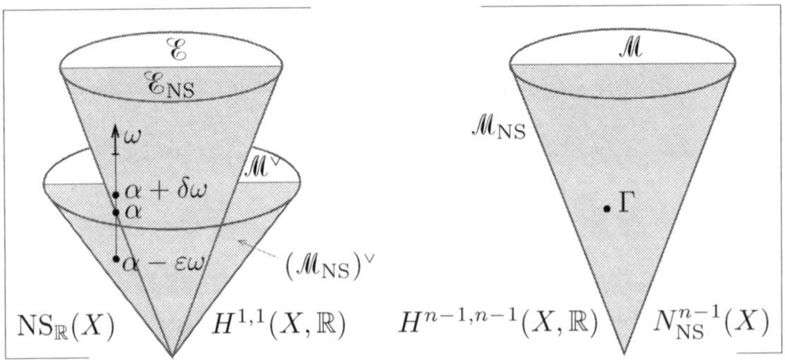

Let $\omega=c_1(H)$ be an ample class. Since $\alpha\in\partial\mathcal{E}_{\mathrm{NS}}$, the class $\alpha+\delta\omega$ is big for every $\delta>0$, and since $\alpha\in((\mathrm{ME}^s(X))^\vee)^\circ$ we still have $\alpha-\varepsilon\omega\in(\mathrm{ME}^s(X))^\vee$ for

$\varepsilon > 0$ small. Therefore

$$\alpha \cdot \Gamma \geqslant \varepsilon\omega \cdot \Gamma \tag{18.24}$$

for every strongly mobile curve Γ, and therefore for every $\Gamma \in \overline{\mathrm{ME}^s(X)}$. We are going to contradict (18.24). Since $\alpha + \delta\omega$ is big, we have an approximate Zariski decomposition

$$\mu_\delta^*(\alpha + \delta\omega) = E_\delta + D_\delta.$$

We pick $\Gamma = (\mu_\delta)_*(D_\delta^{n-1}) \in \overline{\mathrm{ME}^s(X)}$. By the Hovanskii-Teissier concavity inequality

$$\omega \cdot \Gamma \geqslant (\omega^n)^{1/n}(D_\delta^n)^{(n-1)/n}.$$

On the other hand

$$\begin{aligned} \alpha \cdot \Gamma &= \alpha \cdot (\mu_\delta)_*(D_\delta^{n-1}) \\ &= \mu_\delta^*\alpha \cdot D_\delta^{n-1} \leqslant \mu_\delta^*(\alpha + \delta\omega) \cdot D_\delta^{n-1} \\ &= (E_\delta + D_\delta) \cdot D_\delta^{n-1} = D_\delta^n + D_\delta^{n-1} \cdot E_\delta. \end{aligned}$$

By the orthogonality estimate, we find

$$\begin{aligned} \frac{\alpha \cdot \Gamma}{\omega \cdot \Gamma} &\leqslant \frac{D_\delta^n + \big(20(C\omega)^n(\mathrm{Vol}(\alpha + \delta\omega) - D_\delta^n)\big)^{1/2}}{(\omega^n)^{1/n}(D_\delta^n)^{(n-1)/n}} \\ &\leqslant C'(D_\delta^n)^{1/n} + C''\frac{(\mathrm{Vol}(\alpha + \delta\omega) - D_\delta^n)^{1/2}}{(D_\delta^n)^{(n-1)/n}}. \end{aligned}$$

However, since $\alpha \in \partial\mathscr{E}_{\mathrm{NS}}$, the class α cannot be big so

$$\lim_{\delta \to 0} D_\delta^n = \mathrm{Vol}(\alpha) = 0.$$

We can also take D_δ to approximate $\mathrm{Vol}(\alpha + \delta\omega)$ in such a way that $(\mathrm{Vol}(\alpha + \delta\omega) - D_\delta^n)^{1/2}$ tends to 0 much faster than D_δ^n. Notice that $D_\delta^n \geqslant \delta^n\omega^n$, so in fact it is enough to take

$$\mathrm{Vol}(\alpha + \delta\omega) - D_\delta^n \leqslant \delta^{2n},$$

which gives $(\alpha \cdot \Gamma)/(\omega \cdot \Gamma) \leqslant (C' + C'')\delta$. This contradicts (18.24) for δ small. □

(18.25) Conjecture. The Kähler analogue should be :
For an arbitrary compact Kähler manifold X, the cones $\mathscr{E}$ and $\mathscr{M}$ are dual.

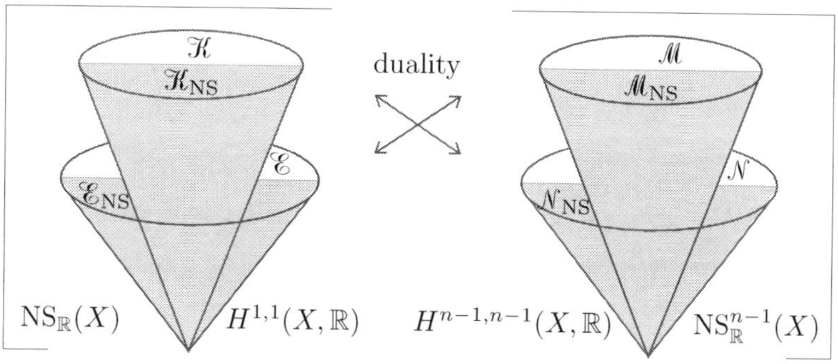

If holomorphic Morse inequalities were known also in the Kähler case (cf. Conjecture 8.21), we would infer by the same proof that "α not pseudo-effective" implies the existence of a modification $\mu : \widetilde{X} \to X$ and a Kähler metric $\widetilde{\omega}$ on $\widetilde{X}$ such that $\alpha \cdot \mu_*(\widetilde{\omega})^{n-1} < 0$. In the special case when $\alpha = K_X$ is not pseudo-effective, we would expect the Kähler manifold X to be covered by rational curves. The main trouble is that characteristic p techniques are no longer available. On the other hand it is tempting to approach the question via techniques of symplectic geometry :

(18.26) Question. *Let (M, ω) be a compact real symplectic manifold. Fix an almost complex structure J compatible with ω, and assume that $c_1(M, J) \cdot \omega^{n-1} > 0$ (the condition does not depend on the choice of J, but only on ω). Does it follow that M is covered by rational J-pseudoholomorphic curves ?*

The relation between the various cones of mobile curves and currents in Definitions 18.1 and 18.2 is now a rather direct consequence of Theorem 18.23. In fact, using ideas hinted in [DPS96], one can say a little bit more. Given an irreducible curve $C \subset X$, we consider its normal "bundle" $N_C = \mathrm{Hom}(\mathcal{I}/\mathcal{I}^2, \mathcal{O}_C)$, where $\mathcal{I}$ is the ideal sheaf of C. If C is a general member of a covering family (C_t), then N_C is nef. By [DPS96], the dual cone of the pseudo-effective cone of X contains the closed cone spanned by curves with nef normal bundle, which in turn contains the cone of mobile curves. In this way we get :

(18.27) Theorem. *Let X be a projective manifold. Then the following cones coincide:*

(a) *the cone* $\mathcal{M}_{\mathrm{NS}} = \mathcal{M} \cap \mathrm{NS}^{n-1}_{\mathbb{R}}(X)$;

(b) *the cone* $\mathcal{M}^s_{\mathrm{NS}} = \mathcal{M}^s \cap \mathrm{NS}^{n-1}_{\mathbb{R}}(X)$;

(c) *the closed cone* $\overline{\mathrm{ME}^s(X)}$ *of strongly mobile curves*;

(d) *the closed cone* $\overline{\mathrm{ME}(X)}$ *of mobile curves*;

(e) *the closed cone* $\overline{\mathrm{ME}_{\mathrm{nef}}(X)}$ *of curves with nef normal bundle.*

Proof. We have already seen that

$$\mathrm{ME}^s(X) \subset \mathrm{ME}(X) \subset \mathrm{ME}_{\mathrm{nef}}(X) \subset (\mathcal{E}_{\mathrm{NS}})^\vee$$

and

$$\mathrm{ME}^s(X) \subset \mathcal{M}^s_{\mathrm{NS}}(X) \subset \mathcal{M}_{\mathrm{NS}} \subset (\mathcal{E}_{\mathrm{NS}})^\vee$$

by 18.4 (c). Now Theorem 18.23 implies $(\mathcal{M}_{\mathrm{NS}})^\vee = \overline{\mathrm{ME}^s(X)}$, and 18.27 follows. □

(18.28) Corollary. *Let X be a projective manifold and L a line bundle on X.*

(a) *L is pseudo-effective if and only if $L \cdot C \geqslant 0$ for all curves C with nef normal sheaf N_C.*

(b) *If L is big, then $L \cdot C > 0$ for all curves C with nef normal sheaf N_C.*

Corollary 18.28 (a) strenghthens results from [PSS99]. It is however not yet clear whether $\mathcal{M}_{\mathrm{NS}}$ is equal to the closed cone of curves with *ample* normal bundle (although we certainly expect this to be true). An important special case of Theorem 18.23 is

(18.29) Theorem. *If X is a projective manifold, then K_X is pseudo-effective (i.e. $K_X \in \mathcal{E}_{\mathrm{NS}}$), if and only if X is not uniruled (i.e. not covered by rational curves).*

Proof. If X is covered by rational curves C_t, then it is well-known that the normal bundle N_{C_t} is nef for a general member C_t, thus

$$K_X \cdot C_t = K_{C_t} \cdot C_t - N_{C_t} \cdot C_t \leqslant -2,$$

and K_X cannot be pseudo-effective. Conversely, if $K_X \notin \mathcal{E}_{\mathrm{NS}}$, Theorem 18.23 shows that there is a mobile curve C_t such that $K_X \cdot C_t < 0$. The standard "bend-and-break" lemma of Mori theory then produces a covering family Γ_t of rational curves with $K_X \cdot \Gamma_t < 0$, so X is uniruled. □

The generalized abundance conjecture 18.16 would then yield the stronger result :

(18.30) Conjecture. *Let X be a projective manifold. If X is not uniruled, then K_X is a $\mathbb{Q}$-effective divisor and $\kappa(X) = \mathrm{nd}(K_X) \geqslant 0$.*

18.E. A Volume Formula for Algebraic (1,1)-Classes on Projective Surfaces

As a further application, we give here a positive answer to Question 8.20, in the case where X is a projective surface.

(18.31) Theorem. *Let (X, ω) be a compact complex n-fold. Then for every class $\alpha \in H^{1,1}_{\mathrm{BC}}(X, \mathbb{R})$ we have*

(a) $\mathrm{Vol}(\alpha) = \sup_{T\in\alpha} \int_{X\smallsetminus \mathrm{Sing}(T)} T^n \leqslant \widehat{h}^0_{\mathrm{tr}}(X, \alpha) = \inf_{u\in\alpha} \int_{X(u,0)} u^n,$

where the supremum runs over all Kähler currents $T \in \alpha$ with analytic singularities, and the infimum runs over all smooth closed $(1,1)$-forms $u \in \alpha$.

(b) *Equality holds in* (a) *if X is a projective surface and $\alpha \in \mathrm{NS}_{\mathbb{R}}(X)$.*

The following result has been proved in [BmD09] (cf. Theorem 1.4 and Corollary 2.5); we will not give details here – the proof relies again in an essential way on regularization techniques ([Dem94a] especially).

(18.32) Lemma. *Assume that $\alpha \in H^{1,1}(X, \mathbb{R}$ contains a Kähler current T_0, and let Z_0 be the analytic set of poles of T_0. Fix a smooth representative $u \in \alpha$ and* *consider a current with minimal singularities*

$$T = u + \mathrm{i}\partial\overline{\partial}\varphi \geqslant 0$$

where

$$\varphi(x) := \sup\{\psi(x)\,;\ \psi \leqslant 0 \text{ and } u + \mathrm{i}\partial\overline{\partial}\psi \geqslant 0 \text{ on } X\}, \tag{18.33}$$

and the supremum is taken over all quasi-psh functions ψ satisfying the conditions $\psi \leqslant 0$ and $u + \mathrm{i}\partial\overline{\partial}\psi \geqslant 0$. Then φ is continuous with locally bounded second derivatives $\partial^2/\partial z_j\partial\overline{z}_k$ on $X \smallsetminus Z_0$. Moreover, if S is the set of points $z \in X \smallsetminus Z_0$ where $\varphi(z) = 0$, then $S \subset \{z\,;u(z) \geqslant 0\}$ and

$$\mathrm{Vol}(\alpha) = \int_S u^n = \int_{X\smallsetminus Z_0} (u + \mathrm{i}\partial\overline{\partial}\varphi)^n. \tag{18.34}$$

Since $S \subset \{z\,;u(z) \geqslant 0\}$, we immediately conclude from these equalities that

$$\mathrm{Vol}(\alpha) \leqslant \int_{\{z\,;u(z)\geqslant 0\}} u^n = \int_{\{z\,;u(z)>0\}} u^n = \int_{X(u,0)} u^n,$$

and Theorem 18.31 (a) follows.

Proof of 18.31 (b). Let $T \in \alpha$ be a Kähler current. Take a blow-up $\mu : \widetilde{X} \to X$ such that $\mu^*T = [E] + \beta$ where E is a normal crossing divisor on $\widetilde{X}$ and $\beta \geqslant 0$ smooth. Until now, this is valid for an arbitrary compact complex manifold X. If moreover X is projective and $\alpha \in \mathrm{NS}_{\mathbb{R}}(X)$, the orthogonality estimate (18.18) yields

$$[E]\cdot\beta^{n-1} = \int_E \beta^{n-1} \leqslant C\big(\mathrm{Vol}(\alpha) - \beta^n\big)^{1/2}. \tag{18.35}$$

Our method consists of approaching $[E] + \beta$ by smooth closed $(1,1)$-forms u_ε in the same $\partial\overline{\partial}$-cohomology class as $[E] + \beta$, in such a way that

$$\int_{\widetilde{X}(u_\varepsilon,0)} u_\varepsilon^n$$

will not be substantially larger than the volume $\int_{\widetilde{X}} \beta^n$. For this, we select a Hermitian metric h on $\mathcal{O}(E)$ and put

$$u_\varepsilon = \frac{\mathrm{i}}{2\pi}\partial\overline{\partial}\log(|\sigma_E|_h^2 + \varepsilon^2) + \Theta_{\mathcal{O}(E),h} + \beta \tag{18.36}$$

where $\sigma_E \in H^0(\widetilde{X}, \mathcal{O}(E))$ is the canonical section and $\Theta_{\mathcal{O}(E),h}$ the Chern curvature form. Clearly,by the Lelong-Poincaré equation, u_ε converges to $[E]+\beta$ in the weak topology as $\varepsilon \to 0$. Straightforward calculations yield

$$u_\varepsilon = \frac{\mathrm{i}}{2\pi}\,\frac{\varepsilon^2 D_h^{1,0}\sigma_E \wedge \overline{D_h^{1,0}\sigma_E}}{(\varepsilon^2 + |\sigma_E|^2)^2} + \frac{\varepsilon^2}{\varepsilon^2 + |\sigma_E|^2}\Theta_{E,h} + \beta.$$

The first term converges to $[E]$ in the weak topology, while the second, which is close to $\Theta_{E,h}$ near E, converges pointwise everywhere to 0 on $\widetilde{X} \smallsetminus E$. A simple asymptotic analysis shows that

$$\Big(\frac{\mathrm{i}}{2\pi}\frac{\varepsilon^2 D_h^{1,0}\sigma_E \wedge \overline{D_h^{1,0}\sigma_E}}{(\varepsilon^2+|\sigma_E|^2)^2} + \frac{\varepsilon^2}{\varepsilon^2+|\sigma_E|^2}\Theta_{E,h}\Big)^p \to [E]\wedge\Theta_{E,h}^{p-1}$$

in the weak topology for $p \geqslant 1$, hence

$$\lim_{\varepsilon\to 0} u_\varepsilon^n = \beta^n + \sum_{p=1}^{n}\binom{n}{p}[E]\wedge\Theta_{E,h}^{p-1}\wedge\beta^{n-p}. \tag{18.37}$$

In arbitrary dimension, the signature of u_ε is hard to evaluate, and it is also non trivial to decide the sign of the limiting measure $\lim u_\varepsilon^n$. However, when $n = 2$, we get the simpler formula

$$\lim_{\varepsilon\to 0} u_\varepsilon^2 = \beta^2 + 2[E]\wedge\beta + [E]\wedge\Theta_{E,h}.$$

In this case, E can be assumed to be an exceptional divisor (otherwise some part of it would be nef and could be removed from the poles of T). Hence the matrix $(E_j \cdot E_k)$ is negative definite and we can find a Hermitian metric h on $\mathcal{O}(E)$ such that $(\Theta_{E,h})_{|E} < 0$. Then $[E]\wedge\Theta_{E,h}$, which is the limit of the product of the first two terms in u_ε^2, contributes negatively to the limit; all other terms are nonnegative or have a mass converging to 0. From this, one can easily infer by (18.35) that

$$\limsup_{\varepsilon\to 0}\int_{\widetilde{X}(u_\varepsilon,0)} u_\varepsilon^2 \leqslant \int_{\widetilde{X}}\beta^2 + 2[E]\wedge\beta \leqslant \mathrm{Vol}(\alpha) + 2C(\mathrm{Vol}(\alpha)-\beta^2)^{1/2}.$$

This is arbitrary close to $\mathrm{Vol}(\alpha)$ when β^2 approaches the volume, and so Theorem 18.31 (b) is proved in dimension 2. Obviously the n-dimensional case would require a deeper analysis of "higher order" orthogonality relations. □

Chapter 19

Super-canonical Metrics and Abundance

A very fundamental fact of the theory of compact Riemann surfaces is the existence of metrics with constant curvature, which is in this case a consequence of the uniformization theorem. In general, "invariant" or "canonical" metrics, such as the Kobayashi and Kobayashi-Eisenman metrics, play an important role in analytic geometry. We introduce here still another way of constructing such "canonical" metrics, following ideas of Narasimhan-Simha [NS68] which have been recently generalized by Tsuji [Tsu07a, 07b].

19.A. Construction of Super-canonical Metrics

Let X be a compact complex manifold and $(L, h_{L,\gamma})$ a holomorphic line bundle over X equipped with a singular Hermitian metric $h_{L,\gamma} = e^{-\gamma} h_L$ with satisfies $\int e^{-\gamma} < +\infty$ locally on X, where h_L is a smooth metric on L. In fact, we can more generally consider the case where $(L, h_{L,\gamma})$ is a "Hermitian $\mathbb{R}$-line bundle"; by this we mean that we have chosen a smooth real d-closed $(1,1)$ form α_L on X (whose dd^c cohomology class is equal to $c_1(L)$), and a specific current $T_{L,\gamma}$ representing it, namely $T_{L,\gamma} = \alpha_L + dd^c\gamma$, such that γ is a locally integrable function satisfying $\int e^{-\gamma} < +\infty$. An important special case is obtained by considering a klt (Kawamata log terminal) effective divisor Δ. In this situation $\Delta = \sum c_j \Delta_j$ with $c_j \in \mathbb{R}$, and if g_j is a local generator of the ideal sheaf $\mathcal{O}(-\Delta_j)$ identifying it to the trivial invertible sheaf $g_j\mathcal{O}$, we take $\gamma = \sum c_j \log |g_j|^2$, $T_{L,\gamma} = \sum c_j [\Delta_j]$ (current of integration on Δ) and α_L given by any smooth representative of the same dd^c-cohomology class; the klt condition precisely means that

$$\int_V e^{-\gamma} = \int_V \prod |g_j|^{-2c_j} < +\infty \tag{19.1}$$

on a small neighborhood V of any point in the support $|\Delta| = \bigcup \Delta_j$ (condition (19.1) implies $c_j < 1$ for every j, and this in turn is sufficient to imply Δ klt if Δ is a normal crossing divisor; the line bundle L is then the real line bundle $\mathcal{O}(\Delta)$, which makes sens as a genuine line bundle only if $c_j \in \mathbb{Z}$). For each klt pair (X, Δ) such that $K_X + \Delta$ is pseudo-effective, H. Tsuji [Tsu07a, 07b] has introduced a "super-canonical metric" which generalizes the metric introduced by Narasimhan and Simha [NS68] for projective algebraic varieties with ample canonical divisor. We take the opportunity to present here a simpler, more direct and more general approach.

We assume from now on that $K_X + L$ is *pseudo-effective*, i.e. that the class $c_1(K_X)+\{\alpha_L\}$ is pseudo-effective, and under this condition, we are going to define a "super-canonical metric" on $K_X + L$. Select an arbitrary smooth Hermitian metric ω on X. We then find induced Hermitian metrics h_{K_X} on K_X and $h_{K_X+L} = h_{K_X}h_L$ on $K_X + L$, whose curvature is the smooth real $(1,1)$-form

$$\alpha = \Theta_{K_X+L,h_{K_X+L}} = \Theta_{K_X,\omega} + \alpha_L.$$

A singular Hermitian metric on $K_X + L$ is a metric of the form $h_{K_X+L,\varphi} = e^{-\varphi}h_{K_X+L}$ where φ is locally integrable, and by the pseudo-effectivity assumption, we can find quasi-psh functions φ such that $\alpha + dd^c\varphi \geqslant 0$. The metrics on L and $K_X + L$ can now be "subtracted" to give rise to a metric

$$h_{L,\gamma}h^{-1}_{K_X+L,\varphi} = e^{\varphi-\gamma}h_L h^{-1}_{K_X+L} = e^{\varphi-\gamma}h^{-1}_{K_X} = e^{\varphi-\gamma}dV_\omega$$

on $K_X^{-1} = \Lambda^n T_X$, since $h^{-1}_{K_X} = dV_\omega$ is just the Hermitian (n,n) volume form on X. Therefore the integral $\int_X h_{L,\gamma}h^{-1}_{K_X+L,\varphi}$ has an intrinsic meaning, and it makes sense to require that

$$\text{(19.2)} \qquad \int_X h_{L,\gamma}h^{-1}_{K_X+L,\varphi} = \int_X e^{\varphi-\gamma}dV_\omega \leqslant 1$$

in view of the fact that φ is locally bounded from above and of the assumption

$$\int e^{-\gamma} < +\infty.$$

Observe that condition (19.2) can always be achieved by subtracting a constant to φ. Now, we can generalize Tsuji's super-canonical metrics on klt pairs (cf. [Tsu07b]) as follows.

(19.3) Definition. *Let X be a compact complex manifold and let (L, h_L) be a Hermitian $\mathbb{R}$-line bundle on X associated with a smooth real closed $(1,1)$ form α_L. Assume that $K_X + L$ is pseudo-effective and that L is equipped with a singular Hermitian metric $h_{L,\gamma} = e^{-\gamma}h_L$ such that $\int e^{-\gamma} < +\infty$ locally on X. Take a Hermitian metric ω on X and define $\alpha = \Theta_{K_X+L,h_{K_X+L}} = \Theta_{K_X,\omega} + \alpha_L$. Then we define the super-canonical metric $h_{\rm can}$ of $K_X + L$ to be*

$$h_{K_X+L,\rm can} = \inf_\varphi h_{K_X+L,\varphi} \quad \textit{i.e.} \quad h_{K_X+L,\rm can} = e^{-\varphi_{\rm can}}h_{K_X+L}, \quad \textit{where}$$

$$\varphi_{\rm can}(x) = \sup_\varphi \varphi(x) \ \textit{ for all } \varphi \textit{ with } \ \alpha + dd^c\varphi \geqslant 0, \quad \int_X e^{\varphi-\gamma}dV_\omega \leqslant 1.$$

In particular, this gives a definition of the super-canonical metric on $K_X + \Delta$ for every klt pair (X, Δ) such that $K_X + \Delta$ is pseudo-effective, and as an even more special case, a super-canonical metric on K_X when K_X is pseudo-effective.

In the sequel, we assume that γ has analytic singularities, otherwise not much can be said. The mean value inequality then immediately shows that the quasi-psh functions φ involved in Definition 19.3 are globally uniformly bounded outside of the poles of γ, and therefore everywhere on X, hence the envelopes $\varphi_{\rm can} = \sup_\varphi \varphi$ are indeed well defined and bounded above. As a consequence, we get a "super-canonical" current $T_{\rm can} = \alpha + dd^c\varphi_{\rm can} \geqslant 0$ and $h_{K_X+L,{\rm can}}$ satisfies

$$\int_X h_{L,\gamma} h^{-1}_{K_X+L,{\rm can}} = \int_X e^{\varphi_{\rm can}-\gamma} dV_\omega < +\infty. \tag{19.4}$$

It is easy to see that in Definition 19.3 the supremum is a maximum and that $\varphi_{\rm can} = (\varphi_{\rm can})^*$ everywhere, so that taking the upper semicontinuous regularization is not needed. In fact if $x_0 \in X$ is given and we write

$$(\varphi_{\rm can})^*(x_0) = \limsup_{x\to x_0} \varphi_{\rm can}(x) = \lim_{\nu\to+\infty} \varphi_{\rm can}(x_\nu) = \lim_{\nu\to+\infty} \varphi_\nu(x_\nu)$$

with suitable sequences $x_\nu \to x_0$ and (φ_ν) such that $\int_X e^{\varphi_\nu-\gamma}dV_\omega \leqslant 1$, the well-known weak compactness properties of quasi-psh functions in L^1 topology imply the existence of a subsequence of (φ_ν) converging in L^1 and almost everywhere to a quasi-psh limit φ. Since $\int_X e^{\varphi_\nu-\gamma}dV_\omega \leqslant 1$ holds true for every ν, Fatou's lemma implies that we have $\int_X e^{\varphi-\gamma}dV_\omega \leqslant 1$ in the limit. By taking a subsequence, we can assume that $\varphi_\nu \to \varphi$ in $L^1(X)$. Then for every $\varepsilon > 0$ the mean value $⨍_{B(x_\nu,\varepsilon)} \varphi_\nu$ satisfies

$$⨍_{B(x_0,\varepsilon)} \varphi = \lim_{\nu\to+\infty} ⨍_{B(x_\nu,\varepsilon)} \varphi_\nu \geqslant \lim_{\nu\to+\infty} \varphi_\nu(x_\nu) = (\varphi_{\rm can})^*(x_0),$$

hence we get $\varphi(x_0) = \lim_{\varepsilon\to 0} ⨍_{B(x_0,\varepsilon)} \varphi \geqslant (\varphi_{\rm can})^*(x_0) \geqslant \varphi_{\rm can}(x_0)$, and therefore the sup is a maximum and $\varphi_{\rm can} = \varphi^*_{\rm can}$. By elaborating on this argument, one can infer certain regularity properties of the envelope.

(19.5) Theorem ([BmD09]). *Let X be a compact complex manifold and (L, h_L) a holomorphic $\mathbb{R}$-line bundle such that $K_X + L$ is big. Assume that L is equipped with a singular Hermitian metric $h_{L,\gamma} = e^{-\gamma}h_L$ with analytic singularities such that $\int e^{-\gamma} < +\infty$ (klt condition). Denote by Z_0 the set of poles of a singular metric $h_0 = e^{-\psi_0}h_{K_X+L}$ with analytic singularities on $K_X + L$ and by Z_γ the poles of γ (assumed analytic). Then the associated super-canonical metric $h_{\rm can}$ is continuous on $X \smallsetminus (Z_0 \cup Z_\gamma)$.*

In fact, using the regularization techniques of [Dem94a], it is shown in [BmD09] that $h_{\rm can}$ possesses some computable logarithmic modulus of continuity. In order to shorten the exposition, we will only give a proof of the continuity in the algebraic case, using approximation by pluri-canonical sections.

(19.6) Algebraic Version of the Super-canonical Metric. Since the klt condition is open and $K_X + L$ is assumed to be big, we can always perturb L a little bit, and after blowing-up X, assume that X is projective and that $(L, h_{L,\gamma})$ is obtained as a sum of $\mathbb{Q}$-divisors

$$L = G + \Delta$$

where Δ is klt and G is equipped with a smooth metric h_G (from which $h_{L,\gamma}$ is inferred, with Δ as its poles, so that $\Theta_{L,h_{L,\gamma}} = \Theta_{G,L_G} + [\Delta]$). Clearly this situation is "dense" in what we have been considering before, just as $\mathbb{Q}$ is dense in $\mathbb{R}$. In this case, it is possible to give a more algebraic definition of the super-canonical metric φ_{can}, following the original idea of Narasimhan-Simha [NS68] (see also H. Tsuji [Tsu07a]) — the case considered by these authors is the special situation where $G = 0$, $h_G = 1$ (and moreover $\Delta = 0$ and K_X ample, for [NS68]). In fact, if m is a large integer which is a multiple of the denominators involved in G and Δ, we can consider sections

$$\sigma \in H^0(X, m(K_X + G + \Delta)).$$

We view them rather as sections of $m(K_X + G)$ with poles along the support $|\Delta|$ of our divisor. Then $(\sigma \wedge \overline{\sigma})^{1/m} h_G$ is a volume form with integrable poles along $|\Delta|$ (this is the klt condition for Δ). Therefore one can normalize σ by requiring that

$$\int_X (\sigma \wedge \overline{\sigma})^{1/m} h_G = 1.$$

Each of these sections defines a singular Hermitian metric on $K_X + L = K_X + G + \Delta$, and we can take the regularized upper envelope

$$\varphi_{\mathrm{can}}^{\mathrm{alg}} = \Big(\sup_{m,\sigma} \frac{1}{m} \log |\sigma|^2_{h^m_{K_X+L}} \Big)^* \tag{19.7}$$

of the weights associated with a smooth metric h_{K_X+L}. It is clear that $\varphi_{\mathrm{can}}^{\mathrm{alg}} \leqslant \varphi_{\mathrm{can}}$ since the supremum is taken on the smaller set of weights $\varphi = \frac{1}{m} \log |\sigma|^2_{h^m_{K_X+L}}$, and the equalities

$$e^{\varphi-\gamma} dV_\omega = |\sigma|^{2/m}_{h^m_{K_X+L}} e^{-\gamma} dV_\omega = (\sigma \wedge \overline{\sigma})^{1/m} e^{-\gamma} h_L = (\sigma \wedge \overline{\sigma})^{1/m} h_{L,\gamma} = (\sigma \wedge \overline{\sigma})^{1/m} h_G$$

imply $\int_X e^{\varphi-\gamma} dV_\omega \leqslant 1$. We claim that the inequality $\varphi_{\mathrm{can}}^{\mathrm{alg}} \leqslant \varphi_{\mathrm{can}}$ is an equality. The proof is an immediate consequence of the following statement based in turn on the Ohsawa-Takegoshi theorem and the approximation technique of [Dem92].

(19.8) Proposition. *With $L = G + \Delta$, ω, $\alpha = \Theta_{K_X+L, h_{K_X+L}}$, γ as above and $K_X + L$ assumed to be big, fix a singular Hermitian metric $e^{-\varphi} h_{K_X+L}$ of curvature $\alpha + dd^c\varphi \geqslant 0$, such that $\int_X e^{\varphi-\gamma} dV_\omega \leqslant 1$. Then φ is equal to a regularized limit*

$$\varphi = \Big(\limsup_{m\to+\infty} \frac{1}{m} \log |\sigma_m|^2_{h^m_{K_X+L}} \Big)^*$$

for a suitable sequence $\sigma_m \in H^0(X, m(K_X + G + \Delta))$ with $\int_X (\sigma_m \wedge \overline{\sigma}_m)^{1/m} h_G \leqslant 1$.

Proof. By our assumption, there exists a quasi-psh function ψ_0 with analytic singularity set Z_0 such that

$$\alpha + dd^c\psi_0 \geqslant \varepsilon_0\omega > 0$$

and we can assume $\int_C e^{\psi_0-\gamma}dV_\omega < 1$ (the strict inequality will be useful later). For $m \geqslant p \geqslant 1$, this defines a singular metric $\exp(-(m-p)\varphi - p\psi_0)h^m_{K_X+L}$ on $m(K_X+L)$ with curvature $\geqslant p\varepsilon_0\omega$, and therefore a singular metric

$$h_{L'} = \exp(-(m-p)\varphi - p\psi_0)h^m_{K_X+L}h^{-1}_{K_X}$$

on $L' = (m-1)K_X + mL$, whose curvature $\Theta_{L',h_{L'}} \geqslant (p\varepsilon_0 - C_0)\omega$ is arbitrary large if p is large enough. Let us fix a finite covering of X by coordinate balls. Pick a point x_0 and one of the coordinate balls B containing x_0. By the Ohsawa-Takegoshi extension theorem applied on the ball B, we can find a section σ_B of $K_X + L' = m(K_X+L)$ which has norm 1 at x_0 with respect to the metric $h_{K_X+L'}$ and $\int_B |\sigma_B|^2_{h_{K_X+L'}}dV_\omega \leqslant C_1$ for some uniform constant C_1 depending on the finite covering, but independent of m, p, x_0 . Now, we use a cut-off function $\theta(x)$ with $\theta(x) = 1$ near x_0 to truncate σ_B and solve a $\overline{\partial}$-equation for $(n,1)$-forms with values in L to get a global section σ on X with $|\sigma(x_0)|_{h_{K_X+L'}} = 1$. For this we need to multiply our metric by a truncated factor $\exp(-2n\theta(x)\log|x-x_0|)$ so as to get solutions of $\overline{\partial}$ vanishing at x_0. However, this perturbs the curvature by bounded terms and we can absorb them again by taking p larger. In this way we obtain

$$\text{(19.9)} \qquad \int_X |\sigma|^2_{h_{K_X+L'}}dV_\omega = \int_X |\sigma|^2_{h^m_{K_X+L}}e^{-(m-p)\varphi-p\psi_0}dV_\omega \leqslant C_2.$$

Taking $p > 1$, the Hölder inequality for congugate exponents m, $\frac{m}{m-1}$ implies

$$\begin{aligned}
\int_X (\sigma\wedge\overline{\sigma})^{\frac{1}{m}}h_G &= \int_X |\sigma|^{2/m}_{h^m_{K_X+L}}e^{-\gamma}dV_\omega \\
&= \int_X \Big(|\sigma|^2_{h^m_{K_X+L}}e^{-(m-p)\varphi-p\psi_0}\Big)^{\frac{1}{m}}\Big(e^{(1-\frac{p}{m})\varphi+\frac{p}{m}\psi_0-\gamma}\Big)dV_\omega \\
&\leqslant C_2^{\frac{1}{m}}\bigg(\int_X \Big(e^{(1-\frac{p}{m})\varphi+\frac{p}{m}\psi_0-\gamma}\Big)^{\frac{m}{m-1}}dV_\omega\bigg)^{\frac{m-1}{m}} \\
&\leqslant C_2^{\frac{1}{m}}\bigg(\int_X \big(e^{\varphi-\gamma}\big)^{\frac{m-p}{m-1}}\Big(e^{\frac{p}{p-1}(\psi_0-\gamma)}\Big)^{\frac{p-1}{m-1}}dV_\omega\bigg)^{\frac{m-1}{m}} \\
&\leqslant C_2^{\frac{1}{m}}\bigg(\int_X e^{\frac{p}{p-1}(\psi_0-\gamma)}dV_\omega\bigg)^{\frac{p-1}{m}}
\end{aligned}$$

using the hypothesis $\int_X e^{\varphi-\gamma}dV_\omega \leqslant 1$ and another application of Hölder's inequality. Since klt is an open condition and

$$\lim_{p\to+\infty}\int_X e^{\frac{p}{p-1}(\psi_0-\gamma)}dV_\omega = \int_X e^{\psi_0-\gamma}dV_\omega < 1,$$

we can take p large enough to ensure that

$$\int_X e^{\frac{p}{p-1}(\psi_0-\gamma)}dV_\omega \leqslant C_3 < 1.$$

Therefore, we see that

$$\int_X (\sigma\wedge\overline{\sigma})^{\frac{1}{m}} h_G \leqslant C_2^{\frac{1}{m}} C_3^{\frac{p-1}{m}} \leqslant 1$$

for p large enough. On the other hand

$$|\sigma(x_0)|^2_{h_{K_X+L'}} = |\sigma(x_0)|^2_{h^m_{K_X+L}} e^{-(m-p)\varphi(x_0)-p\psi_0(x_0)} = 1,$$

thus

$$\frac{1}{m}\log|\sigma(x_0)|^2_{h^m_{K_X+L}} = \Big(1-\frac{p}{m}\Big)\varphi(x_0)+\frac{p}{m}\psi_0(x_0) \tag{19.10}$$

and, as a consequence

$$\frac{1}{m}\log|\sigma(x_0)|^2_{h^m_{K_X+L}} \longrightarrow \varphi(x_0)$$

whenever $m\to+\infty$, $\frac{p}{m}\to 0$, as long as $\psi_0(x_0) > -\infty$. In the above argument, we can in fact interpolate in finitely many points $x_1, x_2, \ldots, x_q$ provided that $p \geqslant C_4 q$. Therefore if we take a suitable dense subset $\{x_q\}$ and a "diagonal" sequence associated with sections $\sigma_m \in H^0(X, m(K_X+L))$ with $m \gg p = p_m \gg q = q_m \to +\infty$, we infer that

$$\Big(\limsup_{m\to+\infty}\frac{1}{m}\log|\sigma_m(x)|^2_{h^m_{K_X+L}}\Big)^* \geqslant \limsup_{x_q\to x}\ \varphi(x_q) = \varphi(x) \tag{19.11}$$

(the latter equality occurring if $\{x_q\}$ is suitably chosen with respect to φ). In the other direction, (19.9) implies a mean value estimate

$$\frac{1}{\pi^n r^{2n}/n!}\int_{B(x,r)} |\sigma(z)|^2_{h^m_{K_X+L}}\,dz \leqslant \frac{C_5}{r^{2n}}\sup_{B(x,r)} e^{(m-p)\varphi+p\psi_0}$$

on every coordinate ball $B(x,r)\subset X$. The function $|\sigma_m|^2_{h^m_{K_X+L}}$ is plurisubharmonic after we correct the non necessarily positively curved smooth metric h_{K_X+L} by a factor of the form $\exp(C_6|z-x|^2)$, hence the mean value inequality shows that

$$\frac{1}{m}\log|\sigma_m(x)|^2_{h^m_{K_X+L}} \leqslant \frac{1}{m}\log\frac{C_5}{r^{2n}} + C_6 r^2 + \sup_{B(x,r)}\Big(1-\frac{p_m}{m}\Big)\varphi + \frac{p_m}{m}\psi_0.$$

By taking in particular $r = 1/m$ and letting $m\to+\infty$, $p_m/m\to 0$, we see that the opposite of inequality (19.9) also holds. □

(19.12) Remark. We can rephrase our results in slightly different terms. In fact, let us put

$$\varphi_m^{\rm alg} = \sup_\sigma \frac{1}{m}\log|\sigma|^2_{h^m_{K_X+L}}, \qquad \sigma\in H^0(X, m(K_X+G+\Delta)),$$

with normalized sections σ such that $\int_X(\sigma\wedge\overline{\sigma})^{1/m}h_G = 1$. Then $\varphi_m^{\rm alg}$ is quasi-psh (the supremum is taken over a compact set in a finite dimensional vector space) and by passing to the regularized supremum over all σ and all φ in (19.10) we get

$$\varphi_{\rm can} \geqslant \varphi_m^{\rm alg} \geqslant \Big(1-\frac{p}{m}\Big)\varphi_{\rm can}(x) + \frac{p}{m}\psi_0(x).$$

As $\varphi_{\rm can}$ is bounded from above, we find in particular

$$0 \leqslant \varphi_{\rm can} - \varphi_m^{\rm alg} \leqslant \frac{C}{m}(|\psi_0(x)|+1).$$

This implies that $(\varphi_m^{\rm alg})$ converges uniformly to $\varphi_{\rm can}$ on every compact subset of $X \subset Z_0$, and in this way we infer again (in a purely qualitative manner) that $\varphi_{\rm can}$ is continuous on $X \smallsetminus Z_0$. Moreover, we also see that in (19.7) the upper semicontinuous regularization is not needed on $X \smallsetminus Z_0$; in case K_X+L is ample, it is not needed at all and we have uniform convergence of $(\varphi_m^{\rm alg})$ towards $\varphi_{\rm can}$ on the whole of X. Obtaining such a uniform convergence when K_X+L is just big looks like a more delicate question, related e.g. to abundance of K_X+L on those subvarieties Y where the restriction $(K_X+L)_{|Y}$ would be e.g. nef but not big.

(19.13) Generalization. In the general case where L is a $\mathbb{R}$-line bundle and K_X+L is merely pseudo-effective, a similar algebraic approximation can be obtained. We take instead sections

$$\sigma \in H^0(X, mK_X + \lfloor mG\rfloor + \lfloor m\Delta\rfloor + p_m A)$$

where (A, h_A) is a positive line bundle, $\Theta_{A,h_A} \geqslant \varepsilon_0\omega$, and replace the definition of $\varphi_{\rm can}^{\rm alg}$ by

$$\varphi_{\rm can}^{\rm alg} = \Big(\limsup_{m\to+\infty}\sup_\sigma \frac{1}{m}\log|\sigma|^2_{h_{mK_X+\lfloor mG\rfloor+p_mA}}\Big)^*, \tag{19.14}$$

$$\int_X (\sigma\wedge\overline{\sigma})^{\frac{2}{m}} h^{\frac{1}{m}}_{\lfloor mG\rfloor + p_m A} \leqslant 1, \tag{19.15}$$

where $m \gg p_m \gg 1$ and $h^{1/m}_{\lfloor mG\rfloor}$ is chosen to converge uniformly to h_G.

We then find again $\varphi_{\rm can} = \varphi_{\rm can}^{\rm alg}$, with an almost identical proof – though we no longer have a sup in the envelope, but just a lim sup. The analogue of Proposition (19.8) also holds true in this context, with an appropriate sequence of sections $\sigma_m \in H^0(X, mK_X + \lfloor mG\rfloor + \lfloor m\Delta\rfloor + p_m A)$.

(19.16) Remark. It would be nice to have a better understanding of the super-canonical metrics. In case X is a curve, this should be easier. In fact X then has a Hermitian metric ω with constant curvature, which we normalize by requiring that $\int_X \omega = 1$, and we can also suppose $\int_X e^{-\gamma}\omega = 1$. The class $\lambda = c_1(K_X + L) \geqslant 0$ is a number and we take $\alpha = \lambda\omega$. Our envelope is $\varphi_{\mathrm{can}} = \sup \varphi$ where $\lambda\omega + dd^c\varphi \geqslant 0$ and $\int_X e^{\varphi-\gamma}\omega \leqslant 1$. If $\lambda = 0$ then φ must be constant and clearly $\varphi_{\mathrm{can}} = 0$. Otherwise, if $G(z,a)$ denotes the Green function such that $\int_X G(z,a)\,\omega(z) = 0$ and $dd^c G(z,a) = \delta_a - \omega(z)$, we find

$$\varphi_{\mathrm{can}}(z) \geqslant \sup_{a\in X} \left(\lambda G(z,a) - \log \int_{z\in X} e^{\lambda G(z,a)-\gamma(z)}\omega(z) \right)$$

by taking already the envelope over $\varphi(z) = \lambda G(z,a) - \mathrm{Const}$. It is natural to ask whether this is always an equality, i.e. whether the extremal functions are always given by one of the Green functions, especially when $\gamma = 0$.

19.B. Invariance of Plurigenera and Positivity of Curvature of Super-canonical Metrics

The concept of super-canonical metric can be used to give a very interesting result on the positivity of relative pluricanonical divisors, which itself can be seen to imply the invariance of plurigenera. The main idea is due to H. Tsuji [Tsu07a], and some important details were fixed by Berndtsson and Păun [BnP09], using techniques inspired from their results on positivity of direct images [Bnd06; BnP08].

(19.17) Theorem. *Let $\pi : \mathcal{X} \to S$ be a deformation of projective algebraic manifolds over some irreducible complex space S (π being assumed locally projective over S). Let $\mathcal{L} \to \mathcal{X}$ be a holomorphic line bundle equipped with a Hermitian metric $h_{\mathcal{L},\gamma}$ of weight γ such that $\mathrm{i}\Theta_{\mathcal{L},h_{\mathcal{L},\gamma}} \geqslant 0$ (i.e. γ is plurisubharmonic), and $\int_{X_t} e^{-\gamma} < +\infty$, i.e. we assume the metric to be klt over all fibers $X_t = \pi^{-1}(t)$. Then the metric defined on $K_{\mathcal{X}} + \mathcal{L}$ as the fiberwise super-canonical metric has semi-positive curvature over $\mathcal{X}$. In particular, $t \mapsto h^0(X_t, m(K_{X_t} + \mathcal{L}_{\restriction X_t}))$ is constant for all $m > 0$.*

Once the metric is known to have a plurisuharmonic weight on the total space of $\mathcal{X}$, the Ohsawa-Takegoshi theorem can be used exactly as at the end of the proof of Lemma 16.3. Therefore the final statement is just an easy consequence. The cases when $\mathcal{L} = \mathcal{O}_{\mathcal{X}}$ is trivial or when $\mathcal{L}_{\restriction X_t} = \mathcal{O}(\Delta_t)$ for a family of klt $\mathbb{Q}$-divisors are especially interesting.

Proof (Sketch). By our assumptions, there exists (at least locally over S) a relatively ample line bundle $\mathcal{A}$ over $\mathcal{X}$. We have to show that the weight of the global super-canonical metric is plurisubharmonic, and for this, it is enough to look at analytic disks $\Delta \to S$. We may thus as well assume that $S = \Delta$ is the unit disk. Consider the super-canonical metric $h_{\mathrm{can},0}$ over the fiber X_0. The approximation argument seen above (see (19.9) and Remark (19.12)) show that $h_{\mathrm{can},0}$ has a weight $\varphi_{\mathrm{can},0}$ which is a regularized upper limit

$$\varphi^{\mathrm{alg}}_{\mathrm{can},0} = \Big(\limsup_{m\to+\infty} \frac{1}{m}\log|\sigma_m|^2\Big)^*$$

defined by sections $\sigma_m \in H^0(X_0, m(K_{X_0} + \mathscr{L}_{\restriction X_0}) + p_m\mathscr{A}_{\restriction X_0})$ such that

$$\int_{X_0} |\sigma|^2 e^{-(m-p_m)\varphi_{\mathrm{can},0} - p_m\psi_0} dV_\omega \leqslant C_2.$$

with the suitable weights. Now, by the proof of the invariance of plurigenera (Chapter 16), these sections extend to sections $\widetilde{\sigma}_m$ defined on the whole family $\mathscr{X}$, satisfying a similar L^2 estimate (possibly with a slightly larger constant C_2' under control). If we set

$$\Phi = \Big(\limsup_{m\to+\infty} \frac{1}{m}\log|\widetilde{\sigma}_m|^2\Big)^*,$$

then Φ is plurisubharmonic by construction, and $\varphi_{\mathrm{can}} \geqslant \Phi$ by the defining property of the super-canonical metric. Finally, we also have $\varphi_{\mathrm{can},0} = \Phi_{\restriction X_0}$ from the approximation technique. It follows easily that φ_{can} satisfies the mean value inequality with respect to any disk centered on the central fiber X_0. Since we can consider arbitrary analytic disks $\Delta \to S$, the plurisubharmonicity of φ_{can} follows. □

19.C. Tsuji's Strategy for Studying Abundance

H. Tsuji [Tsu07c] has recently proposed the following interesting prospective approach of the abundance conjecture.

(19.18) Conjecture/question. *Let (X,Δ) be a klt pair such that $K_X+\Delta$ is pseudoeffective and has numerical dimension $\mathrm{nd}(K_X+\Delta) > 0$. Then for every point $x \in X$ there exists a closed positive current $T_x \in c_1(K_X+\Delta)$ such that the Lelong number at x satisfies $\nu(T_x, x) > 0$.*

It would be quite tempting to try to produce such currents e.g. by a suitable modification of the construction of super-canonical metrics, trying to enforce singularities of the metric at any prescribed point $x \in X$. A related procedure would be to enforce enough vanishing of sections of $A + m(K_X+\Delta)$ at point x, where A is a sufficiently ample line bundle. The number of these sections grows as cm^p where $p = \mathrm{nd}(K_X+\Delta)$. Hence, by an easy linear algebra argument, one can prescribe a vanishing order $s \sim c'm^{p/n}$ of such a section σ, whence a Lelong number $\sim c'm^{\frac{p}{n}-1}$ for the corresponding rescaled current of integration $T = \frac{1}{m}[Z_\sigma]$ on the zero divisor. Unfortunately, this tends to 0 as $m \to +\infty$ whenever $p < n$. Therefore, one should use a more clever argument which takes into account the fact that, most probably, all directions do not behave in an "isotropic way", and vanishing should be prescribed only in certain directions.

Assuming that (19.18) holds true, a simple semi-continuity argument would imply that there exists a small number $c > 0$ such that the analytic set $Z_x = E_c(T_x)$ contains x, and one would expect conjecturally that these sets can be reorganized as the generic fibers of a reduction map $f : X \dashrightarrow Y$, together with a klt divisor Δ' on Y such that (in first approximation, and maybe only after replacing X, Y by suitable blow-ups), one has $K_X + \Delta = f^*(K_Y + \Delta' + R_f) + \beta$ where R_f is a suitable orbifold divisor (in the sense of Campana [Cam04]) and β a suitable pseudo-effective class. The expectation is that $\dim Y = p = \mathrm{nd}(K_X + \Delta)$ and that (Y, Δ') is of general type, i.e. $\mathrm{nd}(K_Y + \Delta') = p$.

Chapter 20

Siu's Analytic Approach and Păun's Non Vanishing Theorem

We describe here briefly some recent developments without giving much detail about proofs. Recall that given a pair (X, Δ) where X is a normal projective variety and Δ an effective $\mathbb{R}$-divisor, the transform of (X, Δ) by a birational morphism $\mu : \widetilde{X} \to X$ of normal varieties is the unique pair $(\widetilde{X}, \widetilde{\Delta})$ such that $K_{\widetilde{X}} + \widetilde{\Delta} = \mu^*(K_X + \Delta) + E$ where E is an effective μ-exceptional divisor (we assume here that $K_X + \Delta$ and $K_{\widetilde{X}} + \widetilde{\Delta}$ are $\mathbb{R}$-Cartier divisors).

In [BCHM06], Birkar, Cascini, Hacon and McKernan proved several fundamental conjectures which had been expected for more than two decades, concerning the existence of minimal models and the finiteness of the canonical ring for arbitrary projective varieties. The latter result was also announced independently by Siu in [Siu06]. The main results can be summarized in the following statement.

(20.1) Theorem. *Let (X, Δ) be a klt pair where Δ is big.*

(a) *If $K_X + \Delta$ is pseudo-effective, (X, Δ) has a log-minimal model, i.e. there is a birational transformation $(\widetilde{X}, \widetilde{\Delta})$ with $\widetilde{X}$ $\mathbb{Q}$-factorial, such that $K_{\widetilde{X}} + \widetilde{\Delta}$ is nef and satisfies additionally strict inequalities for the discrepancies of μ-exceptional divisors.*

(b) *If $K_X + \Delta$ is not pseudo-effective, then (X, Δ) has a Mori fiber space, i.e. there exists a birational transformation $(\widetilde{X}, \widetilde{\Delta})$ and a morphism $\varphi : \widetilde{X} \to Y$ such that $-(K_{\widetilde{X}} + \widetilde{\Delta})$ is φ-ample.*

(c) *If moreover Δ is a $\mathbb{Q}$-divisor, the log-canonical ring $\bigoplus_{m\geqslant 0} H^0(X, m(K_X+\Delta))$ is finitely generated.*

The proof, for which we can only refer to [BCHM06], is an extremely subtle induction on dimension involving finiteness of flips (a certain class of birational transforms improving positivity of $K_X + \Delta$ step by step), and a generalization of Shokurov's non vanishing theorem [Sho85]. The original proof of this non vanishing result was itself based on an induction on dimension, using the existence of minimal models in dimension $n-1$. Independently, Y.T. Siu [Siu06] announced an analytic proof of the finiteness of canonical rings $\bigoplus_{m\geqslant 0} H^0(X, mK_X)$, along with an analytic variant of Shokurov's non vanishing theorem; in his approach, multiplier ideals and Skoda's division theorem are used in crucial ways. Let us mention a basic statement in this direction which illustrates the connection with Skoda's result, and is interesting for two reasons : i) it does not require any strict positivity assumption, ii) it shows that it is enough to have a sufficiently good approximation of the minimal singularity metric $h_{\min}$ by sections of sufficiently large linear systems $|pK_X|$.

(20.2) Proposition. *Let X be a projective n-dimensional manifold with K_X pseudo-effective. Let $h_{\min} = e^{-\varphi_{\min}}$ be a metric with minimal singularity on K_X (e.g. the super-canonical metric), and let $c_0 > 0$ be the log canonical threshold of $\varphi_{\min}$, i.e. $h_{\min}^{c_0-\delta} = e^{-(c_0-\delta)\varphi_{\min}} \in L^1$ for $\delta > 0$ small. Assume that there exists an integer $p > 0$ so that the linear system $|pK_X|$ provides a weight $\psi_p = \frac{1}{p}\log\sum|\sigma_j|^2$ whose singularity approximates $\varphi_{\min}$ sufficiently well, namely*

$$\psi_p \geqslant \Big(1 + \frac{1+c_0-\delta}{pn}\Big)\varphi_{\min} + O(1) \qquad \textit{for some } \delta > 0.$$

Then $\bigoplus_{m\geqslant 0} H^0(X, mK_X)$ is finitely generated, and a set of generators is actually provided by a basis of sections of $\bigoplus_{0\leqslant m\leqslant np+1} H^0(X, mK_X)$.

Proof. A simple argument based on the curve selection lemma (see e.g. [Dem01], Lemma 11.16) shows that one can extract a system $g = (g_1, \cdots, g_n)$ of at most n sections from (σ_j) in such a way that the singularities are unchanged, i.e. $C_1 \log|\sigma| \leqslant \log|g| \leqslant C_2 \log|\sigma|$. We apply Skoda's division (11.8), (11.12) with $E = \mathcal{O}_X^{\oplus n}$, $Q = \mathcal{O}(pK_X)$ and $L = \mathcal{O}((m-p-1)K_X)$ [so that $K_X \otimes Q \otimes L = \mathcal{O}_X(mK_X)$], and with the metric induced by $h_{\min}$ on K_X. By definition of a metric with minimal singularities, every section f in $H^0(X, mK_X) = H^0(X, K_X \otimes Q \otimes L)$ is such that $|f|^2 \leqslant Ce^{m\varphi_{\min}}$. The weight of the metric on $Q \otimes L$ is $(m-1)\varphi_{\min}$. Accordingly, we find

$$\begin{aligned} |f|^2 |g|_{h_{min}}^{-2n-2\varepsilon} e^{-(m-1)\varphi_{\min}} &\leqslant C \exp\Big(m\varphi_{\min} - p(n+\varepsilon)(\psi_p - \varphi_{\min}) - (m-1)\varphi_{\min}\Big) \\ &\leqslant C' \exp\big(-(c_0 - \delta/2)\varphi_{\min}\big) \end{aligned}$$

for $\varepsilon > 0$ small, thus the left hand side is in L^1. Skoda's theorem implies that we can write $f = g \cdot h = \sum g_j h_j$ with $h_j \in H^0(X, K_X \otimes L) = H^0(X, (m-p)K_X)$. The argument holds as soon as the curvature condition $m - p - 1 \geqslant (n - 1 + \varepsilon)p$ is satisfied, i.e. $m \geqslant np + 2$. Therefore all multiples $m \geqslant np + 2$ are generated by sections of lower degree $m - p$, and the result follows. □

Recently, Păun [Pau08] has been able to provide a very strong Shokurov-type analytic non vanishing statement, and in the vein of Siu's approach [Siu06], he gave a very detailed independent proof which does not require any intricate induction on dimension (i.e. not involving the existence of minimal models).

(20.3) Theorem (Păun [Pau08]). *Let X be a projective manifold, and let $\alpha_L \in \mathrm{NS}_{\mathbb{R}}(X)$ be a cohomology class in the real Neron-Severi space of X, such that :*

(a) *The adjoint class $c_1(K_X) + \alpha_L$ is pseudoeffective, i.e. there exist a closed positive current*

$$\Theta_{K_X+L} \in c_1(K_X) + \alpha_L;$$

(b) *The class α_L contains a Kähler current Θ_L (so that α_L is big), such that the respective potentials φ_L of Θ_L and φ_{K_X+L} of Θ_{K_X+L} satisfy*

$$e^{(1+\varepsilon)(\varphi_{K_X+L}-\varphi_L)} \in L^1_{\mathrm{loc}}$$

where ε is a positive real number.

Then the adjoint class $c_1(K_X) + \alpha_L$ contains an effective $\mathbb{R}$-divisor.

The proof is a clever application of the Kawamata-Viehweg-Nadel vanishing theorem, combined with a perturbation trick of Shokurov [Sho85] and with diophantine approximation to reduce the situation to the case of $\mathbb{Q}$-divisors. Shokurov's trick allows to single out components of the divisors involved, so as to be able to take restrictions and apply induction on dimension. One should notice that the poles of φ_L may help in achieving condition 20.3 (b), so one obtains a stronger condition by requiring (b′) $\exp((1+\varepsilon)\varphi_{K_X+L}) \in L^1_{\mathrm{loc}}$ for $\varepsilon > 0$ small, namely that $c_1(K_X) + \alpha_L$ is klt. The resulting weaker statement then makes sense in a pure algebraic setting. In [BrP09], Birkar and Păun announced a relative version of Theorem 20.3, and they showed that this can be used to reprove a relative version of Theorem 20.1. The notes of Mihai Păun [Pau09] give a fairly precise account of these techniques, and incorporate as well some ideas of Ein-Lazarsfeld-Mustaţă-Nakamaye-Popa [E-P06] and of A. Corti and V. Lazić (see [Lzc09]). A similar purely algebraic approach has been described by C. Hacon in his Oberwolfach lectures [Hac08].

References

[AN54] **Y. Akizuki, S. Nakano**. — *Note on Kodaira-Spencer's proof of Lefschetz theorems*, Proc. Jap. Acad., **30** (1954), 266–272.

[AG62] **A. Andreotti, H. Grauert**. — *Théorèmes de finitude pour la cohomologie des espaces complexes*, Bull. Soc. Math. France, **90** (1962), 193–259.

[AS95] **U. Angehrn, Y.T. Siu**. — *Effective freeness and point separation for adjoint bundles*, Invent. Math., **122** (1995), 291–308.

[AV65] **A. Andreotti, E. Vesentini**. — *Carleman estimates for the Laplace-Beltrami equation in complex manifolds*, Publ. Math. I.H.E.S., **25** (1965), 81–130.

[Aub78] **T. Aubin**. — *Equations du type Monge-Ampère sur les variétés kählériennes compactes*, C.R. Acad. Sci. Paris Ser. A, **283** (1976), 119–121 ; Bull. Sci. Math, **102** (1978), 63–95.

[BT76] **E. Bedford, B.A. Taylor**. — *The Dirichlet problem for the complex Monge-Ampère equation*, Invent. Math., **37** (1976), 1–44.

[BT82] **E. Bedford, B.A. Taylor**. — *A new capacity for plurisubharmonic functions*, Acta Math., **149** (1982), 1–41.

[BBS89] **M. Beltrametti, A. Biancofiore, A.J. Sommese**. — *Projective n-folds of log-general type, I*, Trans. Amer. Math. Soc., **314** (1989), 825–849.

[BeS93] **M. Beltrametti, A.J. Sommese**. — *On k-jet ampleness*, Complex Analysis and Geometry, Univ. Series in Math., edited by V. Ancona and A. Silva, Plenum Press, New-York, 1993, 355–376.

[BFS89] **M. Beltrametti, P. Francia, A.J. Sommese**. — *On Reider's method and higher order embeddings*, Duke Math. J., **58** (1989), 425–439.

[BmD09] **R. Berman, J.-P. Demailly**. — *Regularity of plurisubharmonic upper envelopes in big cohomology classes*, arXiv:math.CV/0905.1246v1.

[Bnd06] **B. Berndtsson**. — *Subharmonicity properties of the Bergman kernel and some other functions associated to pseudoconvex domains*, Ann. Inst. Fourier (Grenoble), **56** (2006), 1633–1662.

[BnP08] **B. Berndtsson M. Păun**. — *Bergman kernels and the pseudoeffectivity of relative canonical bundles*, Duke Math. J, **145** (2008), 341–378.

[BnP09] **B. Berndtsson M. Păun**. — *Canonical metrics for projective families*, Privately communicated manuscript, February 2009.

[BCHM06] **C. Birkar, P. Cascini, C.D. Hacon, J. McKernan**. — *Existence of minimal models for varieties of log general type*, arXiv:math.AG/0610203v2.

[BrP09] **C. Birkar, M. Păun**. — *Minimal models, flips and finite generation : a tribute to V.V. Shokurov and Y.-T. Siu*, arXiv:math.AG/0904.2936.

[Boc48] **S. Bochner**. — *Curvature and Betti numbers (I) and (II)*, Ann. of Math., **49** (1948), 379–390 ; **50** (1949), 77–93.

[Bom70] **E. Bombieri**. — *Algebraic values of meromorphic maps*, Invent. Math., **10** (1970), 267–287 and *Addendum*, Invent. Math., **11** (1970), 163–166.

[Bom73] **E. Bombieri**. — *Canonical models of surfaces of general type*, Publ. Math. IHES, **42** (1973), 171–219.

[Bon93] **L. Bonavero**. — *Inégalités de Morse holomorphes singulières*, Prépublication Université de Grenoble I, Institut Fourier, n° 259, 1993.

[Bou02] **S. Boucksom**. — *Cônes positifs des variétés complexes compactes*, Thesis, Grenoble 2002.

[BDPP04] **S. Boucksom, J.-P. Demailly, M. Păun, Th. Peternell**. — *The pseudo-effective cone of a compact Kähler manifold and varieties of negative Kodaira dimension*, manuscript May 2004, math.AG/0405285.

[BGK05] **Ch. P. Boyer, K. Galicki, J. Kollár**. — *Einstein Metrics on Spheres*, Annals of Math, **162** (2005), 557–580.

[BSk74] **J. Briançon, H. Skoda**. — *Sur la clôture intégrale d'un idéal de germes de fonctions holomorphes en un point de $\mathbb{C}^n$*, C. R. Acad. Sc. Paris, Sér. A, **278** (1974), 949–951.

[Buc99] **N. Buchdahl**. — *On compact Kähler surfaces*, Ann. Inst. Fourier, **49** (1999), 287–302.

[Buc00] **N. Buchdahl**. — *A Nakai-Moishezon criterion for non-Kähler surfaces*, Ann. Inst. Fourier, **50** (2000), 1533–1538.

[Cam92] **F. Campana**. — *Connexité rationnelle des variétés de Fano*, Ann. Sci. Ecole Norm. Sup., **25** (1992), 539–545.

[Cam04] **F. Campana**. — *Orbifolds, special varieties and classification theory*, Ann. Inst. Fourier (Grenoble), **54** (2004), 499–630.

[Cat88] **F. Catanese**. — *Footnotes to a theorem of I. Reider*, Proc. Intern. Conf. on Algebraic Geometry (L'Aquila, June 1988), Lecture Notes in Math., Vol. 1417, Springer-Verlag, Berlin, 1990, 67–74.

[Cla07] **B. Claudon**. — *Invariance of plurigenera for multiples of the twisted canonical bundle*, Ann. Inst. Fourier, **57** (2007), 289–300.

[CLN69] **S.S. Chern, H.I. Levine, L. Nirenberg**. — *Intrinsic norms on a complex manifold*, Global Analysis (papers in honor of K.Kodaira), p. 119–139, Univ. of Tokyo Press, Tokyo, 1969.

[CP04] **F. Campana, Th. Peternell**. — *Geometric stability of the cotangent bundle and the universal cover of a projective manifold*, arXiv:math.AG/0405093.

[DeI87] **P. Deligne, L. Illusie**. — *Relèvements modulo p^2 et décomposition du complexe de De Rham*, Invent. Math., **89** (1987), 247–270.

[Dem82a] **J.-P. Demailly**. — *Sur les nombres de Lelong associés à l'image directe d'un courant positif fermé*, Ann. Inst. Fourier (Grenoble), **32** (1982), 37–66.

[Dem82b] **J.-P. Demailly**. — *Estimations L^2 pour l'opérateur $\overline{\partial}$ d'un fibré vectoriel holomorphe semi-positif au dessus d'une variété kählérienne complète*, Ann. Sci. Ec. Norm. Sup., **15** (1982), 457–511.

[Dem82c] **J.-P. Demailly**. — *Courants positifs extrêmaux et conjecture de Hodge*, Invent. Math., **69** (1982), 347-374.

[Dem85a] **J.-P. Demailly**. — *Mesures de Monge-Ampère et caractérisation géométrique des variétés algébriques affines*, Mém. Soc. Math. France (N.S.), **19** (1985), 1–124.

[Dem85b] **J.-P. Demailly**. — *Champs magnétiques et inégalités de Morse pour la d''-cohomologie*, Ann. Inst. Fourier (Grenoble), **35** (1985), 189–229.

[Dem87] **J.-P. Demailly**. — *Nombres de Lelong généralisés, théorèmes d'intégralité et d'analyticité*, Acta Math., **159** (1987), 153–169.

[Dem89] **J.-P. Demailly**. — *Transcendental proof of a generalized Kawamata-Viehweg vanishing theorem*, C. R. Acad. Sci. Paris Sér. I Math., **309** (1989), 123–126 and Proceedings of the Conference "Geometrical and algebraical aspects in several complex variables" held at Cetraro, Univ. della Calabria, June 1989.

[Dem90] **J.-P. Demailly**. — *Singular Hermitian metrics on positive line bundles*, Proc. Conf. Complex algebraic varieties (Bayreuth, April 2–6, 1990), edited by K. Hulek,

T. Peternell, M. Schneider, F. Schreyer, Lecture Notes in Math., Vol. 1507, Springer-Verlag, Berlin, 1992.

[Dem91] **J.-P. Demailly**. — *Holomorphic Morse inequalities*, Lectures given at the AMS Summer Institute on Complex Analysis held in Santa Cruz, July 1989, Proceedings of Symposia in Pure Mathematics, Vol. **52**, Part 2 (1991), 93–114.

[Dem92] **J.-P. Demailly**. — *Regularization of closed positive currents and Intersection Theory*, J. Alg. Geom., **1** (1992), 361–409.

[Dem93a] **J.-P. Demailly**. — *Monge-Ampère operators, Lelong numbers and intersection theory*, Complex Analysis and Geometry, Univ. Series in Math., edited by V. Ancona and A. Silva, Plenum Press, New-York, 1993, 115–193.

[Dem93b] **J.-P. Demailly**. — *A numerical criterion for very ample line bundles*, J. Differential Geom., **37** (1993), 323–374.

[Dem94a] **J.-P. Demailly**. — *Regularization of closed positive currents of type* $(1,1)$ *by the flow of a Chern connection*, Actes du Colloque en l'honneur de P. Dolbeault (Juin 1992), édité par H. Skoda et J.-M. Trépreau, Aspects of Mathematics, Vol. E 26, Vieweg, 1994, 105–126.

[Dem94b] **J.-P. Demailly**. — L^2 *vanishing theorems for positive line bundles and adjunction theory*, Lecture Notes of the CIME Session "Transcendental methods in Algebraic Geometry", Cetraro, Italy, July 1994, Ed. F. Catanese, C. Ciliberto, Lecture Notes in Math., Vol. 1646, 1–97.

[Dem96] **J.-P. Demailly**. — *Effective bounds for very ample line bundles*, Invent. Math., **124** (1996), 243–261.

[Dem98] **J.-P. Demailly**. — *Méthodes* L^2 *et résultats effectifs en géométrie algébrique*, Sém. Bourbaki, Novembre 1998, exposé n°852.

[Dem00] **J.-P. Demailly**. — *On the Ohsawa-Takegoshi-Manivel* L^2 *extension theorem*, Proceedings of the Conference in honour of the 85th birthday of Pierre Lelong, Paris, September 1997, éd. P. Dolbeault, Progress in Mathematics, Birkhäuser, Vol. **188** (2000) 47-82.

[Dem01] **J.-P. Demailly**. — *Multiplier ideal sheaves and analytic methods in algebraic geometry*, Lecture Notes, School on "Vanishing theorems and effective results in Algebraic Geometry, ICTP Trieste, April 2000, Publications of ICTP, 2001.

[Dem08] **J.-P. Demailly**. — *Appendix to I. Cheltsov and C. Shramov's article "Log canonical thresholds of smooth Fano threefolds": On Tian's invariant and log canonical thresholds*, arXiv: math.AG 0806.2107.

[Dem09] **J.-P. Demailly**. — *Regularity of plurisubharmonic upper envelopes in big cohomology classes*, Preprint May 2009, arXiv:math.CV/0905.1246, to appear in the Proceedings volume in honor of Oleg Viro, ed. by Burglind Jöricke and Mikael Passare.

[DEL00] **J.-P. Demailly, L. Ein, R. Lazarsfeld**. — *A subadditivity property of multiplier ideals*, Michigan Math. J., 2000, to appear.

[DK01] **J.-P. Demailly, J. Kollár**. — *Semicontinuity of complex singularity exponents and Kähler-Einstein metrics on Fano orbifolds*, math.AG/9910118 ; Ann. Ec. Norm. Sup., **34** (2001), 525–556.

[DP04] **J.-P. Demailly, M. Păun**. — *Numerical characterization of the Kähler cone of a compact Kähler manifold*, arXiv: math.AG/0105176, Annals of Math, **159** (2004), 1247–1274.

[DPS94] **J.-P. Demailly, Th. Peternell, M. Schneider**. — *Compact complex manifolds with numerically effective tangent bundles*, J. Algebraic Geometry, **3** (1994), 295–345.

[DPS00] **J.-P. Demailly, Th. Peternell, M. Schneider**. — *Pseudo-effective line bundles on compact Kähler manifolds*, June 2000, Algebraic Geometry e-prints math.AG/0006205.

[DSk79] **J.-P. Demailly, H. Skoda**. — *Relations entre les notions de positivité de P.A. Griffiths et de S. Nakano*, Séminaire P. Lelong-H. Skoda (Analyse), année 1978/79, Lecture notes in Math. no **822**, Springer-Verlag, Berlin (1980) 304–309.

[DF83] **H. Donnelly, C. Fefferman**. — *L^2-cohomology and index theorem for the Bergman metric*, Ann. Math., **118** (1983), 593–618.

[DX84] **H. Donnelly, F. Xavier**. — *On the differential form spectrum of negatively curved Riemann manifolds*, Amer. J. Math., **106** (1984), 169–185.

[Ein94] **L. Ein**. — *Note on higher dimensional adjoint linear systems*, preprint 1994, personal communication to the author.

[EKL95] **L. Ein, O. Küchle R. Lazarsfeld**. — *Local positivity of ample line bundles*, J. Differ. Geom., **42** (1995), 193–219.

[EL92] **L. Ein, R. Lazarsfeld**. — *Seshadri constants on smooth surfaces*, Journées de Géométrie Algébrique d'Orsay, July 1992, Astérisque **282**, 1993, 177–186.

[EL93] **L. Ein, R. Lazarsfeld**. — *Global generation of pluricanonical and adjoint linear series on smooth projective threefolds*, Jour. of Am. Math. Soc., **6** (1993), 875–903.

[ELM95] **L. Ein, R. Lazarsfeld, V. Masek**. — *Global generation of linear series on terminal threefolds*, Intern. J. Math., **6** (1995), 1–18.

[ELN96] **L. Ein, R. Lazarsfeld, M. Nakamaye**. — *Zero estimates, intersection theory, and a theorem of Demailly*, in Higher Dimensional Complex Varieties (Trento, 1994), de Gruyter (1996), 183–207.

[E-P06] **L. Ein, R. Lazarsfeld, M. Mustaţă, M. Nakamaye, M. Popa**. — *Asymptotic invariants of base loci* , Ann. Inst. Fourier (Grenoble) **56** (2006), no. 6, 1701–1734.

[EM84] **H. El Mir**. — *Sur le prolongement des courants positifs fermés*, Acta Math., **153** (1984), 1–45.

[Eno93] **I. Enoki**. — *Kawamata-Viehweg vanishing theorem for compact Kähler manifolds*, Einstein metrics and Yang-Mills connections (ed. T. Mabuchi, S. Mukai), Marcel Dekker, 1993, 59–68.

[EV86] **H. Esnault, E. Viehweg**. — *Logarithmic De Rham complexes and vanishing theorems*, Invent. Math., **86** (1986), 161–194.

[EV92] **H. Esnault, E. Viehweg**. — *Lectures on vanishing theorems*, DMV Seminar, Band **20**, Birkhäuser Verlag, 1992.

[Fed69] **H. Federer**. — *Geometric measure theory*, Vol. 153, Springer Verlag, Berlin, Heidelberg, New-York, 1969.

[FdB94] **G. Fernández del Busto**. — *UCLA Thesis*, 1994.

[FdB95] **G. Fernández del Busto**. — *Bogomolov instability and Kawamata-Viehweg vanishing*, J. Algebr. Geom., **4** (1995), 693–700.

[FdB96] **G. Fernandez del Busto**. — *A Matsusaka-type theorem on surfaces*, J. Algebr. Geom., **5** (1996), 513–520.

[FKL07] **T. de Fernex, A. Küronya, R. Lazarsfeld**. — *Higher cohomology of divisors on a projective variety*, Math. Ann., **337** (2007), 443–455.

[Fuj83] **T. Fujita**. — *Semipositive line bundles*, J. Fac. Sci. Univ. of Tokyo, **30** (1983), 353–378.

[Fuj87] **T. Fujita**. — *On polarized manifolds whose adjoint bundles are not semi-positive*, Algebraic Geometry, Sendai, 1985, Adv. Stud. in Pure Math., 10(1987), 167–178. North Holland, T. Oda (ed.).

[Fuj88] **T. Fujita**. — *Problem list*, Conference held at the Taniguchi Foundation, Katata, Japan, August 1988.

[Fuj93] **T. Fujita**. — *Remarks on Ein-Lazarsfeld criterion of spannedness of adjoint bundles of polarized threefolds*, Preprint 1993.

[Fuj94] **T. Fujita**. — *Approximating Zariski decomposition of big line bundles*, Kodai Math. J., **17** (1994), 1–3.

[Gir76] **J. Girbau**. — *Sur le théorème de Le Potier d'annulation de la cohomologie*, C. R. Acad. Sci. Paris, série A, **283** (1976), 355–358.

[Gra60] **H. Grauert**. — *Ein Theorem der analytischen Garbentheorie und die Modulräume komplexer Strukturen*, Publ. Math. I.H.E.S., **5** (1960), 233–292.

[GR84] **H. Grauert, R. Remmert**. — *Coherent analytic sheaves*, Grundlehren der math. Wissenschaften **265**, Springer-Verlag, Berlin, 1984.

[Gri69] **P.A. Griffiths**. — *Hermitian differential geometry, Chern classes and positive vector bundles*, Global Analysis, papers in honor of K. Kodaira, Princeton Univ. Press, Princeton, 1969, 181–251.

[GrH78] **P.A. Griffiths, J. Harris**. — *Principles of algebraic geometry*, Wiley, New York, 1978.

[Hac08] **C. Hacon**. — *Extension theorems and the existence of flips*, Series of 8 lectures given at Oberwolfach Math. Institute, Oct. 12-18, 2008.

[HaMK07] **C. Hacon, J. McKernan**. — *Extension theorems and the existence of flips*, In "flips for 3-folds and 4-folds" (ed. A. Corti), Oxford lecture series in mathematics and its applications **35**, Oxford University Press, 2007.

[HaMK08] **C. Hacon, J. McKernan**. — *Existence of minimal models for varieties of general type, II*, (Existence of pl-flips), arXiv:math.AG/0808.1929v1.

[Har70] **R. Hartshorne**. — *Ample subvarieties of algebraic varieties*, Lecture Notes in Math., Vol. 156, Springer-Verlag, Berlin (1970).

[Hel97] **S. Helmke**. — *On Fujita's conjecture*, Duke Math. J., **88** (1997), 201–231.

[Hel98] **S. Helmke**. — *Global generation of adjoint linear systems*, Communication to the Oberwolfach Conference "Komplexe Analysis", September 1998.

[Hir64] **H. Hironaka**. — *Resolution of singularities of an algebraic variety over a field of characteristic zero*, Ann. of Math., **79** (1964), 109–326.

[Hör65] **L. Hörmander**. — *L^2 estimates and existence theorems for the $\overline{\partial}$ operator*, Acta Math., **113** (1965), 89–152.

[Hör66] **L. Hörmander**. — *An introduction to Complex Analysis in several variables*, 1966, 3rd edition, North-Holland Math. Libr., vol.**7**, Amsterdam, London, 1990.

[Hov79] **A.G. Hovanski**. — *Geometry of convex bodies and algebraic geometry*, Uspehi Mat. Nauk, **34** (4) (1979), 160–161.

[Huy99] **D. Huybrechts**. — *Compact Hyperkähler Manifolds: Basic Results*, Invent. Math., **135** (1999), 63–113.

[Huy01] **D. Huybrechts**. — *The Projectivity Criterion for Hyperkähler manifolds as a Consequence of the Demailly-Păun Theorem*, personal communication, manuscript Köln University, May 2001, 3 p, to appear in Invent. Math.

[JK01] **J.M. Johnson, J. Kollár**. — *Kähler-Einstein metrics on log del Pezzo surfaces in weighted projective 3-spaces*, Ann. Inst. Fourier, **51** (2001), 69–79.

[Kaw82] **Y. Kawamata**. — *A generalization of Kodaira-Ramanujam's vanishing theorem*, Math. Ann., **261** (1982), 43–46.

[Kaw84] **Y. Kawamata**. — *The cone of curves of algebraic varieties*, Ann. of Math., **119** (1984), 603–633.

[Kaw85] **Y. Kawamata**. — *Pluricanonical systems of minimal algebraic varieties*, Invent. Math., **79** (1985), 567–588.

[Kaw97] **Y. Kawamata**. — *On Fujita's freeness conjecture for 3-folds and 4-folds*, Math. Ann., **308** (1997), 491–505.

[Kaw98] **Y. Kawamata**. — *On the extension problem of pluricanonical forms*, Preprint Univ. of Tokyo at Komaba, September 1998.

[Kaw99] **Y. Kawamata**. — *Deformation of canonical singularities*, J. Amer. Math. Soc., **12** (1999), 85–92.

[KMM87] **Y. Kawamata, K. Matsuda, K. Matsuki**. — *Introduction to the minimal problem*, Algebraic Geometry, Sendai, 1985, Adv. Stud. in Pure Math., Vol. 10, T. Oda (ed.), North Holland, Amsterdam, 1987, 283–360.

[Kis78] **C.O. Kiselman**. — *The partial Legendre transformation for plurisubharmonic functions*, Invent. Math., **49** (1978), 137–148.

[Kis79] **C.O. Kiselman**. — *Densité des fonctions plurisousharmoniques*, Bull. Soc. Math. France, **107** (1979), 295–304.

[Kis84] **C.O. Kiselman**. — *Sur la définition de l'opérateur de Monge-Ampère complexe*, Analyse Complexe, Proceedings of the Journées Fermat (SMF), Toulouse, 1983, Lecture Notes in Math., **1094** (1984), 139–150. Springer-Verlag Berlin.

[KobO73] **S. Kobayashi, T. Ochiai**. — *Characterisations of complex projective spaces and hyperquadrics*, J. Math. Kyoto Univ., **13** (1973), 31–47.

[Kod53] **K. Kodaira**. — *On a differential geometric method in the theory of analytic stacks*, Proc. Nat. Acad. Sci. USA, **39** (1953), 1268–1273.

[Kod54] **K. Kodaira**. — *On Kähler varieties of restricted type*, Ann. of Math., **60** (1954), 28–48.

[Koh63] **J.J. Kohn**. — *Harmonic integrals on strongly pseudo-convex manifolds I*, Ann. Math., **78** (1963), 206–213.

[Koh64] **J.J. Kohn**. — *Harmonic integrals on strongly pseudo-convex manifolds II*, Ann. Math., **79** (1964), 450–472.

[Kol85] **J. Kollár**. — *Vanishing theorems for cohomology groups*, Algebraic Geometry, Bowdoin 1985, Proc. Symp. Pure Math., Vol. 46, 1987, 233–243.

[Kol92] **J. Kollár**. — *Effective basepoint freeness*, Math. Ann., **296** (1993), 595–605.

[KoM83] **J. Kollár** and **T. Matsusaka**. — *Riemann-Roch type inequalities*, Amer. J. of Math., **105** (1983), 229–252.

[KoMM92] **J. Kollár, Y. Miyaoka, S. Mori**. — *Rational connectedness and boundedness of Fano manifolds*, J. Differential Geom., **36** (1992), 765–779.

[KoM98] **J. Kollár, S. Mori**. — *Birational geometry of algebraic varieties*, With the collaboration of C.H. Clemens and A. Corti. Cambridge Tracts in mathematics, **134**, Cambridge University press, 1998.

[Kur06] **A. Küronya**. — *Asymptotic cohomological functions on projective varieties*, Amer. J. Math., **128** (2006), 1475–1519.

[Lae02] **L. Laeng**. — *Estimations spectrales asymptotiques en géométrie hermitienne*, Thèse de Doctorat de l'Université de Grenoble I, octobre 2002, http://www-fourier.ujf-grenoble.fr/THESE/ps/laeng.ps.gz and http://tel.archives-ouvertes.fr/tel-00002098/en/.

[Lam99a] **A. Lamari**. — *Courants kählériens et surfaces compactes*, Ann. Inst. Fourier, **49** (1999), 263–285.

[Lam99b] **A. Lamari**. — *Le cone Kählérien d'une surface*, J. Math. Pures Appl., **78** (1999), 249–263.

[Laz97] **R. Lazarsfeld,** with the assistance of **G. Fernández del Busto**. — *Lectures on linear series*, Complex Algebraic Geometry (Park City, UT, 1993), Amer. Math. Soc., Providence, RI (1997), 161–219.

[Laz99] **R. Lazarsfeld**. — *Private communication*, University of Ann Arbor, 1999.

[Lzc09] **V. Lazić**. — *Adjoint rings are finitely generated,* arXiv: 0905.2707.

[Lel57] **P. Lelong**. — *Intégration sur un ensemble analytique complexe*, Bull. Soc. Math. France, **85** (1957), 239–262.

[Lel67] **P. Lelong**. — *Fonctionnelles analytiques et fonctions entières (n variables)*, Sém. de Math. Supérieures, 6e session, été 1967, Presses Univ. Montréal 1968.

[Lel72] **P. Lelong.** — *Eléments extrêmaux sur le cône des courants positifs fermés*, Séminaire P. Lelong (Analyse), année 1971/72, Lecture Notes in Math., Vol. 332, Springer-Verlag, Berlin, (1972), 112–131.

[Lel69] **P. Lelong.** — *Plurisubharmonic functions and positive differential forms*, Gordon and Breach, New-York, and Dunod, Paris, 1969.

[LP75] **J. Le Potier.** — *Annulation de la cohomologie à valeurs dans un fibré vectoriel holomorphe de rang quelconque*, Math. Ann., **218** (1975), 35–53.

[Man93] **L. Manivel.** — *Un théorème de prolongement L^2 de sections holomorphes d'un fibré vectoriel*, Math. Zeitschrift, **212** (1993), 107–122.

[Mat72] **T. Matsusaka.** — *Polarized varieties with a given Hilbert polynomial*, Amer. J. of Math., **94** (1972), 1027–1077.

[Mor82] **S. Mori.** — *Threefolds whose canonical bundles are not numerically effective*, Ann. of Math., **116** (1982), 133–176.

[Mou99] **Ch. Mourougane.** — *Théorèmes d'annulation génériques pour les fibrés vectoriels semi-négatifs*, Bull. Soc. Math. Fr., **127** (1999), 115–133.

[Nad89] **A.M. Nadel.** — *Multiplier ideal sheaves and Kähler-Einstein metrics of positive scalar curvature*, Proc. Nat. Acad. Sci. U.S.A., **86** (1989), 7299–7300 and Annals of Math., **132** (1990), 549–596.

[Nak55] **S. Nakano.** — *On complex analytic vector bundles*, J. Math. Soc. Japan, **7** (1955), 1–12.

[Ohs88] **T. Ohsawa.** — *On the extension of L^2 holomorphic functions, II*, Publ. RIMS, Kyoto Univ., **24** (1988), 265–275.

[Ohs94] **T. Ohsawa.** — *On the extension of L^2 holomorphic functions, IV: A new density concept*, Mabuchi, T. (ed.) et al., Geometry and analysis on complex manifolds. Festschrift for Professor S. Kobayashi's 60th birthday. Singapore: World Scientific, (1994) 157–170.

[Ohs95] **T. Ohsawa.** — *On the extension of L^2 holomorphic functions, III: negligible weights*, Math. Zeitschrift, **219** (1995), 215–225.

[Ohs02] **T. Ohsawa.** — *A precise L^2 division theorem*, Complex geometry (Göttingen, 2000), 185–191, Springer, Berlin, 2002.

[Ohs04] **T. Ohsawa.** — *Generalization of a precise L^2 division theorem*, Complex analysis in several variables – Memorial Conference of Kiyoshi Oka's Centennial Birthday, 249–261, Adv. Stud. Pure Math., 42, Math. Soc. Japan, Tokyo, 2004.

[OT87] **T. Ohsawa, K. Takegoshi.** — *On the extension of L^2 holomorphic functions*, Math. Zeitschrift, **195** (1987), 197–204.

[Pau98a] **M. Păun.** — *Sur l'effectivité numérique des images inverses de fibrés en droites*, Math. Ann., **310** (1998), 411–421.

[Pau98b] **M. Păun.** — *Fibrés en droites numériquement effectifs et variétés Kählériennes compactes à courbure de Ricci nef*, Thèse, Université de Grenoble, **1**, (1998), 80 .

[Pau00] **M. Păun.** — *Semipositive $(1,1)$-cohomology classes on projective manifolds*, Preprint de l'I.R.M.A., Strasbourg.

[Pau07] **M. Păun.** — *Siu's invariance of plurigenera: a one-tower proof*, J. Differential Geom., **76** (2007), 485–493.

[Pau08] **M. Păun.** — *Relative critical exponents, non-vanishing and metrics with minimal singularities*, arXiv:math.CV/0807.3109.

[Pau09] **M. Păun.** — *Lecture notes on rational polytopes and finite generation*, Preprint Université de Nancy I, October 2009, to appear in the volume of the 2008 Park City Summer School, PCMI.

[Pop08] **D. Popovici.** — *Regularization of currents with mass control and singular Morse inequalities*, J. Differential Geom., **80** (2008), 281–326.

[Rei88] **I. Reider**. — *Vector bundles of rank 2 and linear systems on algebraic surfaces*, Ann. of Math., **127** (1988), 309–316.

[Ric68] **R. Richberg**. — *Stetige streng pseudokonvexe Funktionen*, Math. Ann., **175** (1968), 257–286.

[Sak88] **F. Sakai**. — *Reider-Serrano's method on normal surfaces*, Proc. Intern. Conf. on Algebraic Geometry (L'Aquila, June 1988), Lecture Notes in Math., **1417** (1990), 301–319. Springer-Verlag, Berlin.

[Sch74] **M. Schneider**. — *Ein einfacher Beweis des Verschwindungssatzes für positive holomorphe Vektorraumbündel*, Manuscripta Math., **11** (1974), 95–101.

[Ser54] **J.-P. Serre**. — *Fonctions automorphes: quelques majorations dans le cas où X/G est compact*, Sém. Cartan (1953-54) 2–1 à 2–9.

[Ser55] **J.-P. Serre**. — *Un théorème de dualité*, Comment. Math., **29** (1955), 9–26.

[Ser56] **J.-P. Serre**. — *Géométrie algébrique et géométrie analytique*, Ann. Inst. Fourier (Grenoble), **6** (1956), 1–42.

[ShSo85] **B. Shiffman, A.J. Sommese**. — *Vanishing Theorems on Complex Manifolds*, Progress in Math. **56**, (1985), Birkhäuser.

[Sho85] **V.V. Shokurov**. — *A nonvanishing theorem*, Izv. Akad. Nauk SSSR Ser. Mat, **49** (1985), 635–651.

[Sib85] **N. Sibony**. — *Quelques problèmes de prolongement de courants en analyse complexe*, Duke Math. J., **52** (1985), 157–197.

[Sie55] **C.L. Siegel**. — *Meromorphic Funktionen auf kompakten Mannigfaltigkeiten*, Nachrichten der Akademie der Wissenschaften in Göttingen, Math.-Phys. Klasse, **4** (1955), 71–77.

[Siu74] **Y.T. Siu**. — *Analyticity of sets associated to Lelong numbers and the extension of closed positive currents*, Invent. Math., **27** (1974), 53–156.

[Siu88] **Y.T. Siu**. — *The existence of Kähler-Einstein metrics on manifolds with positive anticanonical line bundle and a suitable finite symmetry group*, Ann. of Math, **127** (1988), 585–627.

[Siu93] **Y.T. Siu**. — *An effective Matsusaka big theorem*, Ann. Inst. Fourier, **43** (1993), 1387–1405.

[Siu95] **Y.T. Siu**. — *Very ampleness criterion of double adjoint of ample line bundles*, Princeton Conf. in honor of Robert C. Gunning and Joseph J. Kohn, Princeton University, NJ, USA, Mar. 16-20, 1992; Ann. Math. Stud. **137**, (1995), 291–318.

[Siu96] **Y.T. Siu**. — *Effective Very Ampleness*, Invent. Math., **124** (1996), 563–571.

[Siu98] **Y.T. Siu**. — *Invariance of Plurigenera*, Invent. Math., **134** (1998), 661–673.

[Siu00] **Y.T. Siu**. — *Extension of twisted pluricanonical sections with plurisubharmonic weight and invariance of semi-positively twisted plurigenera for manifolds not necessarily of general type*, Complex Geometry (Göttingen, 2000), Springer, Berlin, 2002, 223–277.

[Siu06] **Y.-T. Siu**. — *A General Non-Vanishing Theorem and an Analytic Proof of the Finite Generation of the Canonical Ring*, arXiv:math.AG/0610740.

[Siu08] **Y.-T. Siu**. — *Finite Generation of Canonical Ring by Analytic Method*, arXiv: math.CV/0803.2454.

[Sko72a] **H. Skoda**. — *Sous-ensembles analytiques d'ordre fini ou infini dans $\mathbb{C}^n$*, Bull. Soc. Math. France, **100** (1972), 353–408.

[Sko72b] **H. Skoda**. — *Applications des techniques L^2 à la théorie des idéaux d'une algèbre de fonctions holomorphes avec poids*, Ann. Scient. Ec. Norm. Sup. 4e Série, **5** (1972), 545–579.

[Sko75] **H. Skoda**. — *Estimations L^2 pour l'opérateur $\overline{\partial}$ et applications arithmétiques*, Séminaire P. Lelong (Analyse), année 1975/76, Lecture Notes in Math., **538** **(1977)** (314–323. Springer-Verlag), Berlin.

[Sko78] **H. Skoda**. — *Morphismes surjectifs de fibrés vectoriels semi-positifs*, Ann. Sci. Ec. Norm. Sup., **11** (1978), 577–611.

[Sko80] **H. Skoda**. — *Relèvement des sections globales dans les fibrés semi-positifs*, Sém. P. Lelong-H. Skoda (Analyse), 19e année, 1978-1979, Lecture Notes in Math. 822 (1980), 259–303.

[Sko82] **H. Skoda**. — *Prolongement des courants positifs fermés de masse finie*, Invent. Math., **66** (1982), 361–376.

[Som78] **A.J. Sommese**. — *Submanifolds of abelian varieties*, Math. Ann., **233** (1978), 229–256.

[Sug88] **K.-I. Sugiyama**. — *A geometry of Kähler cones*, Math. Ann., **281** (1988), 135–144.

[Taka06] **S. Takayama**. — *Pluricanonical systems on algebraic varieties of general type*, Invent. Math., **165** (2006), 551–587.

[Taka07] **S. Takayama**. — *On the invariance and lower semi-continuity of plurigenera of algebraic varieties*, J. Algebraic Geom., **16** (2007), 1–18.

[Take97] **K. Takegoshi**. — *On cohomology groups of nef line bundles tensorized with multiplier ideal sheaves on compact Kähler manifolds*, Osaka J. Math., **34** (1997), 783–802.

[Tei79] **B. Teissier**. — *Du théorème de l'index de Hodge aux inégalités isopérimétriques*, C. R. Acad. Sc. Paris, sér. A, **288** (29 Janvier 1979), 287–289.

[Tei82] **B. Teissier**. — *Bonnesen-type inequalities in algebraic geometry*, Sem. on Diff. Geom., edited by S.T. Yau, 1982, Princeton Univ. Press, 85–105.

[Thi67] **P. Thie**. — *The Lelong number of a point of a complex analytic set*, Math. Annalen, **172** (1967), 269–312.

[Tia87] **G. Tian**. — *On Kähler-Einstein metrics on certain Kähler manifolds with $c_1(M) > 0$*, Invent. Math., **89** (1987), 225–246.

[Tra95] **S. Trapani**. — *Numerical criteria for the positivity of the difference of ample divisors*, Math. Zeitschrift, **219** (1995), 387–401.

[Tsu05] **H. Tsuji**. — *Extension of log pluricanonical forms from subvarieties*, arXiv: math.CV/0511342.

[Tsu07a] **H. Tsuji**. — *Canonical singular Hermitian metrics on relative canonical bundles*, prepublication Sophia University, Japan, arXiv:math.AG/0704.0566.

[Tsu07b] **H. Tsuji**. — *Canonical volume forms on compact Kähler manifolds*, prepublication Sophia University, Japan, arXiv: math.AG/0707.0111.

[Tsu07c] **H. Tsuji**. — *Talk and personal communication*, MSRI Hot-Topics Workshop "Minimal and canonical models in algebraic geometry" held at Berkeley in April 2007.

[Vie82] **E. Viehweg**. — *Vanishing theorems*, J. Reine Angew. Math., **335** (1982), 1–8.

[Yau78] **S.T. Yau**. — *On the Ricci curvature of a complex Kähler manifold and the complex Monge-Ampère equation I*, Comm. Pure and Appl. Math., **31** (1978), 339–411.

Made in the USA
Monee, IL
08 November 2026

68144969R20133